2.

Benchmark Papers in Ecology

Series Editor: Frank B. Golley
University of Georgia

Benchmark Papers
in Ecology / 8

A BENCHMARK® Books Series

PATTERNS OF
PRIMARY PRODUCTION
IN THE BIOSPHERE

Edited by

HELMUT F. H. LIETH

Universität Osnabrück

Dowden, Hutchinson & Ross, Inc.

STROUDSBURG, PENNSYLVANIA

LIBRARY OF CONGRESS CATALOGING IN PUBLICATION DATA
Main entry under title:
Patterns of primary production in the biosphere.
   (Benchmark papers in ecology ; 8)
   Includes indexes.
   1. Primary productivity (Biology)—Addresses, essays, lectures. I.
Lieth, Helmut.
QH541.3.P37   581.5   78-18691
ISBN 0-87933-327-8

Distributed world wide by Academic Press,
a subsidiary of Harcourt Brace Jovanovich,
Publishers.

SERIES EDITOR'S FOREWORD

Ecology—the study of interactions and relationships between living systems and environment—is an extremely active and dynamic field of science. The great variety of possible interactions in even the most simple ecological system makes the study of ecology compelling but difficult to discuss in simple terms. Further, living systems include individual organisms, populations, communities, and ultimately the entire biosphere; there are thus numerous subspecialties in ecology. Some ecologists are interested in wildlife and natural history, others are intrigued by the complexity and apparently intractable problems of ecological systems, and still others apply ecological principles to the problems of man and the environment. This means that a Benchmark Series in Ecology could be subdivided into innumerable subvolumes that represent these diverse interests. However, rather than take this approach, I have tried to focus on general patterns or concepts that are applicable to two particularly important levels of ecological understanding: the population and the community. I have taken the dichotomy between these two concepts as my major organizing concept in the series.

In a field that is rapidly changing and evolving, it is often difficult to chart the transition of single ideas into cohesive theories and principles. In addition, it is not easy to make judgments as to the benchmarks of the subject when the theoretical features of a field are relatively young. These twin problems—the relationship between interweaving ideas and the elucidation of theory, and the youth of the subject itself—make development of a Benchmark series in the field of ecology difficult. Each of the volume editors has recognized this inherent problem, and each has acted to solve it in his or her unique way. Their collective efforts will, we anticipate, provide a survey of the most important concepts in the field.

The Benchmark series is especially designed for libraries of colleges, universities, and research organizations that cannot purchase the older literature of ecology because of costs, lack of staff to select from the hundreds of thousands of journals and volumes, or from the unavailability of the reference materials. For example, in developing countries where a science library must be developed *de novo* I have seen where the Benchmark series can provide the only background literature available to the students and staff. Thus, the intent of the series is to provide an authoritative selection of literature, which can be read in the original form, but that is cast in a matrix of thought provided by the editor. The volumes are designed to explore the historical development of a concept in ecology and point the way toward new developments, without being a historical study. We hope that even though the Benchmark Series in Ecology is a library-oriented series and bears an appropriate cost it will also be a sufficient utility so that many professionals will place it in their personal libraries. In a few cases the volumes have even been used as textbooks for advanced courses. Thus we expect that the Benchmark Series in Ecology will be useful not only to the student who seeks an authoritative selection of original literature but also to the professional who wants to quickly and efficiently expand his or her background in an area of ecology outside his special competence.

Helmut Lieth, the editor of this Benchmark volume, is presently professor of biology at the University of Osnabrück, Germany, having recently moved from a professorship at the University of North Carolina. Dr. Lieth is an authority of long standing on primary production, with considerable personal experience with field measurement of production and with modelling. His training in the German tradition of biogeography and plant ecology is reflected by his focus on the historical beginnings of the study of botany and on the global features of production. Primary production is a central process in ecosystem function and as such merits special treatment in the Benchmark series. The present volume is complementary to previous Benchmark books on cycles of essential elements and on ecological energetics.

F. B. GOLLEY

PREFACE

When Dr. Golley invited me to compile this Benchmark book about global net primary productivity pattern, it was for me one of the many invitations during my lifetime to work on this subject matter.

The farmhouse in Germany I was born in was built at the time when Priestley, Ingenhousz, and their contemporaries found out the constitution of plant matter. My grandfather, who had a small country dealership for organic and inorganic fertilizer for the local co-op, taught me to distinguish between good and bad land before I went to grammar school. The village of Steeg where our farmhouse is located is about 25 kilometers from the city of Cologne (KÖLN), where Albertus Magnus worked about 700 years earlier on the scholastic understanding of plants. Before I went to school, I raised plants and learned their basic needs.

As an undergraduate student I learned analytical chemistry from Walter Noddak, who in the thirties had written his essays on the global carbon cycle. I would never have met him if the postwar conditions in Germany had not by chance brought the two of us together in the same college, the Philosophical Theological College of Bamberg/Bavaria. Far ahead of his time, he urged us in 1948 to work on the global production levels and pattern of plants because we would run out of oil by the end of the century and out of coal in a few hundred years more. Plant matter was in his opinion the most precious future raw material for energy as well as for food.

As a graduate student I went back home to the University of Cologne, the one Albertus Magnus originally founded. Soon after I received my Ph.D., I got my first job in Stuttgart Hohenheim, at the Agricultural University, where Dr. Wolff had worked 100 years earlier on the chemical composition of food and feed and Dr. Schwertz, as a contemporary of Liebig, had fought a decade-long battle with Liebig about the validity of humus for plant production.

The department of botany in Hohenheim was chaired at the time I taught there (1955–66) by Dr. Heinrich Walter who had contributed to production ecology of the African savannah, although he mostly worked on plant-water relationships. One of his earlier predecessors had been Dr. Schröder who in 1919 wrote a standard paper on the global primary productivity. Schröder's entire personal library was still in Hohenheim, and I was the assistant in charge of the library. That position brought me in contact with the works of Liebig and Ebermayer, Vernadsky and Boussingault, Boysen Jensen and Lundegårdh, the Blackmans and Mitscherlich. I learned at that time that one good table, procedure, or equation out of thousands of pages of books and papers written by these men was all that remained of them after 50 to 200 years.

In the late fifties, my interest in productivity research was further enforced when I met Dr. V. G. Blackman in Oxford and Drs. Eugene Odum and Frank Golley in Athens. I worked on the light compensation point of plants at that time, a key problem in primary productivity research. When I embarked on a large program on primary productivity pattern in southern Germany, I organized a symposium on primary productivity in Hohenheim in 1960, which was probably the first of its kind and very timely. Among the knowledgeable men who came from all over Europe were the successor of Boysen Jensen, Dr. D. Müller from Copenhagen, and Dr. P. Duvigneaud, from Brussels. The symposium was timely insofar as soon thereafter the International Biological Program was started, with net primary productivity as the central theme.

We had just found out how difficult it is to assess net primary productivity in terrestrial ecosystems with sufficient accuracy when the IBP called everybody together who knew how to do such work and who was able to come. We worked together: Drs. Ovington, Ellenberg, Müller, Duvigneaud, Tamm, Monsi, and Satoo and many younger colleagues who became known for their later work on plant productivity. The work on this program carried me to many countries, and I have not only met almost everyone who has worked in the IBP within the last ten years but have been fortunate enough to see a great number of the sites of productivity research.

In the early sixties a number of symposia were organized by UNESCO as well as other international organizations. Participation in these gave me the opportunity to meet more people engaged in productivity research. Among them were Rodin from Russia; Lemee and Bourliere from France; Kira, Ogawa, and Numata from Japan; Petrusewicz from Poland; and many others. The most remarkable event during that time for me was a geophysical symposium organized (1963) by Dr. Junge in Utrecht. He asked me to address the convention on the possible impact of the vegetation on the annual fluctuation of the CO_2 content of the atmosphere. The latter topic had just been studied by Bolin and Keeling and the geophysicists were searching for an explanation. During the meeting Dr. Junge and one of his students encouraged me to construct a world map of the annual carbon production pattern and a seasonal flux model of the CO_2 input and output for the conditions in Germany. I agreed and as a result produced the first global productivity map with which Junge and Czeplak were able to simulate Bolin and Keeling's annual CO_2 flux model. Soon thereafter the map by Rodin and Bazilevitch appeared, and now we have many others available indicating that vegetation mapping has now an alternative—the functional vegetation maps—to the classical approaches of physiognomy and phytosociology.

When I moved to Chapel Hill in 1967, almost immediately I became involved in the US/IBP. For a number of years I served with Drs. Stanley Auerbach, David Goodall, Stan Gessel, Jerry Brown, George van Dyne, Fred Smith, and later Frank Blair on the development of the program. Among my closest colleagues on the UNC campus was Dr. Howard T. Odum who had just switched at that time from nutrient cycling studies to energy considerations in ecosystems.

My objective after initial organizational work in the deciduous forest biome was to develop a "tropical forest biome" study in the Western hemisphere. This never materialized during the lifetime of the IBP, but recently we do have such a study organized by Ernesto Medina and Rafael Herrera in Venezuela. Medina happened to be a student of mine when I worked in Hohenheim. It seems, therefore, that the great need for productivity work in tropical areas, which I saw early, is finally spreading to the new world tropics with the delay of about one decade.

While the organizational off-campus work for IBP was cumbersome and far from enjoyable at times, the UNC campus offered some rewarding development. The sixties were the time of "logarithmic growth phase" for computer science, and at UNC we were fortunate to have Dr. Brooks who was one of the prime designers of the IBM 360 operational system. Curve fitting and other calculating routines were developed everywhere, and Dana Quade's Nonreg program developed at UNC was of specific help to us. Computer mapping routines were developed at about the same time. We selected the SYMAP routine developed by the Harvard Graduate School of Design and combined all those elements into one procedure. Numerous UNC students picked up one task after another, and within a matter of two years we developed the world computer map and its computation facilities. This is now known as the UNC Biosphere Model. The first presentation of this model in Miami in 1971 resulted in the term "the Miami Model" and a close cooperation with Dr. Whittaker. The result of this association was the joint editorship on the book *Net Primary Productivity of the Biosphere*, from which several readings in this book were taken.

I hope this book will encourage more young people to work on primary productivity in geographical areas where little is known, on better resolutions where such are desirable, on more refined partitioning where required, and last but not least on better models for the management of primary productivity on a global scale. Man is so severely infringing upon the vegetation that only careful planning will prevent disaster.

No book dealing with productivity should be published without the statement that all our work is in vain without the restriction of the further growth of the human population. Although I personally would prefer to see this done by self-discipline of individuals, I must admit that almost every other means to stop the population explosion would be better than the global war that mankind inadvertently will run into in less than two generations.

Many people have helped me in collating this book. I must acknowledge the help of several from the University of North Carolina at Chapel Hill for the work done for Part I: Drs. George Lane and Petrus Tax from the Germanic languages department, Dr. Jack M. Sasson from the department of religion, Dr. P. S. Koda from the rare book room of the main library, Mrs. B. Conway from the chemistry library, and Mrs. B. Zouck from the botany library. My appreciation also goes to Mrs. Fähnrich from the main library of the KFA in Jülich and especially to my daughter Margot. All were instrumental in retrieving literature of various kinds from which I include a few original lines or figures. In the overall work, I must thank my wife Magdalene and my daughter Margot for their loyal support in collecting material, for their secretarial work, and for their continuous criticism. I also thank Ms. Karen Henry for repeatedly typing versions of this manuscript. Mrs. Hildegard Schnitzler, my wife Magdalene, and Mr. A. Suntrup deserve credit for preparation of the subject index. Dr. Golley is to be thanked for his valuable advice in selecting papers and the format of the book. The staff of the publishing company deserves special thanks for their willingness to entertain the many special inconvenient demands of the editor.

HELMUT F. H. LIETH

CONTENTS

Contents

PART III: MODELING PRODUCTIVITY PATTERNS

Contents

PART IV: GLOBAL PRODUCTIVITY PATTERN

CONTENTS BY AUTHOR

PATTERNS OF
PRIMARY PRODUCTION
IN THE BIOSPHERE

INTRODUCTION

The discussion of the patterns of primary productivity on earth must start with the definition of plant matter, since it is the basic output of the primary production process. The nature of plant matter has intrigued man since earliest time and even though it is not usual in the primary production literature it will be interesting to follow man's understanding of plant matter from a historical perspective. Part I demonstrates the change in concepts and levels of understanding during the last 4,000 years or so—the time span from which we have written documents. Collecting the evidence for the statements in Part I was not easy. I will be the first to admit that the presentation is incomplete and that some statements are not supported to the extent that one would like. But in absence of anything better, this compilation is offered as a beginning. Historical treatments of the early philosophies of man are being enriched all the time. One can hope, therefore, that this historical background will be more complete in the future.

Part I begins with the history covering approximately 3,500 years from the ancient Chinese and Indian cultures to the first half of the nineteenth century and concludes with the last 150 years and a look into the future. Collection of the first selections proved to be the more difficult task. The available documents are in strange scripts. I had to find my way through a number of volumes of annotated translations of the scripts into either German, English, or Latin to find a few statements relevant to my task. Some of the older documents would never have been consulted if the first lead had not been provided by casual conversations with friends and colleagues.

The translations sometimes had inherent difficulties. Professional knowledge of science, plants, and environment leads one to interpretations of the original texts quite different in meaning from those presented by the liberal arts scholar. This is especially evident in interpretations of the Rigveda and the writings of Nicholas of Cusa.

The modern scientific study of primary production concerns the quantitative description of plant matter which requires evaluation of methodology. Thus, Part II deals with the measurement of primary productivity. To list the benchmark papers for the development of production-measuring techniques is a difficult task. All methods have been refined several times and in several different ways. Therefore, Part II is an attempt to compile papers that demonstrate the early development together with papers that show the more recent application of any given method. All methods we use today for productivity assessment have their roots in the last century, and yield measurements can even be traced back at least to the Roman times. Also presented is a collection of photographs with productivity data and environmental parameters of eighteen International Biological Program Study sites. The IBP had a special focus on primary production and provided a great increase in valuable data for the student of production ecology.

Part III deals with modeling productivity patterns, an activity that began in the early part of this century and that is going on at this time on all levels. It contributes to the understanding of natural processes and their management needs as well. This part does not deal specifically with model building but it will be obvious how much less could be presented if modeling had been left out of productivity studies.

The last part of this volume is concerned with the global productivity pattern because of its importance to the human population. Primary production estimates are a base line against which we can judge the value of agricultural or forest yield. Ultimately man depends upon the global input of energy captured by plant production. A good estimate of global productivity is now available. How reliable it is the future must tell. I received a pleasant surprise at a recent (1974) conference when a young colleague entered a discussion circle without knowing me personally. "I don't understand," he said, "what is all that fuss about primary productivity. Such old stuff; we have global productivity maps in our textbook." Having most likely drawn the ones he referred to I wished I could share his confidence.

Anyway, primary productivity assessment has never been an end in itself in my estimation. I want it to be a baseline for future probes into the laws that govern ecosystems processes and evolution. At several points in this collection of papers scientific problems are presented that can be solved now since new methods or an extended data base are available from current literature. Any young reader may see this and exploit it profitably for his or her own progress. With this in mind, the collection of historical papers may become a stimulant for new research in this field of crucial significance for mankind.

2

Part I

WHAT IS PLANT MATTER?

Editor's Comments
on Papers 1 Through 11

MYTH AND SEARCH FROM SHEN-NUNG TO LIEBIG

Man's understanding of the nature of plant matter throughout early history is poorly documented, and still not well treated. Most of the early English treatments of the history of botany start with Sachs' (1890) viewpoint that was later continued by Green (1909). However, since botanists now realize that Sachs' handling of the subject is anything but accurate (Harvey-Gibson 1919, Reed 1942), Green (1909) might better have based his work on *Botanik der Gegenwart und Vorzeit* by Karl F. W. Jessen (1864). With regard to the topic of plant matter and primary production, Jessen's book is by far superior to any other of its time. It contains an excellent selection of references, enumeration of sources, and philosophical treatment. During the last hundred years, however, new discoveries of old manuscripts have provided additional information, some of which is included in this Benchmark volume. For example, Paper 1 from Reed (1942) contains the oldest available reports of agriculture in China during the time of Emperor Shen-Nung (2700 B.C.).

In his book, Jessen (1864) did include reports about the ancient Babylonic knowledge of plant production and general agriculture that originated from the translation of Chaldaic scripts into Arabic by Abu Bekr Ahmed around 900 A.D., and he cited a further translation into Russian by Chwolson (1859).[1] While the Russian scholar's reports, which credited agricultural essays to Janbushad (around 1800 B.C.) and Dhagrit (around 2000 B.C.), were hardly credible in Jessen's time, today we do have ample documentation of Assyrian and Sumerian skill in botany, agriculture, and medicine (see Thompson, 1949).

Although many of these older books reflect a sound knowledge about plant yield on the part of various early people, no information exists about their beliefs concerning the real nature of the plant substances they ate. As Liebig (1862) notes in discussing the approach of the Chinese people to plant production, ". . . they were masterly in developing recipes for improvements but apparently lacking any interest in analyzing the nature of the substances they were dealing with."

The philosophies of the ancient East Indian cultures, including their views on plant substances, are known today to the extent that they are reflected in old Sanskrit scripts such as the Rigveda (circa 3000 B.C.). The translation by Geldner (1951), contains several verses that allow us to deduce that people regarded plants as entities on the same level as water, sky, mountains, oceans, and so forth. Plants (probably herbaceous plants) are described as being different from trees, and they

---

[1] Cited by Jessen without bibliographic details.

were obviously held as transitory manifestations between water and Agni (the god of fire). This philosophy seems very reasonable for a people living in a monsoon climate and experiencing frequent or at least annual brush fires. For our purposes the most significant of the verses in the Rigveda translation by Geldner—song 10-91—is presented here as Paper 2 of this volume together with a reprint of the Sanscrit version of the song edited by Müller.

The philosophic treatment of plant substances in the book of Genesis ascribed to Moses offers more precise statements than the Rigveda about the material nature of plants. However, the sequence of creation (Gen.1:11-13, plants [see Paper 3], and 14ff, the sun and the moon) has confused philosophers for the 2,000 years following its authorship. According to Moses, plants were part of the solid earth created on the third day, and they existed before the sun and moon were created on the fourth day.

The cosmology of the Germanic nations, before Christianity swept the European continent, also includes in the song of the creation the origin of plants from the hairs of a giant. This song is part of the collection of Germanic sagas, *The Edda* (see Paper 4). *The Edda* as we know it today was recorded by Saemund Sigfusson and Snorri Sturluson in Iceland around 1100 to 1200 A.D.

As Western civilization evolved, intellectual thought developed along many lines, and the scientific historians of the nineteenth century later saw Theophrastus Eresii (371-286 B.C.) or Aristotle (384-322 B.C.) as the fathers of all of the botanical philosophies that evolved through the Middle Ages. Jessen (1864) covered the entire development of thought on the origin and nature of plants from the time of Theophrastus to that of Nicholas of Cusa (1450 A.D.). Paper 5 presents extracts from Jessen on the writings of Theophrastus Eresii and Aristotle that show the relationship between plants, soil, and water in Greek philosophy. The first extract is an evaluation of a passage on the element system that we now assume to be from the writings of Theophrastus Eresii, although Jessen and his contemporaries believed all Greek books on plants were written by Aristotle. (For a detailed discussion of the controversy over Greek authorship, see Green 1909; for the gradual development of the element concept in the sequence of Greek philosophers, see Solmsen 1960.) Later, Greek philosophy was amalgamated with the Hebrew heritage of Genesis 1 by the early patriarchs of the Christian church: Moses had supplied the story of the creation and Theophrastus had elaborated on the relationship of the elements; St. Basil (315-379 A.D.) put both together into a functional unit (see Paper 6). His teachings became prominent in the early medieval literature when they were rediscovered by Albertus Magnus (1193-1280), who also revived Theophrastus for the Christian culture

Moses (around 1225 B.C.) provided in *Genesis 1* the statement that plants come from the earth.

Aristotle (384–322 B.C.) culminated the Greek philosophic tradition of the four elements—solid, liquid, gas, fire— and plants primarily belong to the solids. (Source: Mägdefrau 1973)

Theophrastus (370–287 B.C.) applied Aristotelian categories to plants; that is, mostly solid, some liquid, and some warmth. (Source: drawing by Marion Seiler after figure 2 in Mägdefrau 1973)

7

in Europe. In Albertus's time, the writings of the Greek scholars had reached central Europe by way of Alexandria, Morocco, and Spain, and Aristotle and Theophrastus were believed to be the same source.

The following extract on the nature of plants from Albertus's work shows his scholastic approach to problems and his reliance on biblical text, Augustinus's (354–430) evaluations of it, and Aristotelian philosophy.

> ... Solution of the third link:
> After this one must say that plants are part (of the earth) and not (its) ornament. It is sound logic which Augustinus cites for this: because they are solidly attached to the soil. Also it is true, as the philosopher (Aristotle) said, that each body stems from the same material of which it is nourished. The plant has roots, which correspond to the mouth as the philosopher says in his second book about the soul. Now, they would not have fastened their roots, which are in analogy their mouth if they were not naturally originating from earth. Also says the philosopher in his second book about plants (this is now Theophrastus): the plant has three powers, and these it gains from the soil. And the philosopher Abrutallus (or Protagoras) said that in all such plants the earth (soil) is the mother and sun is the principle. . . [translated from Jessen 1864, p. 147].

While the philosophers of the early medieval period seemed to uphold the concept that plants were primarily part of the soil, other members of society seemed to have different interpretations, as reflected in one verse of the *Parzival* ballad (The story of a knight in search of the Holy Grail). The poem was composed (circa 1200) by Wolfram von Eschenbach and stanza 817, 25 ff. reads: "From water trees derive their sap. Water fructifies all created things . . ." (Mustard and Passarge 1961).

This emphasis on water is almost repetitious of some Sanskrit verses of the Rigveda of about 3,000 years earlier. While there may have been connections between Eastern and Western cultures, the more probable source in Europe for the idea that water generated plants may exist in the visions of a famous nurse of the twelfth century, St. Hildegardis of Bingen. This extraordinary woman founded the Benedictine cloister and hospital, Rupertsberg, near Bingen on the Rhine. Although she was illiterate, she became the abbess of the cloister and found aides[2] to help put her pharmaceutical experiences and visions into script. Following are her ideas on the origin of plants:

> . . . The soil bringeth forth sweat, steam and juice. The sweat of the soil

[2] "Among the aides (in writing) was Probst Volmer of Rupertsberg by far the most important. His name reappears in the newly discovered material after it was lost for some time behind the name of the monk Gottfried who was put in his place by Tritenius . . ." (Liebeschütz 1930.)

Albertus Magnus (1193-1280 A.D.) revived Aristotelian and Theophrastian philosophies for the medieval scholar with the script of *Genesis 1* that indicates that plants are part of the earth. (Source: Mägdefrau 1973)

St. Hildegardis of Bingen (1098-1199) wrote that plants are the manifestations of different liquids excreted from the earth. No connection to the similar Brahman views of then 4000 years ago is known. (Source: Mägdefrau 1973)

Nicholas of Cusa (1401-1464) stated that plants are made of water. This statement was based on the first reported experiments with plants grown in pots to disprove that plants are basically made of earth. (Source: Haubst 1964)

9

> produces herbs of nuisance. The steam of the earth produces useful
> herbs, the ones that one may eat and those useful to man for other
> things. The juice of the earth generates vineyards and green trees. The
> vegetables, planted or sowed by human labor, and growing without
> delay from day to day . . . lose through the same effort with which
> they were sowed and harvested, the acid and bitterness of their sap
> and their moisture as well. But herbs growing from their seeds without
> human labor and care . . . are rejected by men for food . . . . However,
> these herbs are partially good as medicine and calm detrimental wicked
> moisture within the human body. . . [translated from Jessen 1864,
> p. 125].

The text is as mythical as the Rigveda but it reveals that conceptually
water was put between plants and earth as the medium to generate
certain properties.

This idea persisted in the minds of the people, although the official
dogma of the church was the one presented by Albertus Magnus (1193?–
1280) and Thomas Aquinus (1225–1274) in *Summa theologica*. Three
hundred years later in 1450, however, Nicholas of Cusa (1401?–1464) a
cardinal of the Roman Catholic church from the town of Cues on the
Moselle, tipped the balance in favor of water generating plants (see
Paper 7). He came to the conclusion in his influential treatise *De
experimentis staticis* (the experiments with the balance). In this treatise
Nicholas describes in theory an experiment that was performed again
150 years later by the Dutch physician van Helmont (1577–1644).
Krikorian and Steward (Paper 7) have critically studied the relation
between van Helmont's experiment and Nicholas's essay. The statements by Nicholas in this excerpt are extraordinary inasmuch as they
were not authorized by the Roman Curia. He cleverly circumvented
any trouble by putting heretical innovation into the mouth of the
layman. We are probably safe in assuming that he had actually conducted the experiments that are presented in Paper 7.

The discussion of the relationship between living and nonliving
matter went in a different direction, however. The typical interpretation of the position of plants in the general medieval cosmology is
presented in Figure 1 which contains two excerpts from a treatise by
Carolus Bovillus (see Klibansky 1927). The table and drawing show
how Bovillus applied the Aristotelian four-element system—Terra,
Aqua, Aer, and Ignis—to the array of bodies surrounding and including man. In this context, plants (represented by arbor, a tree) stand
between the mineral and animal level on the basis that they are living
(vivit) but not feeling (sentit). The other entries in the table and drawing compare the sequences of the four categories of matter or being
with human activities that are not directly relevant to the topic of
plant production but subconsciously support the presented position
of plants.

Mineralia	Viventia	Sensibilia	Rationalia
Lapis	Arbor	Bestia	Homo naturalis
Esse	Vivere	Sentire	Intelligere
Infans in utero	Ablactans	Progrediens	Stans
Prima etas	Secunda etas	Tertia etas	Quarta etas
Fetus	Infans	Puer	Vir
Acedia	Gula	Luxuria	Virtus
Piger	Concupiscens	Amans	Studiosus
Terra	Aqua	Aer	Ignis
Scriptura	Vox	Conceptus	Mens

Figure 1. Medieval cosmology representing different stages of being. Plants are placed between stone and animal, analogous to the vital sphere of man, to voice, child, water, etc. (Source: Klibansky 1927, pp. 312 and 306)

11

One basic difference between plants and animals in the table in Figure 1 is the homology of plants with water, and animals with air (rows 2 and 8). Indeed one of the presumptions of the time was that breathing was a basic difference between plants and animals. For this reason, gas exchange information was first gathered by animal physiologists (see the discussion of Hooke 1667 and Mayow 1674 by Transeau (1926.)

About 350 years after Nicholas of Cusa, mankind witnessed the final breakthrough in the analysis of plant matter in the sense that we usually discuss it today, when Priestley, Scheele, Ingenhousz, Senebier, De Saussure, and Liebig unveiled, one after another, the components of plant matter: carbon, oxygen, hydrogen, and minerals. The concept that plants exchanged matter only with earth and/or water had to be dropped. The scientists of the day understood during the first half of

Priestley (1733–1804), above, and **Ingenhousz** (1730–1799), at right, described portions of the gas exchange of plants. (Source: Gibbs 1967 and Mägdefrau 1973, respectively)

De Saussure (1767–1845) wrote the photosynthetic equation. (Source: Mägdefrau 1973)

Justus von Liebig (1803–1873) analyzed the elemental composition of plant matter including mineral nutrients. (Source: Shenstone 1895)

the nineteenth century that plants exchanged matter with the atmosphere, hydrosphere, and lithosphere. (See Rabinovitch 1971 and Lieth 1975 for concentrated reviews of the work conducted in this era.)

Paper 8 contains extracts from Lieth (1975) that summarize the developments from van Helmont's crucial experiment, the growth of a hundred or so pounds of a willow tree in a container with the loss of 16 ounces of soil, around 1600 A.D. to the first half of the nineteenth century. We need to correct here the inclusion of the names of Mayer and Boltzmann as persons first responsible for the discovery of chemical energy generated in the photosynthetic process. The treatise on the correlation and conservation of forces compiled by Youmans (1881), does not indicate either one of them made the discovery.

13

Despite Boltzmann's works on solar radiation, he was certainly not the first one who showed the convertibility of one energy form into another, and J. R. Mayer, although responsible for the clear formulation of the law of conservation of energy, made no specific attempt to explain plant growth in his law. Conservation of energy has little bearing on the fact that plants produce chemical energy; this process falls much more under the context of "conversion" of energy. The name of Benjamin Thompson, Count of Rumford, might more suitably be added to the list for his experiment of converting horse labor into heat via friction, which was described in his *Essays* (1797, pp. 470–493). It was conducted while Sir Rumford served as aide-de-camp and chamberlain in the Bavarian court. The design of the crucial experiment was deduced from his experience that steel drills produce heat when holes were drilled into brass cylinders in the process of making cannons. Sir Rumford concluded that the power of the horses turning the drills was converted into heat, but even this experiment omits the clue for the conversion of solar energy to plant energy. From the papers included in Youmans's book, only that of Helmholtz, originating in

Benjamin Thompson (1753–1814) made the first experiment of the conversion of energy. (Source: Williams 1904, vol. 3)

Helmholtz (1821–1894) described in detail the energy relation between the sun, plants, and sensible heat. (Source: Williams 1904, vol. 3)

1854, contains particular treatment of the conversion of solar energy by plants. An excerpt from Helmholtz's paper appears in this volume as Paper 9. It demonstrates well the clear concept of energy that had rapidly evolved between 1800 and 1850 and also gives credit to a number of people that have helped to elaborate this principle, particularly Carnot, a French worker, for the first discovery of the conservation of energy. Since the people included in Helmholtz' paper (1854) were contemporaries of Liebig, and Liebig himself is also mentioned, it marks a natural transition to a discussion of the rapid developments in dry matter analysis during and after Liebig's time.

By the time Liebig started his work in organic chemistry, about 50 percent of the chemical elements had been identified. The Swedish chemist, Berzelius, had done much of the pioneering along with De Saussure and Lavoisier in France, and Woehler in Germany. The main emphasis at this time was on elemental analysis of substances, and the distinction between organic and inorganic compounds had basically been settled. The term *element* was adopted for the "pure" chemical elements, and some of the former elements of the Aristotelian philosophy lived on as physical states—solid, liquid, gas. Within a few decades in the early nineteenth century, the number of categories of plant material jumped by an order of magnitude to about forty or fifty, and the later papers of Liebig reflect the full command of the major features of plant chemistry that the leading chemists of his time had.

Liebig is probably the anchor man for the development of inorganic fertilizer. However, chemical and agricultural experiments with crop plants were carried out by hundreds of people. Why Liebig, together with a young farmer near Giessen, used chemical compounds that were not water soluble when he made his first fertilizer experiments in the early 1840s is difficult to understand. About ten years later, after he had talked to the plant physiologists of the day, he switched to soluble compounds. They became a huge success and are practically the starting point for the present worldwide fertilizer use. Liebig's failure to realize immediately the water-related uptake of minerals is the reason that some scientists give credit to Wolff (1862) or Sachs (1868) for the understanding of minerals needed as plant nutrients (Sachs 1882 and Transeau 1925). However, the pertinent chapters of his book *Chemistry in Its Application to Agriculture and Physiology* make clear the extent to which Liebig knew the composition and needs of plants. A few pages of this work translated in 1856 by Lyon Playfair appear as Paper 10 of this volume.

Liebig must have left the research scene in the second half of the nineteenth century. Sachs (1882) does not quote him at all but rather cites Wolff (1862) as the source for mineral nutrition analysis of plants. Sachs' textbook mentions the first significant proof of the

Table 1. ppm elements in dry plant tissues. (Source: Bowen 1966, pp. 68 and 69. Copyright © 1966 by Academic Press Inc. (London) Ltd.)

Element	Plankton[1]	Brown algae[2]	Bryophytes	Ferns[3]	Gymnosperms	Angiosperms[4]	Bacteria[5]	Fungi[6]
Ag	0·25 Va	0·28 Bl	0·1 La, Se	0·23 H, La	0·07 La	0·06 C, La		0·15 La
Al	1000 Va	62 V	1400 Se		65 By	550 C	210 P	29 By
As		30 V, Y				0·2 Bu, Sm		
Au		0·012 F				<0·00045 C		
B	15 Va	120 Bl, V	20 La	77 La	63 L	50 C, La	5·5 P	5 La
Ba		31 Bl, Bo	150 Se	8 Bd		14 Bd		
Be			<0·2 Se			<0·1 C		<0·1 Le
Bi			<1 Se			0·06 Sh		
Br						15 N		20 N
C	225000	345000	450000	450000	450000	454000	538000	494000
Ca	8000 V	11500 V	3000 Se	3700 H+	6500 L	18000 L	5100 P	1700 P
Cd	0·4 Va	0·4 V	0·1 La	0·5 La	0·24 La	0·64 Bu, La		4 La
Ce			≤14 Se			≤34 Se		
Cl		4700 V	670 Se	6000 H+		2000 Sp	2300 P	10000 Mc
Co	5 Va	0·7 Bl, V, Y	0·33 La, Se	0·8 La	0·2 C, La	0·48 C, La	0·5 La	0·5 La
Cr	3·5 Va	1·3 Bl	2 La, Se	0·8 La	0·16 C, La	0·23 C, La		1·5 La
Cs		0·067 S				0·2 Ya		
Cu	200 Va	11 Bl, V, Y	7 La, Se	15 La	15 L	14 C, La	42 P	15 La
Eu						0·021 Si		
F		4·5 Y				0·5 Mo		
Fe	3500 Va	690 Bl	1200 M	300 M	130 L	140 L, M	250 P	130 M, P
Ga	1·5 Va	0·5 B	0·1 La, Se	0·23 La	<0·07 La	0·05 Bu		1·5 La
H	46000	41000	55000	55000	55000	55000	74000	55000
Hg		0·03 V				0·015 Bu, Mo		
I	300 V	1500 V, Y	5 Se			0·4 J		
K		52000 V, Y	2400 Ba, Se	18000 H+	6300 L	14000 L	115000 P	22300 L
La		10 B	3 Se			0·085 Bu, Si		
Li		5·4 Bl				0·1 Mi		
Mg	3200 V	5200 V	800 Ba, Se	1800 H+	1300 L	3200 L	7000 P	1500 P
Mn	75 Va	53 Bl, V, Y	290 M	250 M, H+	330 L	630 L, M	30 P	25 M
Mo	1 Va	0·45 Bl, V, Y	0·7 La, Se	0·8 H, La	0·13 C, La	0·9 C, La		1·5 La

	38000	15000	25000	20500	32000	30000	96000	51000
N	38000	15000	25000	20500	32000	30000	96000	51000
Na	6000 Va	33000 V, Y	1100 Ba	1400 H+	340 L	1200 L	4600 P	1500 P
Nb			0.3 Ty		0.3 Ty	0.3 Ty		
Nd			≤6.5 Se			≤24 Se		
Ni	36 Va	3, Bl, V, Y	2.5 La, Se	1.5 H, La	1.8 C, La	2.7 C, La		1.5 La
O	440000	470000	450000	430000	440000	410000	230000	340000
P	4250 Va	2800 V	400 Ba, Se	2000 H+	2900 L	2300 L	30000 P	14000 P
Pb	5 Va	8.4 Bl	3.3 La, Se	2.3 La	1.8 C, La	2.7 C, La		50 La
Ra	4×10^{-7} K	9×10^{-8} V				10^{-9} Tb		
Rb		7.4 S				20 Bu		
Re		0.014 F						
Ru						0.005 Bu		
S	6000 V	12000 V	1100 Ba	1000 Th	1100 Th	3400 Th	5300 P, Sp	4000 P
Sb						0.06 Bu		
Sc						0.008 Bu		
Se		0.84 Ch	≤0.3 Se			0.2 Bn		2 Sf
Si	200000 V	1500 V	2000 Se	5500 H+		200 Sp	180 P	
Sm						0.0055 Si		
Sn	35 Va	1.1 Bl	1 La, Se	2.3 La	<0.24 La	<0.3 C		5 La
Sr	260 Va	1400 Bl, Bo	15 Se	13 Bd, H		26 Bd		320 Le
Tb						<0.0015 Si		
Ti	80 Va	12 Bl, V	65 Se	5.3 H		1 Mi		
Tm						0.0015 Si		
U			<0.35 Mk		≤0.35 Mk	0.038 C		0.25 Le
V	5 Va	2 Bl	2.3 Se	0.13 H	0.69 Ca	1.6 Ca		0.67 Be
W		0.035 F				0.07 Bw		
Y			0.33 La, Se	0.77 La	<0.24 La	<0.6 La		0.5 La
Yb			0.2 Se			<0.0015 Si		
Zn	2600 Va	150 Bl, V, Y	50 La, Se	77 H, La	26 C, La	160 C, La		150 La
Zr	20 Va		0.33 La, Se	2.3 La	0.24 La	0.64 La		5 La

[1] Mainly diatoms.
[2] Figures for Au, Re and W are for red/green algae.
[3] Does not include horsetails or clubmosses.
[4] Figures are for woody species where there are plenty of data.
[5] Figures are for vegetative cells.
[6] Lichens included.
*See page 78 for references.

involvement of chlorophyll for photosynthesis by Adolph Baeyer in 1871. Today we know that practically every chemical element is present in plants with the range being $10^5 - 10^{-20?}$ ppm. Table 1 (pages 16 and 17) gives a survey of the concentration of elements in plant tissues from the compendium of Bowen (1966). This table does not yet include a differentiation into isotopes, which are becoming increasingly important in biology.

The organic chemistry of plant material deals with many more than the hundred entities of inorganic chemistry. A tabulation and evaluation by Morowitz (1968) lists names of simple carbon compounds. The figure of 40,000 that he mentions in his assessment of carbon compounds possible in biological material brings us probably to a correct order of magnitude of $10^5 - 10^6$ different types of compounds in which plant biomass may be distinguished. Thus, in the hundred years since Liebig, plant matter categories have jumped five orders of magnitude, which is a level of refinement not within the grasp of any single person's memory but only manageable by the collective intelligence of a highly skilled scientific society.

Although knowledge of plant matter had grown significantly by Liebig's time, the analyses did not yet include the Aristotelian element of fire as a major property of plants. Today we refer to this "element" as "energy." The early discoveries of the intimate relationship between plant production and energy fixation and solar energy were discussed, as mentioned earlier, by Helmholtz in 1854, but Transeau (1926) gives credit to Becquerel (1868) for the first attempt to assess the percentage of solar energy fixed by plants. The first caloric measurements on organic substances related to plants were made by Favre and Silberman (1852). Berthelot and Vieille (1885) introduced high oxygen pressure bomb calorimetry and published the first value for cotton, 4200 cal/g (p. 552). Since energy is one of the most important properties of primary produced plant matter, Paper 11, which contains excerpts from Cummins and Wuycheck, has been included in this section. This paper summarizes the caloric-content measurements of many plants and plant parts. The table it contains gives a good overview of the variability in energy content of plant material and some of the reasons for it.

The value of plant material as food is another important aspect of plant matter analysis. Although knowledge of the chemical composition of the plant material is required for very detailed studies, for the most part, identifying the major categories of such compounds as carbohydrates (raw fiber and water solubles), fat (ether extract), proteins, and ash is sufficient for nutritional work. Most nutrition analyses distinguish these categories, which have been listed according to their combinations in different ecosystems by Lieth (1975).

In recent times other possible aspects of plant matter have been discussed. Margalef (1956) and later Patten (1959) have considered the information content of plant material. Although this property of plants is not yet adequately defined, it is a challenging area for future research. Interesting in this respect is the calculation by Patten, that ecosystems should contain 10^{28} "bits of information." If "bits of information" become realistic entities at some time, we could certainly assume that a good portion of them would reside in the plant matter.

This consideration invites a summary of the diversification of man's concepts throughout his documented past and foreseeable future. The sequence in Table 2 shows that starting in 3,000 B.C. with vague statements by Aristotle of four categories of material, the number of categories gradually increases one order of magnitude until the great turning point in 1800 A.D. From then on the number of distinguishable categories increases by an order of magnitude every fifty years. Assuming this speed to be constant, we might be well past the year 3000 A.D. before we have full knowledge about the now-dubious concept of "bits of information."

Table 2. The number of categories distinguished in plant material over the last 5,000 years of human evolution and a prognosis for possible future levels of insight.

Year	Distinguished Categories (Information Bits)
Past	
3000 B.C.	10^0–10^2
1800 A.D.	10^1–10^3
1850 A.D.	10^2–10^3
1900 A.D.	10^3
1950 A.D.	10^4
Future*	
2250 A.D.	10^{10}
2700 A.D.	10^{19}
3150 A.D.	10^{28}

*The predicted figures are dependent upon the development speed shown during the period 1800–1950 remaining constant.

19

REFERENCES

Becquerel, E. 1868. Chemical changes by friction, p. 16 in *Traite de l'Electricite*, Vol. 5 (cited after Youmans, E. L., ed. 1881. *The Correlation and Conservation of Natural Forces: A Series of Expositions by Prof. Grove, Prof. Helmholtz, Dr. Mayer, Dr. Faraday, Prof. Liebig, and Dr. Carpenter.* New York: Appleton).

Berthelot, C. L., and M. Vieille. 1885. Nouvelle méthode pour mesurer la chaleur de combustion du charbon et des composés organiques. *Ann. Chim. (Phys.)* 6:546-556.

Bowen, H. J. M. 1966. *Trace Elements in Biochemistry.* London: Academic Press.

Favre, P. A., and J. T. Silberman. 1852. Recherches sur les quantités de chaleur degagées dans les action chimiques et moléculaires. *Ann. Chim. (Phys.)* 34: 357-450.

Geldner, K. F. 1951. *Der Rig-Veda, aus dem Sanskrit in Deutsche übersetzt,* Volume 35, Harvard Oriental Series. Cambridge, Mass.: Harvard University Press.

Gibbs, F. W. 1967. *Joseph Priestley: Revolutions of the Eighteenth Century.* Garden City, N.Y.: Doubleday.

Green, J. R. 1909. *A History of Botany (1860-1900), Being a Continuation of Sachs's "History of Botany (1530-1860)."* Oxford: Clarendon Press.

Harvey-Gibson, R. J. 1919. *Outlines of the History of Botany.* London: A. & C. Black Ltd.

Haubst, R. (ed.) 1964. *Das Cusanus Jubiläum. Mitteilungen und Forschungsbeiträge der Cusanus Gesellschaft 4.* Mainz: Matthias Grunewald Verlag.

Helmholtz, H. L. F. 1854/1881. Interaction of natural forces, translated by John Tyndall, pp. 208-247 in *The Correlation and Conservation of Forces.* E. L. Youmans, ed., New York: Appleton.

Jessen, K. F. W. 1864. *Botanik der Gegenwart und Vorzeit in culturhistorischer Entwicklung. Ein Beitrage zur Geschichte der abendländischen Völker.* Waltham, Mass.: The Chronica Botanica Co., 1948.

Klibansky, R. (ed.) 1927. Caroli Bovilli, liber de sapiente, pp. 299-412 in *Individuum und Cosmos in der Philosophie der Renaissance,* E. Cassirer, ed. Berlin: B. G. Teubner.

Liebeschütz, H. 1930. *Das allegorische Weltbild der heiligen Hildegard von Bingen.* Leipzig: B. G. Teubner.

Liebig, J. von. 1862. *Der chemische Process der Ernährung der Vegetabilien.* Braunschweig: F. Viehweg & Sohn.

Lieth, H. 1975. Historical survey of primary productivity research, pp. 7-16 in *Primary Productivity of the Biosphere,* H. Lieth and R. H. Whittaker, eds. New York: Springer-Verlag.

Mägdefrau, K. 1973. *Geschichte der Botanik.* Stuttgart: Gustav Fischer Verlag.

Margalef, R. 1956. Informacion y diversidad especifica en las communidades de organismos. *Investigacion pesquera* 3:99-106.

Morowitz, H. I. 1968. *Energy Flow in Biology.* New York: Academic Press.

Mustard, H. M., and C. E. Passarge. 1961. *Parzival by Wolfram von Eschenbach.* New York: Vintage Books.

Patten, B. C. 1959. An introduction to the cybernetics of the ecosystem: The trophic-dynamic aspect. *Ecoloy* 40:221-231.

Rabinovitch, E. 1971. An unfolding discovery. *Proc. Natl. Acad. Sci. (USA)* 68:2875-2876.

Reed, H. S. 1942. *A Short History of the Plant Sciences.* Waltham, Mass.: The Chronica Botanica Co.

Rumford, Count (Benjamin Thompson) 1797. *Essays Political, Economical, and Philosophical*, Vol. 2. London: T. Cadell, Jr. and W. Davies.

Sachs, J. von. 1868. *Lehrbuch der Botanik*. I. Auflage. Leipzig: Engelmann Verlag.

———. 1882. *Textbook of Botany*. Edited with an appendix by S. H. Vines, 2nd ed. New York: Macmillan.

———. 1890. *History of Botany (1530–1860)*. Oxford: Clarendon Press.

Shenstone, W. A. 1895. *Justus von Liebig, His Life and Work 1803–1874*. New York: Macmillan.

Solmsen, F. 1960. *Aristotle's System of the Physical World*. Ithaca, N.Y.: Cornell University Press.

Thompson, R. C. 1949. *A Dictionary of Assyrian Botany*. London: The British Academy.

Transeau, E. N. 1926. The accumulation of energy by plants. *Ohio J. Sci.* **26**:1–10.

Williams, H. S. 1904. *A History of Science*, 5 Volumes. New York and London: Harper Brothers.

Wolff, E. 1862. *Vegetationsversuche in wässrigen Lösungen ihrer Nährstoffe. Hohenheimer Jubiläumsschrift*. Stuttgart: Hohenheim.

Youmans, E. L. (ed.) 1881. *The Correlation and Conservation of Forces: A Series of Expositions by Prof. Grove, Prof. Helmholtz, Dr. Mayer, Dr. Faraday, Prof. Liebig, and Dr. Carpenter*. New York: Appleton.

ADDITIONAL READINGS

Carnot, S. 1824. *Reflexions sur la puissance motrice de feu*. Paris: Didot Fréres.

Chardon, C. E. 1953. *Boussingault, Juicio critico del eminente agronomo del sigle XIX su viaje a la gran Colombia. . . sus relaciones con el libertador y Manuelita Saenz*. Cuidad Trujillo, R. D. Editora Montalvo.

Hoffman, E. 1947. *Nikolaus von Cues*. Heidelberg: F. H. Kerle Verlag.

Menzel-Rogner, H. 1967. Der Laie über Versuche mit der Waage, in *Schriften des Nikolaus von Cues*, E. Hoffman, ed. Leipzig: Felix Meiner.

Meuthen, E. 1964. *Nikolaus von Kues*. Muenster: Verlag Aschendorff.

1

Reprinted from page 18 of *A Short History of the Plant Sciences,* by H. S. Reed, Waltham, Mass.: Chronica Botanica Co., 1942, 320pp.

GARDENERS AND HERBALISTS OF ANTIQUITY

H. S. Reed

[*Editor's Note:* In the original, material precedes and follows this excerpt.]

There is a tradition that the Emperor SHEN-NUNG, who reigned about 2700 B.C., is the Father of Agriculture and Medicine. He is supposed to have invented the plow and the plowshare, to have sowed first the five kinds of grain, and to have composed the first treatise on medicinal plants in a work known as Shên Nung pên ts'ao ching, the classical herbal of SHEN-NUNG. The first mention of this work occurs in the bibliographical section of the Chien Han Shu (History of the Former Han dynasty, B.C. 206 - A.D. 24) in which it was said that the herbal of SHEN-NUNG consisted of 20 chapters. It is not known exactly at what time the Shên Nung pên ts'ao ching was first written down, but there can be no doubt that it is one of the most ancient documents of Chinese materia medica. In the centuries which followed, copyists and transcribers sometimes added references to additional plants.

SWINGLE found a valuable Japanese reprint of this ancient Chinese work published about 1625, written by MIU HAI-YUNG, a native of Ch'ang-shu in the Soochow prefecture, entitled Shên Nung pên ts'ao ching su.

2

The original version is reprinted from Rig-Veda-Samhitâ, The Sacred Hymns of the Brâhmans, *2nd ed., Vol. IV, Mandala X, edited by F. M. Müller, London: Henry Frowde, 1892. The German translation is reprinted from page 289 of* Der Rig-Veda, aus dem Sanskrit in Deutsche übersetzt, *by K. F. Geldner, Volume 35, Harvard Oriental Series, Cambridge, Mass.: Harvard University Press, 1951, 422 pp.; copyright © 1951 by the President and Fellows of Harvard College. The English translation was prepared by H. H. Lieth.*

SONG 10-91: AN AGNI

[*Editor's Note:* In the original, material precedes and follows these excerpts.]

तमोषधीर्दधिरे गर्भमृत्वियं तमापो अग्निं जनयंत मातरः ।
तमित्समानं वनिनश्च वीरुधो ऽतर्वतीश्च सुवंते च विश्वहा ॥ ६ ॥
तं । ओषधीः । दधिरे । गर्भे । ऋत्वियं । तं । आपः । अग्निं । जनयंत । मातरः ।
तं । इत् । समानं । वनिनः । च । वीरुधः । अंतःऽवंतीः । च । सुवंते । च । विश्वहा ॥ ६ ॥

6. Ihn empfingen die Pflanzen als rechtzeitige Leibesfrucht; den Agni erzeugten die Gewässer als seine Mütter. Mit ihm gehen gleichmäßig die Bäume und die Gewächse schwanger und gebären ihn allezeit.

VERSE 6

The plants conceived him as timely fetus
The waters as his mothers created Agni
Trees and plants are evenly pregnant with him
and bring him forth at all times.

3

Reprinted from page 1 of *The Holy Scriptures of the Old Testament in Hebrew and English*, Vienna: The British and Foreign Bible Society, 1870, 1384pp.

בראשית
G E N E S I S.
CAPUT I. א

בְּרֵאשִׁית בָּרָא אֱלֹהִים אֵת הַשָּׁמַיִם וְאֵת הָאָרֶץ: וְהָאָרֶץ א 2
הָיְתָה תֹהוּ וָבֹהוּ וְחֹשֶׁךְ עַל־פְּנֵי תְהוֹם וְרוּחַ אֱלֹהִים
מְרַחֶפֶת עַל־פְּנֵי הַמָּיִם: וַיֹּאמֶר אֱלֹהִים יְהִי אוֹר וַיְהִי־ 3
אוֹר: וַיַּרְא אֱלֹהִים אֶת־הָאוֹר כִּי־טוֹב וַיַּבְדֵּל אֱלֹהִים בֵּין 4
הָאוֹר וּבֵין הַחֹשֶׁךְ: וַיִּקְרָא אֱלֹהִים לָאוֹר יוֹם וְלַחֹשֶׁךְ ה
קָרָא לָיְלָה וַיְהִי־עֶרֶב וַיְהִי־בֹקֶר יוֹם אֶחָד:
פ
וַיֹּאמֶר אֱלֹהִים יְהִי רָקִיעַ בְּתוֹךְ הַמָּיִם וִיהִי מַבְדִּיל בֵּין 6
מַיִם לָמָיִם: וַיַּעַשׂ אֱלֹהִים אֶת־הָרָקִיעַ וַיַּבְדֵּל בֵּין הַמַּיִם 7
אֲשֶׁר מִתַּחַת לָרָקִיעַ וּבֵין הַמַּיִם אֲשֶׁר מֵעַל לָרָקִיעַ וַיְהִי־
כֵן: וַיִּקְרָא אֱלֹהִים לָרָקִיעַ שָׁמָיִם וַיְהִי־עֶרֶב וַיְהִי־בֹקֶר 8
יוֹם שֵׁנִי:
פ
וַיֹּאמֶר אֱלֹהִים יִקָּווּ הַמַּיִם מִתַּחַת הַשָּׁמַיִם אֶל־מָקוֹם אֶחָד 9
וְתֵרָאֶה הַיַּבָּשָׁה וַיְהִי־כֵן: וַיִּקְרָא אֱלֹהִים לַיַּבָּשָׁה אֶרֶץ י
וּלְמִקְוֵה הַמַּיִם קָרָא יַמִּים וַיַּרְא אֱלֹהִים כִּי־טוֹב: וַיֹּאמֶר 11
אֱלֹהִים תַּדְשֵׁא הָאָרֶץ דֶּשֶׁא עֵשֶׂב מַזְרִיעַ זֶרַע עֵץ פְּרִי
עֹשֶׂה פְּרִי לְמִינוֹ אֲשֶׁר זַרְעוֹ־בוֹ עַל־הָאָרֶץ וַיְהִי־כֵן:
וַתּוֹצֵא הָאָרֶץ דֶּשֶׁא עֵשֶׂב מַזְרִיעַ זֶרַע לְמִינֵהוּ וְעֵץ עֹשֶׂה 12
פְּרִי אֲשֶׁר זַרְעוֹ־בוֹ לְמִינֵהוּ וַיַּרְא אֱלֹהִים כִּי־טוֹב: וַיְהִי־ 13
עֶרֶב וַיְהִי־בֹקֶר יוֹם שְׁלִישִׁי:
פ
וַיֹּאמֶר אֱלֹהִים יְהִי מְאֹרֹת בִּרְקִיעַ הַשָּׁמַיִם לְהַבְדִּיל בֵּין 14
הַיּוֹם וּבֵין הַלָּיְלָה וְהָיוּ לְאֹתֹת וּלְמוֹעֲדִים וּלְיָמִים וְשָׁנִים:
והיו

THE FIRST BOOK OF MOSES,
CALLED
GENESIS.

CHAPTER I.

IN the beginning God created the heaven and the earth.

2 And the earth was without form, and void; and darkness *was* upon the face of the deep. And the Spirit of God moved upon the face of the waters.

3 And God said, Let there be light: and there was light.

4 And God saw the light, that *it was* good: and God divided the light from the darkness.

5 And God called the light Day, and the darkness he called Night. And the evening and the morning were the first day.

6 ¶ And God said, Let there be a firmament in the midst of the waters, and let it divide the waters from the waters.

7 And God made the firmament, and divided the waters which *were* under the firmament from the waters which *were* above the firmament: and it was so.

8 And God called the firmament Heaven. And the evening and the morning were the second day.

9 ¶ And God said, Let the waters under the heaven be gathered together unto one place, and let the dry *land* appear: and it was so.

10 And God called the dry *land* Earth; and the gathering together of the waters called he Seas: and God saw that *it was* good.

11 And God said, Let the earth bring forth grass, the herb yielding seed, *and* the fruit tree yielding fruit after his kind, whose seed *is* in itself, upon the earth: and it was so.

12 And the earth brought forth grass, *and* herb yielding seed after his kind, and the tree yielding fruit, whose seed *was* in itself, after his kind: and God saw that *it was* good.

13 And the evening and the morning were the third day.

14 ¶ And God said, Let there be lights in the firmament of the heaven to divide the day from the night; and let them be for signs, and for seasons, and for days, and years:

[*Editor's Note:* Material has been omitted at this point.]

24

4

The original version is reprinted from the facsimile reproduction of The Edda *by Snorri Sturluson (circa 1200 A.D.) published as* Codex Wormianus (The Younger Edda), *edited by S. Nordal, Copenhagen: Levin and Munksgard, 1931. The transliterated version is reprinted from page 13 of* Edda Snorra Sturlusonar, Codex Wormianus AM 242, Fol., *edited by The Commission for the Arnamagnaeaske Legat: Finnur Jónsson, J. C. H. R. Steenstrup, S. Larsen, K. Erslev, and V. Dahlrup, Copenhagen: Gyldendalske Boghandel, 1924, 122pp. The English translation was prepared by H. H. Lieth.*

THE CREATION OF THE WORLD

Snorri Sturluson

[*Editor's Note:* In the originals of both of these versions, material precedes and follows these excerpts.]

Or Ymis holldi uar iorð of skoput
enn or sueita siar biorg or beínum baðmr or haarí enn or
hausi hímínn. Enn or hans brám giorðu blid regín Midgarð
manna sonum. enn or hans heila uoru þau enn hardmoðgu
sky ǫll of skopuð).

Of ymir's flesh	From the giant's eyelashes
Was created the earth	Friendly gods
And from blood the sea	Created Midgard
Mountains from bones	For the sons of men
Tree from hair	From the colosses brain
And skys from skull.	Created were clouds . . .

5

GREEK BOTANICAL PHILOSOPHY

K. F. W. Jessen

*These excerpts have been translated expressly for this Benchmark
volume by H. H. Lieth from pages 18–20 of* Botanik der Gegenwart
und Vorzeit in culturhistorischer Entwicklung. Ein Beitrag zur
Geschichte der abendländischen Völker, *by Karl F. W. Jessen,
Waltham, Mass.: The Chronica Botanica Co., 1948, 495pp.
(originally published in 1864 by F. A. Brockhaus, Leipzig).*

. . . Plants find their nourishment in the earth, and since they are sta-
tionary and do not walk about, they do not need to carry their food around
inside their bodies like the animals. They do not have a stomach, but just
as the vessels of animals take up the nourishing juice from the stomach, so
the roots of plants take their nourishing juice from the earth. The leaves
serve as cover for the fruits, the blossoms as ornaments, like the wedding
songs of birds, as a sign of beginning fertility. . . .

. . . About the chemical composition of bodies, Aristotle has his own
opinions differing from those of his predecessors. They deserve all the more
attention, since they are the basis of all chemistry up until the discoveries
of the last centuries. He speaks often and much of the four elements, earth,
water, air, fire, and lets everything be composed of those. But for him the
term "element" does not have the same meaning that the chemical "element" has
to us. His four elements are several conditions of the same material, or
rather of different combinations of four fundamental principles or basic matters, in that every element does represent a basic matter, but is not pure.
Rather it contains a small portion of another (element). Thus the elements
consist of the following principles: fire consists of heat and some dryness;
air of moisture and some heat; water of cold and some moisture; earth of dryness and some cold. All these beliefs, as is easily agreed, are taken directly
from the observation of the effect of the matters on the human body, which at
that time was the only testing and measuring instrument used.
As incomplete [as] these beliefs of Aristotle may be, they do seem to be
important progress when compared to Plato's concepts. He, too, distinguished
four elements, but he viewed them as chemically different matters and explained
their difference. He attributed to them atoms of different compositions and
various sizes. With his closer observations Aristotle soon was convinced that
such views could not persist; for example, water sometimes changed into steam,
that is into air, sometimes into a hard lump of ice, that is into earth, whereby the forms of the atoms had to change. Aristotle explained such transitions
and mixtures of elements in that he connects to each of the same not a pure
main principle alone, but an addition of another principle. He thinks that
the natural bodies are composed of these principles and their elementary form.
Animals contain more water and need more heat. Plants, being the coldest and
dryest, consist mainly of soil. Just as soil, kneaded together with water,

becomes a coherent mass, so all organic bodies are formed mainly from soil and water, and soil particles and water serve as their nourishment. These matters are first of all changed into plant substance through contact with air and fire (heat), of which all organic bodies also possess a portion. Summarizing these views, one can present them best in the following manner: Aristotle has four basic matters with different compositions and effects; each two produce one element, like two basic chemical matters produce an acid or a base, not as a mechanical mixture, but as a chemical binding of the basic matters. In the organic bodies elements combine themselves to salts. In this way Aristotle has, on the one hand, introduced chemical concepts into natural science. On the other hand, he also laid the foundation for the later enthusiastically followed teaching of the equal composition and interchangeability of mineral and organic bodies.

6

HOMILY 5

St. Basil

[*Editor's Note:* In the original, material precedes and follows this excerpt.]

(3) 'Let the earth bring forth vegetation: the plant producing seed of its own kind and likeness.'[4] Even to the present
day, the order in plants testifies to the first orderly arrangement. For, germination is the beginning of every herb and
every plant. If something is produced from the root, coming
out of the protuberance below, like the crocus or dog's-tooth
grass, it must germinate and emerge; or, if it is produced from
seed, even so it is necessary that there be first germination,
then a seedling, then green foliage, and finally the fruit swell-
ing on the stalk, which up to this time was dry and thick.
'Let the earth bring forth vegetation.' Whenever the seed falls
on ground which contains moisture and heat in moderation,
becoming spongy and very porous, it grasps the surrounding
soil and draws to itself all that is proper and suitable for it.
The very light particles of earth falling in and slipping around
it in the pores expand its bulk even more, so that it sends roots
downward and also thrusts shoots upward, producing stalks

---
4 Gen. 1.11 (Septuagint version).

28

equal in number to the roots; and, since the shoot is always
being warmed, moisture, drawn through the roots by the
attraction of the heat, brings from the earth the proper amount
of nourishment, and distributes this to the stalk and the bark,
to the husks and the grain itself, as well as to the beards. Thus,
as the gradual increase continues, each of the plants reaches
its natural proportions, whether it happens to be one of the
grains or legumes or vegetables or shrubs. One grass, even one
blade of grass is sufficient to occupy all your intelligence completely in the consideration of the art which produced it—
how it is that the stalk of wheat is encircled with nodes, so that
they, like some bonds, may bear easily the weight of the ears,
when, full of fruit, they bend down to the earth. For this very
reason, the oat stalk is completely devoid of these, inasmuch
as its head is not made heavy by any weight. But, nature has
strengthened the wheat with these bonds, placing the grain
in a sheath so as not to be easily snatched by grain-picking
birds; and besides, it keeps off any harm from small insects
by the projecting barrier of the needlelike beards.

Copyright © 1968 by the American Institute of Biological Sciences

Reprinted from page 288 of *BioScience* 18(4):286–292 (1968)

WATER SOLUTES IN PLANT NUTRITION: WITH SPECIAL REFERENCE TO VAN HELMONT AND NICHOLAS OF CUSA

A. D. Krikorian and F. C. Steward

[*Editor's Note:* In the original, material precedes and follows this excerpt.]

From the beginnings recorded in *De Staticis Experimentis* derive the ideas which led John Dee (1570) to praise Nicholas of Cusa's quantitative use of the balance and "Archemastrie" or *Experimentall Science* (Euclides, 1651 sig. N1b) and which prompted Santorio Santorio to write in 1614 his *De Statica Medicina* (Santorio, 1703, 1720). In the same tradition, Stephen Hales later wrote his *Vegetable Staticks* (Hales, 1727), later to form a part of his two-volume *Statical Essays* (Hales, 1738) credibly regarded as the first English treatise on plant physiology.

The crucial passage which bears upon the extent to which Nicholas of Cusa anticipated van Helmont is given by the "Idiot" in response to a question posed by the "Orator." This will first be given from the 1650 edition (Nicolaus Cusanus, 1650).

Orator: There is a saying, that no pure element is to be given, how is this prov'd by the Ballance?

Idiot: If a man should put an hundred weight of earth into a great earthen pot, and then should take some Herbs, and Seeds, and weigh them, and then plant or sow them in that pot, and then should let them grow there so long, untili hee had successively by little and little, gotten an hundred weight of them, hee would finde the earth but very little diminished, when he came to weigh it againe: by which he might gather, that all the aforesaid herbs, had their weigh from the water. Therefore the waters being ingrossed (or impregnated) in the earth, attracted a terrestreity, and by the opperation of the Sunne, upon the Herb were condensed (or were condensed into an Herb.) If those Herbs bee then burn't to ashes, mayest not thou guesse by the diversity of the

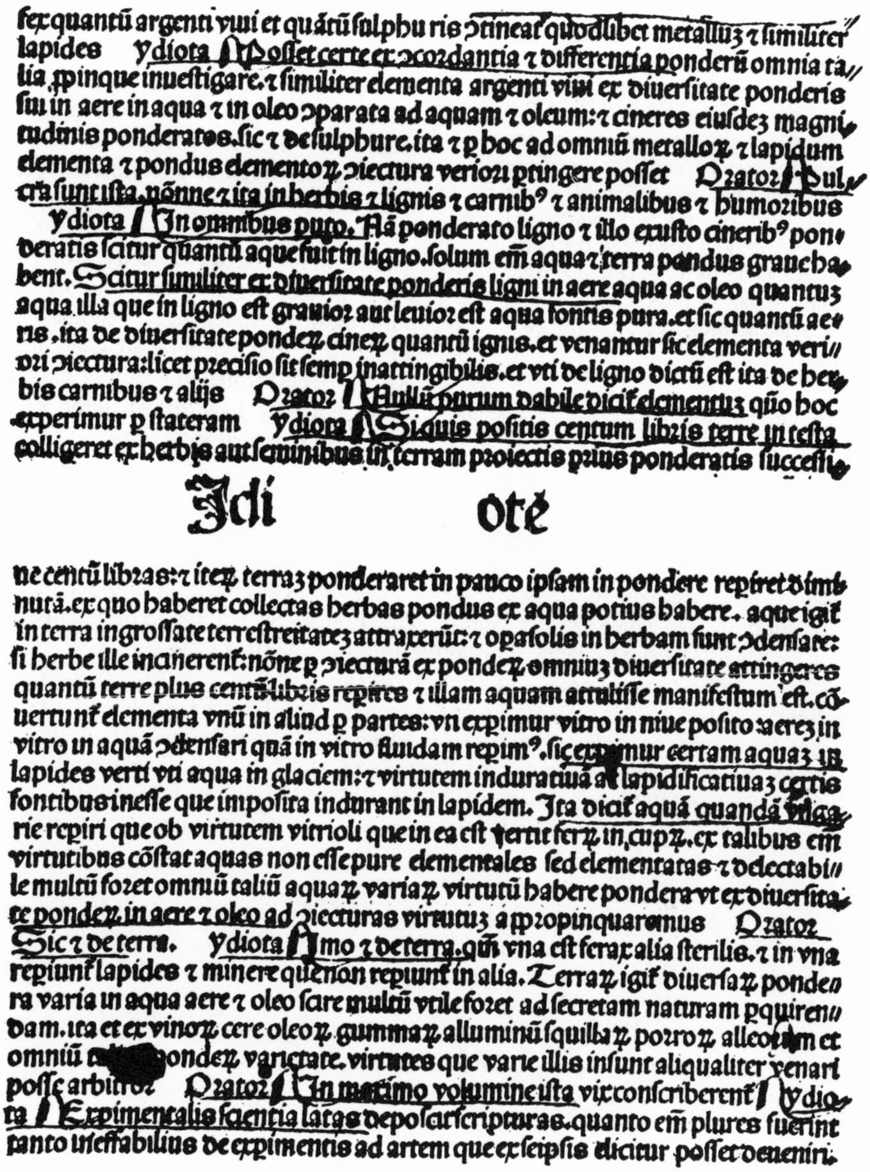

Fig. 2. Dialogue between the Idiot and the Orator about the role of water and solutes in plant nutrition. Photographic reproduction of portions of two pages from *Ydiote de staticis experimentis* in Cusa's Opera Omnia, namely 16 lines at the bottom of one page (sig. riia) and 21 lines from the top of the next (sig. riib) here shown as one figure. Reproduction by kind permission of the Trustees of the British Museum.

weights of all; How much earth though foundest more than the hundred weight, and then conclude that the water brought all that? For the Elements are convertible one into another by parts, as wee finde by a glass put into the snow, where wee shall see the aire condensed into water, and flowing in the glass. So we finde by in experience, that some water is turned into stones, as some is into Ice; . . . (p. 188-189)

The first point is clear. The essence of the van Helmont experiment was indeed stated by Nicholas of Cusa 150 years earlier, namely, that water is transmutable into the stuff of plants. Furthermore, the notion that the balance might record these changes is clearly implied. Although no experimental data were cited, it seems hardly credible that Nicholas of Cusa could have been so specific unless he was already aware of some actual evidence to support his quite positive statements. Another alternative is that Nicholas of Cusa may not have performed the actual experiments but may have drawn upon even earlier sources that may have been lost.

REFERENCES

Cusanus, N., Cardinal. 1650. *The Idiot in Four Books*. The First and Second of Wisdome. The Third of the Minde. The Fourth of Statick Experiments, or Experiments of the Ballance. London. Printed for William Leake.

Dee, John. 1570. (no reference is given in the original article)

Euclides. 1651. Euclides's *Elements of Geometry:* The First VI Books: In a compendious form contracted and demonstrated by Captain Thomas Rudd. Whereunto is added, The Mathematicall Preface of Mr. John Dee. London. Printed by Robert and William Leybourn for Richard Tomlins and Robert Boydell.

Hales, Stephen. 1727. *Vegetable Staticks:* or, An account of some statical experiments on the sap in vegetables: being an essay towards a natural history of vegetation . . . London. W. and J. Innys. 376p.

. 1738. *Statical essays:* containing Vegetable Staticks, etc. Also, Haemastaticks, or an account of hydraulic and hydrostatical experiments made on the Blood and Blood-vessels of Animals, etc. The third edition. 2 vol. London. [1]738.

Santorio, Santorio. 1703. *De Statica Medicina Aphorismorum Sectiones Septem cum Commentario Martini Lister.* 231p. Lugduni Batavorum, apud Cornelium Boutesteyn.

8

HISTORICAL SURVEY OF PRIMARY PRODUCTIVITY RESEARCH

Helmut H. Lieth

[*Editor's Note:* In the original, material precedes and follows this excerpt.]

From Aristotle to Liebig

384–322 B.C.   Aristotle taught that soil, in a manner comparable to that of the intestinal tract of animals, provides predigested food for the plants to take up through their roots. Thus he rightly emphasized the relationship between plant and soil while wrongly interpreting plant nutrition with an idea that was held generally for 1800 years.

1450 A.D.   Nicolai de Cusa expressed the almost revolutionary idea that "the water thickens within the soil, sucks off soil substances and becomes then condensed to herb by the action of the sun."

A reading of the entire paper "Ydiote de staticis experimentis" (the Ydiote here meant is layman, most likely a practitioner with high technical skill) in Nicolaus de Cusa (Cusanus) *Werke* (1967) gives the impression that the "agricultural engineers" of his time held this plant–water relationship as a general consensus. Nicolai's view emphasized this relationship between plant and water. This paper appears to be the design for van Helmont's experiment about 150 years later.

ca. 1600   van Helmont, besides performing odd experiments to find meth-
(1577–1644)   ods of obtaining mice from junk and sawdust, did one rather intelligent experiment. He grew a willow twig weighing 5 lb in a large clay pot containing 300 lb of soil, and irrigated it with rainwater. After 5 years, he harvested a willow tree of 164 lb of wood with a loss of only 2 oz of soil. van Helmont concluded from this that water was condensed to form plants.

1772–1777   Priestley, Scheele, and Ingenhousz were the first to discuss the
or 1779   interaction between plants and air. They spoke about "melioration" and the "spoiling" of the air by plants in light or darkness.

1804   de Saussure studied the gas exchange of plants and gave the correct equation for photosynthesis:

$$\text{Carbon dioxide} + \text{water} = \text{plant matter} + \text{oxygen}$$

Following Rabinovitch's (1971) manner of indicating persons whose work led up to the primary production equation (not the photosynthetic equation), we have added the names of those who were instrumental in first evaluating the importance or necessity or both of each of the elements. Entries from Rabinovitch are in parentheses; our entries are in brackets [ ].

$$CO_2 \quad + \quad H_2O \quad + \quad \text{light} \quad + \quad \begin{array}{c}\text{inorganic}\\\text{nutrients}\end{array}$$

(Senebier)   *(de Saussure)*   *(Ingenhousz)*

[*Priestley–Scheele*]   [*van Helmont*]   [*Liebig*]

YIELD

$$= \quad \text{Oxygen} \quad + \quad \text{organic matter} \quad + \quad \text{chemical energy}$$

(Priestley)   *(Ingenhousz)*   *(Mayer)*

[*van Helmont*]   [*Sir Thompson-Helmholtz*]

Following the development of this equation, plant production was subjected to widespread, serious investigation, although not on the scale of present-day studies. The newly founded Colleges of Agriculture and Forestry dealt with various aspects of such questions.

REFERENCES

Cusanus, N. (Nicolai de Cusa). 1967. *Schriften des-Heft 5. Der Laie über Versuche mit der Waage* (Transl. by Hildegund Menzel-Rogner), Leipzig: Meiner Verlag, 85pp.

Rabinovitch, E. 1971. An unfolding discovery. *Proc. Natl. Acad. Sci. US* 68: 2875–2876.

9

Reprinted from page 240 of *The Correlation and Conservation of Natural Forces: A Series of Expositions by Prof. Grove, Prof. Helmholtz, Dr. Mayer, Dr. Faraday, Prof. Liebig, and Dr. Carpenter,* edited by E. L. Youmans, New York: Appleton, 1881, 438pp.

INTERACTION OF NATURAL FORCES

H. L. F. Helmholtz

[*Editor's Note:* In the original, material precedes and follows this excerpt.]

Hence a certain portion of force disappears from the sunlight, while combustible substances are generated and accumulated in plants; and we can assume it as very probable, that the former is the cause of the latter. I must indeed remark, that we are in possession of no experiments from which we might determine whether the *vis viva* of the sun's rays which have disappeared, corresponds to the chemical forces accumulated during the same time; and as long as these experiments are wanting, we cannot regard the stated relation as a certainty. If this view should prove correct, we derive from it the flattering result, that all force, by means of which our bodies live and move, finds its source in the purest sunlight; and hence we are all, in point of nobility, not behind the race of the great monarch of China, who heretofore alone called himself Son of the Sun. But it must also be conceded that our lower fellow-beings, the frog and leech, share the same ethereal origin, as also the whole vegetable world, and even the fuel which comes to us from the ages past, as well as the youngest offspring of the forest with which we heat our stoves and set our machines in motion.

10

Reprinted from pages 10–23 and 36–43 of *Chemistry in Its Applications to Agriculture and Physiology*, edited by L. Playfair, Phila.: Peterson, 1845, 135pp.

ON THE CHEMICAL PROCESSES IN THE NUTRITION OF VEGETABLES

J. von Liebig

[*Editor's Note:* In the original material precedes this excerpt.]

CHAPTER I.

OF THE CONSTITUENT ELEMENTS OF PLANTS.

THE ultimate constituents of plants are those which form organic matter in general, namely, Carbon, Hydrogen, Nitrogen, and Oxygen. These elements are always present in plants, and produce by their union the various proximate principles of which they consist. It is, therefore, necessary to be acquainted with their individual characters, for it is only by a correct appreciation of these that we are enabled to explain the functions which they perform in the vegetable organization.

Carbon is an elementary substance, endowed with a considerable range of affinity. With oxygen it unites in two proportions, forming the gaseous compounds known under the names of carbonic acid and carbonic oxide. The former of these is emitted in immense quantities from many volcanoes and mineral springs, and is a product of the combustion and decay of organic matter. It is subject to be decomposed by various agencies, and its elements then arrange themselves into new combinations. Carbon is familiarly known as *charcoal*, but in this state it is mixed with several earthy bodies; in a state of absolute purity it constitutes the diamond.

Hydrogen is a very important constituent of vegetable matter. It possesses a special affinity for oxygen, with which it unites and forms water. The whole of the phenomena of decay depend upon the exercise of this affinity, and many of the processes engaged in the nutrition of plants originate in the attempt to gratify it. Hydrogen, when in the state of a gas, is very combustible, and the lightest body known; but it is never found in nature in an isolated condition. Water is the most common combination in which it is presented; and it may be removed by various processes from the oxygen, with which it is united in this body.

Nitrogen is quite opposed in its chemical characters to the two bodies now described. Its principal characteristic is an indifference to all other substances, and an apparent reluctance to enter into combination with them. When forced by peculiar circumstances to do so, it seems to remain in the combination by a *vis inertiæ;* and very slight forces effect the disunion of these feeble compounds.

Yet nitrogen is an invariable constituent of plants, and during their life is subject to the control of the vital powers. But when the mysterious principle of life has ceased to exercise its influence, this element resumes its chemical character, and materially assists in promoting the decay of vegetable matter, by escaping from the compounds of which it formed a constituent.

Oxygen, the only remaining constituent of organic matter, is a gaseous element, which plays a most important part in the economy of nature. It is the agent employed in effecting the union and disunion of a vast number of compounds. It is superior to all other elements in the extensive range of its affinities. The phenomena of combustion and decay are examples of the exercise of its power.

Oxygen is the most generally diffused element on the surface of the earth; for, besides constituting the principal part of the atmosphere which surrounds it, it is a component of almost all the earths and minerals found on its surface. In an isolated state it is a gaseous body, possessed of neither taste nor smell. It is slightly soluble in water, and hence is usually found dissolved in rain and snow, as well as in the water of running streams.

Such are the principal characters of the elements which constitute organic matter; but it remains for us to consider in what form they are united in plants.

The substances which constitute the principal mass of every vegetable are compounds of carbon with oxygen and hydrogen, in the proper relative proportions for forming water. Woody fibre, starch, sugar, and gum, for example, are such compounds of carbon with the elements of water. In another class of substances containing carbon as an element, oxygen and hydrogen are again present; but the proportion of oxygen is greater than would be required for producing water by union with the hydrogen. The numerous organic acids met with in plants belong, with few exceptions, to this class.

A third class of vegetable compounds contains carbon and hydrogen, but no oxygen, or less of that element than would be required to convert all the hydrogen into water. These may be regarded as compounds of carbon with the elements of water, and an excess of hydrogen. Such are the volatile and fixed oils, wax, and the resins. Many of them have acid characters.

The juices of all vegetables contain organic acids, generally combined with the inorganic bases, or metallic oxides; for these metallic oxides exist in every plant, and may be detected in its ashes after incineration.

Nitrogen is an element of vegetable albumen and gluten; it is a constituent of the acid, and of what are termed the " indifferent substances" of plants, as well as of those peculiar vegetable compounds which possess all the properties of metallic oxides, and are known as " organic bases."

Estimated by its proportional weight, nitrogen forms only a very small part of plants; but it is never entirely absent from any part of them. Even when it does not absolutely enter into the composition of a particular part or organ, it is always to be found in the fluids which pervade it.

It follows from the facts thus far detailed, that the development of a plant requires the presence, first, of substances containing carbon and nitrogen, and capable of yielding these elements to the growing organism; secondly, of water and its elements; and lastly, of a soil to furnish the inorganic matters which are likewise essential to vegetable life.

OF THE COMPOSITION OF THE ATMOSPHERE.

In the normal state of growth plants can only derive their nourishment from the atmosphere and the soil. Hence it is of importance to be acquainted with the composition of these, in order that we may be enabled to judge from which of their constituents the nourishment is afforded.

The composition of the atmosphere has been examined by many chemists with great care, and the result of their researches have shown, that its principal constituents are always present in the same proportion. These are the two gases, oxygen and nitrogen, the general properties of which have been already described. One hundred parts, by weight, of atmospheric air contain 23·1 parts of oxygen, and 76·9 parts of nitrogen; or 100 volumes of air contain nearly 21 volumes of oxygen gas. From the extensive range of affinity which this gas possesses, it is obvious, that were it alone to constitute our atmosphere, and left unchecked to exert its powerful effects, all nature would be one scene of universal destruction. It is on this account that nitrogen is present in the air in so large proportion. It is peculiarly adapted for this purpose, as it does not possess any disposition to unite with oxygen, and exerts no action upon the processes proceeding on the earth. These two gases are intimately mixed, by virtue of a property which all gasses possess in common, of diffusing themselves equally through every part of another gas, with which they are placed in contact.

Although oxygen and nitrogen form the principal constituents of the atmosphere, yet they are not the only substances found in it. Watery vapour and carbonic acid gas materially modify its properties. The for-

mer of these falls upon the earth as rain, and brings with it any soluble matter which it meets in its passage through the air.

Carbonic acid gas is discharged in immense quantities from the active volcanoes of America, and from many of the mineral springs which abound in various parts of Europe; it is also generated during the combustion and decay of organic matter. It is not, therefore, surprising that it should have been detected in every part of the atmosphere in which its presence has been looked for. Saussure found it even in the air on the summit of Mont Blanc, which is covered with perpetual snow, and where it could not be produced by the immediate agency of vegetable matter. Carbonic acid gas performs a most important part in the process of vegetable nutrition, the consideration of which belongs to another part of the work.

Carbonic acid, water, and ammonia (a compound of hydrogen and nitrogen) are the final products of the decay of animal and vegetable matter. In an isolated condition, they usually exist in the gaseous form. Hence, on their formation, they must escape into the atmosphere. But ammonia has not hitherto been enumerated among the constituents of the air, although, according to our view, it can never be absent. The reason of this is, that it exists in extremely minute quantity in the amount of air usually subjected to experiment in chemical analysis; it has consequently escaped detection. But rain which falls through a large extent of air, carries down in solution all that remains in suspension in it. Now ammonia always exists in rainwater, and from this fact we must conclude that it is invariably present in the atmosphere. Nor can we be surprised at its presence when we consider that many volcanoes now in activity emit large quantities of it. This subject will, however, be discussed more fully in another part of the work.

Such are the principal constituents of the atmosphere from which plants derive their nourishment; for although other matters are supposed to exist in it in minute quantity, yet they do not exercise any influence on vegetation, nor has even their presence been satisfactorily demonstrated.

OF SOILS.

A soil may be considered a magazine of inorganic matters, which are prepared by the plant to suit the purposes destined for them in its nutrition. The composition and uses of such substances cannot, however, be studied with advantage, until we have considered the manner in which the organic matter is obtained by plants.

Some virgin soils, such as those of America, contain vegetable matter in large proportion; and as these have been found eminently adapted for the cultivation of most plants, the organic matter contained in them has naturally been recognised as the cause of their fertility. To this matter, the term " vegetable mould" or *humus* has been applied. Indeed, this peculiar substance appears to play such an important part in the phenomena of vegetation, that vegetable physiologists have been induced to ascribe the fertility of every soil to its presence. It is believed by many to be the principal nutriment of plants, and is supposed to be extracted by them from the soil in which they grow. It is itself the product of the decay of vegetable matter, and must, therefore, contain many of the constituents which are found in plants during life. Its action will, therefore, be examined in considering whence these constituents are derived.

CHAPTER II.

OF THE ASSIMILATION OF CARBON.

COMPOSITION OF HUMUS.

THE humus, to which allusion has been made, is described by chemists as a brown substance easily soluble in alkalies, but only slightly so in water, and produced during the decomposition of vegetable matters by the action of acids or alkalies. It has, however, received various names according to the different external characters and chemical properties which it presents. Thus, *ulmin, humic acid, coal of humus*, and *humin*, are names applied to modifications of *humus*. They are obtained by treating peat, woody fibre, soot, or brown coal with alkalies; by decomposing sugar, starch, or sugar of milk by means of acids; or by exposing alkaline solutions of tannic and gallic acids to the action of the air.

The modifications of *humus* which are soluble in alkalies, are called *humic acid;* while those which are insoluble have received the designations of *humin* and *coal of humus*.

The names given to these substances might cause it to be supposed that their composition is identical. But a more erroneous notion could not be entertained; since even sugar, acetic acid, and resin do not differ more widely in the proportions of their constituent elements, than do the various modifications of *humus*.

Humic acid formed by the action of hydrate of potash upon sawdust contains, according to the accurate analysis of Peligot, 72 per cent. of carbon, while the humic acid obtained from turf and brown coal contains, according to Sprengel, only 58 per cent.; that prod ced by the action of dilute sulphuric acid upon sugar, 57 per cent. according to Malaguti; and that, lastly, which is obtained from sugar or from starch, by means of muriatic acid, according to the analysis of Stein, 64 per cent. All these analyses have been repeated with care and accuracy,

and the proportion of carbon in the respective cases has been found to agree with the estimates of the different chemists above mentioned; so that there is no reason to ascribe the difference in this respect between the varieties of *humus* to the mere difference in the methods of analysis or degrees of expertness of the operators. Malaguti states, moreover, that *humic acid* contains an equal number of equivalents of oxygen and hydrogen, that is to say, that these elements exist in it in the proportions for forming water; while, according to Sprengel, the oxygen is in excess, and Peligot even estimates the quantity of oxygen at 14 equivalents, and the hydrogen at only 6 equivalents, making the deficiency of hydrogen as great as 8 equivalents. And although Mulder[*] has very recently explained many of these conflicting results, by showing that there are several kinds of humus and humic acids essentially distinct in their characters, and fixed in their composition, yet he has afforded no proof that the definite compounds obtained by him really exist, as such, in the soil. On the contrary, they appear to have been formed by the action of the potash and ammonia, which he employed in their preparation.

It is quite evident, therefore, that chemists have been in the habit of designating all products of the decomposition of organic bodies which had a brown or brownish black colour, by the names of *humic acid* or *humin*, according as they were soluble or insoluble in alkalies; although in their composition and mode of origin, the substances thus confounded might be in no way allied.

Not the slightest ground exists for the belief that one or other of these artificial products of the decomposition of vegetable matters exists in nature in the form and endowed with the properties of the vegetable constituents of mould; there is not the shadow of a proof that one of them exerts any influence on the growth of plants either in the way of nourishment or otherwise.

Vegetable physiologists have, without any apparent reason, imputed the known properties of the *humus* and *humic acids* of chemists to that constituent of mould which has received the same name, and in this way have been led to their theoretical notions respecting the functions of the latter substance in vegetation.

The opinion that the substance called *humus* is extracted from the soil by the roots of plants, and that the carbon entering into its composition serves in some form or other to nourish their tissues, is considered by many as so firmly established that any new argument in its favour has been deemed unnecessary; the obvious difference in the growth of plants according to the known abundance or scarcity of *humus* in the soil,

---

[*] Bulletin des Scienc. Phys. et Natur. de Neerl. 1840, p. 1—102.

seemed to afford incontestable proof of its correctness.*

Yet, this position, when submitted to a strict examination, is found to be untenable, and it becomes evident from most conclusive proofs that *humus* in the form in which it exists in the soil, does not yield the smallest nourishment to plants.

The adherence to the above incorrect opinion has hitherto rendered it impossible for the true theory of the nutritive process in vegetables to become known, and has thus deprived us of our best guide to a rational practice in agriculture. Any great improvement in that most important of all arts is inconceivable without a deeper and more perfect acquaintance with the substances which nourish plants, and with the sources whence they are derived; and no other cause can be discovered to account for the fluctuating and uncertain state of our knowledge on this subject up to the present time, than that modern physiology has not kept pace with the rapid progress of chemistry.

In the following inquiry we shall suppose the *humus* of vegetable physiologists to be really endowed with the properties recognised by chemists in the brownish black deposits which they obtain by precipitating an alkaline decoction of mould or peat by means of acids, and which they name *humic acid*.

Humic acid, when first precipitated, is a flocculent substance, is soluble in 2500 times its weight of water, and combines with alkalies, lime and magnesia, forming compounds of the same degree of solubility. (Sprengel.)

Vegetable physiologists agree in the supposition that by the aid of water *humus* is rendered capable of being absorbed by the roots of plants. But according to the observation of chemists, humic acid is soluble only when newly precipitated, and becomes completely insoluble when dried in the air, or when exposed in the moist state to the freezing temperature. (Sprengel.)

Both the cold of winter and the heat of summer, therefore, are destructive of the solubility of humic acid, and at the same time of its capability of being assimilated by plants. So that, if it is absorbed by plants, it must be in some altered form.

The correctness of these observations is easily demonstrated by treating a portion of good mould with cold water. The fluid remains colourless, and is found to have dissolved less than 100,000 part of its weight of organic matters, and to contain merely the salts which are present in rainwater.

Decayed oak wood, likewise, of which humic acid is the principal constituent, was found by Berzelius to yield to cold water

only slight traces of soluble materials; and I have myself verified this observation on the decayed wood of beech and fir.

These facts, which show that humic, in its unaltered condition, cannot serve for the nourishment of plants, have not escaped the notice of physiologists; and hence they have assumed that the lime or the different alkalies found in the ashes of vegetables render soluble the humic acid and fit it for the process of assimilation.

Alkalies and alkaline earths do exist in the different kinds of soil in sufficient quantity to form such soluble compounds with the humic acid.

Now, let us suppose that humic acid is absorbed by plants in the form of that salt which contains the largest proportion of humic acid, namely, in the form of humate of lime, and then from the known quantity of the alkaline bases contained in the ashes of plants, let us calculate the amount of humic acid which might be assimulated in this manner. Let us admit, likewise, that potash, soda, and the oxides of iron and manganese have the same capacity of saturation as lime with respect to humic acid, and then we may take as the basis of our calculation the analysis of M. Berthier, who found that 1000 lbs. of dry fir wood yielded 4 lbs. of ashes, and that in every 100 lbs. of these ashes, after the chloride of potassium and sulphate of potash were extracted, 53 lbs. consisted of the basic metallic oxides, potash, soda, lime, magnesia, iron, and manganese.

One Hessian acre* of woodland yields annually, according to Dr. Heyer, on an average, 2920 lbs. of dry fir wood, which contain 6.17 lbs. of metallic oxides.

Now, according to the estimates of Malaguti and Sprengel, 1 lb. of lime combines chemically with 12 lbs. of humic acid; 6.17 lbs. of the metallic oxides would accordingly introduce into the trees 67 lbs. of humic acid, which, admitting humic acid to contain 58 per cent. of carbon, would correspond to 100 lbs. of dry wood. But we have seen that 2920 lbs. of fir wood are really produced.

Again, if the quantity of humic acid which might be introduced into wheat in the form of humates is calculated from the known proportion of metallic oxides existing in wheat straw, (the sulphates and chlorides also contained in the ashes of the straw not being included, it will be found that the wheat growing on 1 Hessian acre would receive in that way 63 lbs. of humic acid, corresponding to 93.6 lbs. of woody fibre. But the extent of land just mentioned produces, independently of the roots and grain, 1961 lbs. of straw, the composition of which is the same as that of woody fibre.

It has been taken for granted in these cal-

---

* This remark applies more to German than to English botanists and physiologists. In England, the idea that humus, as such, affords nourishment to plants is by no means general; but on the Continent, the views of Berzelius on this subject have been almost universally adopted.—ED.

* One Hessian acre is equal to 40,000 square feet, Hessian, or 26,910 square feet, English measure.

culations that the basic metallic oxides which have served to introduce humic acid into the plants do not return to the soil, since it is certain that they remain fixed in the parts newly formed during the process of growth.

Let us now calculate the quantity of humic acid which plants can receive under the most favourable circumstances, viz. the agency of rainwater.

The quantity of rain which falls at Erfurt, one of the most fertile districts of Germany, during the months of April, May, June, and July, is stated by Schubler to be 19.3 lbs. over every square foot of surface; 1 Hessian acre, or 26,910 square feet, consequently receive 771,000 lbs. of rainwater.

If, now, we suppose that the whole quantity of this rain is taken up by the roots of a summer plant, which ripens four months after it is planted, so that not a pound of this water evaporates except from the leaves of the plant; and if we farther assume that the water thus absorbed is saturated with humate of lime (the most soluble of the humates, and that which contains the largest proportion of humic acid;) then the plants thus nourished would not receive more than 330 lbs. of humic acid, since one part of humate of lime requires 2500 parts of water for solution.

But the extent of land which we have mentioned produces 2843 lbs. of corn (in grain and straw, the roots not included,) or 22,000 lbs. of beet root (without the leaves and small radicle fibres.) It is quite evident that the 330 lbs. of humic acid, supposed to be absorbed, cannot account for the quantity of carbon contained in the roots and leaves alone, even if the supposition were correct, that the whole of the rainwater was absorbed by the plants. But since it is known that only a small portion of the rainwater which falls upon the surface of the earth evaporates through plants, the quantity of carbon which can be conveyed into them in any conceivable manner by means of humic acid must be extremely trifling, in comparison with that actually produced in vegetation.

Other considerations of a higher nature confute the common view respecting the nutritive office of humic acid, in a manner so clear and conclusive that it is difficult to conceive how it could have been so generally adopted.

Fertile land produces carbon in the form of wood, hay, grain, and other kinds of growth, the masses of which differ in a remarkable degree.

2920 lbs. of firs, pines, beeches, &c. grow as wood upon one Hessian acre of forest land with an average soil. The same superfices yields 2755 lbs. of hay.

A similar surface of corn land gives from 19,000 to 22,000 lbs. of beet root, or 881 lbs. of rye, and 1961 lbs. of straw, 160 sheaves of 15.4 lbs. each,—in all, 2843 lbs.

One hundred parts of dry fir wood con-

tain 38 parts of carbon; therefore, 2920 lbs. contain 1109 lbs. of carbon.

One hundred parts of hay,* dried in air, contain 44.31 parts carbon. Accordingly, 2755 lbs. of hay contain 1111 lbs. of carbon.

Beet roots contain from 89 to 89.5 parts water, and from 10.5 to 11 parts solid matter, which consists of from 8 to 9 per cent. sugar, and from 2 to $2\frac{1}{2}$ per cent. cellular tissue. Sugar contains 42.4 per cent; cellular tissue, 47 per cent. of carbon.

22,000 lbs. of beet root, therefore, if they contain 9 per cent. of sugar, and 2 per cent. of cellular tissue, would yield 1032 lbs. of carbon, of which 833 lbs. would be due to the sugar, and 198 lbs. to the cellular tissue; the carbon of the leaves and small roots not being included in the calculation.

One hundred parts of straw,† dried in air contain 38 per cent. of carbon; therefore, 1961 lbs. of straw contain 745 lbs. of carbon. One hundred parts of corn contain 43 parts of carbon; 882 lbs. must, therefore, contain 379 lbs.—in all, 1124 lbs. of carbon.

26,910 square feet of wood and meadow land produce, consequently, 1109 lbs. of carbon; while the same extent of arable land yields in beet root, without leaves, 1032 lbs., or in corn, 1124 lbs.

It must be concluded from these incontestable facts, that equal surfaces of cultivated land of an average fertility produce equal quantities of carbon; yet, how unlike have been the different conditions of the growth of the plants from which this has been deduced!

Let us now inquire whence the grass in a meadow, or the wood in a forest, receives its carbon, since there no manure—no carbon—has been given to it as nourishment? and how it happens, that the soil, thus exhausted, instead of becoming poorer, becomes every year richer in this element?

A certain quantity of carbon is taken every year from the forest or meadow, in the form of wood or hay, and, in spite of this, the quantity of carbon in the soil augments; it becomes richer in humus.

It is said that in fields and orchards all the carbon which may have been taken away as herbs, as straw, as seeds, or as fruit, is replaced by means of manure; and yet this soil produces no more carbon than that of the forest or meadow, where it is never replaced. It cannot be conceived that the laws for the nutrition of plants are changed by culture,—that the sources of

---

* 100 parts of hay, dried at 100° C. (212° F.) and burned with oxide of copper in a stream of oxygen gas, yielded 51.93 water, 165.8 carbonic acid, and 6.82 of ashes. This gives 45.87 carbon, 5.76 hydrogen, 31.55 oxygen, and 6.82 ashes. Hay, dried in the air, loses 11.2 p. c. water at 100° C. (212 F.)—(Dr. Will.)

† Straw analyzed in the same manner, and dried at 100° C., gave 46.37 p. c. of carbon, 5.68 p. c. of hydrogen, 43.93 p. c. of oxygen, and 4.02 p. c. of ashes. Straw dried in the air at 100° C. lost 18 p. c. of water.—Dr. Will.

carbon for fruit or grain, and for grass or trees, are different.

It is not denied that manure exercises an influence upon the development of plants; but it may be affirmed with positive certainty, that it neither serves for the production of the carbon, nor has any influence upon it, because we find that the quantity of carbon produced by manured lands is not greater than that yielded by lands which are not manured. The discussion as to the manner in which manure acts has nothing to do with the present question, which is, the origin of the carbon. The carbon must be derived from other sources; and as the soil does not yield it, it can only be extracted from the atmosphere.

In attempting to explain the origin of carbon in plants, it has never been considered that the question is intimately connected with that of the origin of humus. It is universally admitted that humas arises from the decay of plants. No primitive humus, therefore, can have existed—for plants must have preceded the humus.

Now, whence did the first vegetables derive their carbon? and in what form is the carbon contained in the atmosphere?

These two questions involve the consideration of two most remarkable natural phenomena, which by their reciprocal and uninterrupted influence maintain the life of the individual animals and vegetables, and the continued existence of both kingdoms of organic nature.

One of these questions is connected with the invariable condition of the air with respect to oxygen. One hundred volumes of air have been found, at every period and in every climate, to contain 21 volumes of oxygen, with such small deviations that they must be ascribed to errors of observation.

Although the absolute quantity of oxygen contained in the atmosphere appears very great when represented by numbers, yet it is not inexhaustible. One man consumes by respiration 25 cubic feet of oxygen in 22 hours; 10 cwt. of charcoal consume 32,066 cubic feet of oxygen during its combustion; and a small town, like Giessen, (with about 7000 inhabitants) extracts yearly from the air, by the wood employed as fuel, more than 551 millions of cubic feet of this gas.

When we consider facts such as these, our former statement, that the quantity of oxygen in the atmosphere does not diminish in the course of ages[*]—that the air at the present day, for example, does not contain less oxygen than that found in jars buried for 1800 years in Pompeii—appears quite incomprehensible, unless some source exists whence the oxygen abstracted is replaced. How does it happen, then, that the proportion of oxygen in the atmosphere is thus invariable?

The answer to this question depends upon another; namely, what becomes of the carbonic acid, which is produced during the respiration of animals, and by the process of combustion? A cubic foot of oxygen gas, by uniting with carbon so as to form carbonic acid, does not change its volume. The billions of cubic feet of oxygen extracted from the atmosphere, produce the same number of billions of cubic feet of carbonic acid, which immediately supply its place.

The most exact and most recent experiments of De Saussure, made in every season for a space of three years, have shown, that the air contains on an average 0·000415 of its own volume of carbonic acid gas; so that, allowing for the inaccuracies of the experiments, which must diminish the quantity obtained, the proportion of carbonic acid in the atmosphere may be regarded as nearly equal to 1-1000 part of its weight. The quantity varies according to the seasons; but the yearly average remains continually the same.

We have no reason to believe that this proportion was less in past ages; and nevertheless, the immense masses of carbonic acid which annually flow into the atmosphere from so many causes, ought perceptibly to increase its quantity from year to year. But we find that all earlier observers describe its volume as from one-half to ten times greater than that which it has at the present time; so that we can hence at most conclude that it has diminished.

It is quite evident that the quantities of carbonic acid and oxygen in the atmosphere, which remain unchanged by lapse of time, must stand in some fixed relation to one another; a cause must exist which prevents the increase of carbonic acid by removing that which is constantly forming; and there

Volume of atmosphere = 9,307,500 cubic miles.
= cube of 210·4 miles.
Volume of oxygen = 1,954,578 cubic miles.
= cube of 125 miles.
Vol. of carbonic acid = 3,862·7 cubic miles.
= cube of 15·7 miles.

The maximum of the carbonic acid contained in the atmosphere has not here been adopted, but the mean, which is equal to 0·000415.

A man daily consumes 45,000 cubic inches (Parisian.) A man yearly consumes 9505·2 cubic feet. 100 million men yearly consume 9,505,-200,000,000 cubic feet.

Hence a thousand million men yearly consume 0·79745 cubic miles of oxygen. But the air is rendered incapable of supporting the process of respiration, when the quantity of its oxygen is decreased 12 per cent.; so that a thousand million men would make the air unfit for respiration in a million years. The consumption of oxygen by animals, and by the process of combustion, is not introduced into the calculation.

---

[*] If the atmosphere possessed, in its whole extent, the same density as it does on the surface of the sea, it would have a height of 24,555 Parisian feet; but it contains the vapour of water, so that we may assume its height to be one geographical mile = 22,843 Parisian feet. Now the radius of the earth is equal to 860 geographical miles; hence the

must be some means of replacing the oxygen, which is removed from the air by the processes of combustion and putrefaction, as well as by the respiration of anmials.

Both these causes are united in the process of vegetable life.

The facts which we have stated in the preceding pages prove that the carbon of plants must be derived exclusively from the atmosphere. Now, carbon exists in the atmosphere only in the form of carbonic acid, and therefore, in a state of combination with oxygen.

It has been already mentioned likewise, that carbon and the elements of water form the principal constituents of vegetables; the quantity of the substances which do not possess this composition being in a very small proportion. Now, the relative quantity of oxygen in the whole mass is less than in carbonic acid; for the latter contains two equivalents of oxygen, while one only is required to unite with hydrogen in the proportion to form water. The vegetable products which contain oxygen in larger proportion than this, are, comparatively, few in number; indeed, in many the hydrogen is in great excess. It is obvious, that when the hydrogen of water is assimilated by a plant, the oxygen in combination with it must be liberated, and will afford a quantity of this element sufficient for the wants of the plant. If this be the case, the oxygen contained in the carbonic acid is quite unnecessary in the process of vegetable nutrition, and it will consequently escape into the atmosphere in a gaseous form. It is, therefore, certain, that plants must possess the power of decomposing carbonic acid, since they appropriate its carbon for their own use. The formation of their principal component substances must necessarily be attended with the separation of the carbon of the carbonic acid from the oxygen, which must be returned to the atmosphere, while the carbon enters into combination with water or its elements. The atmosphere must thus receive a volume of oxygen for every volume of carbonic acid which has been decomposed.

This remarkable property of plants has been demonstrated in the most certain manner, and it is in the power of every person to convince himself of its existence. The leaves and other green parts of a plant absorb carbonic acid, and emit an equal volume of oxygen. They possess this property quite independently of the plant; for if, after being separated from the stem, they are placed in water containing carbonic acid, and exposed in that condition to the sun's light, the carbonic acid is, after a time, found to have disappeared entirely from the water. If the experiment is conducted under a glass receiver filled with water, the oxygen emitted from the plant may be collected and examined. When no more oxygen gas is evolved, it is a sign that all the dissolved carbonic acid is decomposed; but

the operation recommences if a new portion of it is added.

Plants do not emit gas when placed in water which either is free from carbonic acid, or contains an alkali that protects it from assimilation.

These observations were first made by Priestly and Sennebier. The excellent experiments of De Saussure have farther shown, that plants increase in weight during the decomposition of carbonic acid and separation of oxygen. This increase in weight is greater than can be accounted for by the quantity of carbon assimilated; a fact which confirms the view, that the elements of water are assimilated at the same time.

The life of plants is closely connected with that of animals, in a most simple manner, and for a wise and sublime purpose.

The presence of a rich and luxuriant vegetation may be conceived without the concurrence of animal life, but the existence of animals is undoubtedly dependent upon the life and development of plants.

Plants not only afford the means of nutrition for the growth and continuance of animal organization, but they likewise furnish that which is essential for the support of the important vital process of respiration; for, besides separating all noxious matters from the atmosphere, they are an inexhaustible source of pure oxygen, which supplies the loss which the air is constantly sustaining. Animals on the other hand expire carbon, which plants inspire; and thus the composition of the medium in which both exist, namely, the atmosphere, is maintained constantly unchanged.

It may be asked—is the quantity of carbonic acid in the atmosphere, which scarcely amounts to 1-10th per cent., sufficient for the wants of the whole vegetation on the surface of the earth,—is it possible that the carbon of plants has its origin from the air alone? This question is very easily answered. It is known, that a column of air of 2441 lbs. weight rests upon every square Hessian foot (=0.567 square foot English) of the surface of the earth; the diameter of the earth and its superficies are likewise known, so that the weight of the atmosphere can be calculated with the greatest exactness. The thousandth part of this is carbonic acid, which contains upwards of 27 per cent. carbon. By this calculation it can be shown, that the atmosphere contains 3306 billion lbs. of carbon; a quantity which amounts to more than the weight of all the plants, and of all the strata of mineral and brown coal, which exist upon the earth. This carbon is, therefore, more than adequate to all the purposes for which it is required. The quantity of carbon contained in seawater is proportionally still greater.

If, for the sake of argument, we suppose the superficies of the leaves and other green parts of plants, by which the absorption of carbonic acid is effected, to be double that of the soil upon which they grow, a supposition which is much under the truth in the case of woods, meadows, and corn fields; and if we farther suppose that carbonic acid equal to 0.00067 of the volume of the air, or 1-1000th of its weight is abstracted from it during every second of time, for eight hours daily, by a field of 53,814 square feet (= 2 Hessian acres;) then those leaves would receive 1102 lbs. of carbon in two hundred days.*

But it is inconceivable, that the functions of the organs of a plant can cease for any one moment during its life. The roots and other parts of it, which possess the same power, absorb constantly water and carbonic acid. This power is independent of solar light. During the day, when the plants are in the shade, and during the night, carbonic acid is accumulated in all parts of their structure; and the assimilation of the carbon and the exhalation of oxygen commence from the instant that the rays of the sun strike them. As soon as a young plant breaks through the surface of the ground, it begins to acquire colour from the top downwards; and the true formation of woody tissue commences at the same time.

The proper, constant, and inexhaustible sources of oxygen gas are the tropics and warm climates, where a sky, seldom clouded, permits the glowing rays of the sun to shine upon an immeasurably luxuriant vegetation. The temperate and cold zones, where artificial warmth must replace deficient heat of the sun, produce, on the contrary, carbonic acid in superabundance, which is expended in the nutrition of the tropical plants. The same stream of air, which moves by the revolution of the earth from the equator to the poles, brings to us in its passage from the equator, the oxygen generated there, and carries away the carbonic acid formed during our winter.

The experiments of De Saussure have

* The quantity of carbonic acid which can be extracted from the air in a given time, is shown by the following calculation. During the whitewashing of a small chamber, the superficies of the walls and roof of which we will suppose to be 105 square metres, and which receives six coats of lime in four days, carbonic acid is abstracted from the air, and the lime is consequently converted, on the surface, into a carbonate. It has been accurately determined that one square decimetre receives in this way, a coating of carbonate of lime which weighs 0.732 grammes. Upon the 105 square metres already mentioned there must accordingly be formed 7686 grammes of carbonate of lime, which contain 4325.6 grammes of carbonic acid. The weight of one cubic decimetre of carbonic acid being calculated at two grammes. (more accurately 1.97978.) the above mentioned surface must absorb in four days 2.163 cubic metres of carbonic acid. 2500 square metres (one Hessian acre) would absorb, under a similar treatment, 51½ cubic metres= 1818 cubic feet of carbonic acid in four days. In 200 days it would absorb 2575 cubic metres= 904.401 cubic feet, which contain 11,353 lbs. of carbonic acid, of which 3304 lbs. are carbon, a quantity three times as great as that which is assimilated by the leaves and roots growing upon the same space.

proved, that the upper strata of the air contain more carbonic acid than the lower, which are in contact with plants; and that the quantity is greater by night than by day, when it undergoes decomposition.

Plants thus improve the air by the removal of carbonic acid, and by the renewal of oxygen, which is immediately applied to the use of man and animals. The horizontal currents of the atmosphere bring with them as much as they carry away, and the interchange of air between the upper and lower strata, which their difference of temperature causes, is extremely trifling when compared with the horizontal movements of the winds. Thus vegetable culture heightens the healthy state of a country, and a previously healthy country would be rendered quite uninhabitable by the cessation of all cultivation.

The various layers of wood and mineral coal, as well as peat, form the remains of a primeval vegetation. The carbon which they contain must have been originally in the atmosphere as carbonic acid in which form it was assimilated by the plants which constitute these formations. It follows from this, that the atmosphere must be richer in oxygen at the present time than in former periods of the earth's history. The increase must be exactly proportional to the quantity of carbon and hydrogen contained in these carboniferous deposits. Thus, during the formation of 353 cubic feet of Newcastle splint coal, the atmosphere must have received 643 cubic feet of oxygen produced from the carbonic acid assimilated, and also 158 cubic feet of the same gas resulting from the decomposition of water. In former ages, therefore, the atmosphere must have contained less oxygen, but a much larger proportion of carbonic acid, than it does at the present time, a circumstance which accounts for the richness and luxuriance of the earlier vegetation.

But a certain period must have arrived in which the quantity of carbonic acid contained in the air experienced neither increase nor diminution in any appreciable quantity. For if it received an additional quantity to its usual proportion, an increased vegetation would be the natural consequence, and the excess would thus be speedily removed. And, on the other hand, if the gas was less than the normal quantity, the progress of vegetation would be retarded, and the proportion would soon attain its proper standard.

The most important function in the life of plants, or, in other words, in their assimilation of carbon, is the separation, we might almost say the generation, of oxygen. No matter can be considered as nutritious, or as necessary to the growth of plants, which possesses a composition either similar to or identical with theirs, and the assimilation of which, therefore, could take place without exercising this function. The reverse is the case in the nutrition of animals. Hence such substances as sugar, starch, and gum, which are themselves products of plants, cannot be adopted for assimilation. And this is rendered certain by the experiments of vegetable physiologists, who have shown that aqueous solutions of these bodies are imbibed by the roots of plants, and carried to all parts of their structure, but are not assimilated, they cannot, therefore, be employed in their nutrition. We could scarcely conceive a form more convenient for assimilation than that of gum, starch, and sugar, for they all contain the elements of woody fibre, and nearly in the same proportions.

In the second part of the work we shall adduce satisfactory proofs that decayed woody fibre (*humus*) contains carbon and the elements of water, without an excess of oxygen; its composition differing from that of woody fibre in its being richer in carbon.

Misled by this simplicity in its constitution, physiologists found no difficulty in discovering the mode of the formation of woody fibre; for they say,* humus has only to enter into combination with water, in order to effect the formation of woody fibre, and other substances similarly composed, such as sugar, starch, and gum. But they forget that their own experiments have sufficiently demonstrated the inaptitude of these substances for assimilation.

All the erroneous opinions concerning the *modus operandi* of humus have their origin in the false notions entertained respecting the most important vital functions of plants; analogy, that fertile source of error, having, unfortunately, led to the very unapt comparison of the vital functions of plants with those of animals.

Substances, such as sugar, starch, &c., which contain carbon and the elements of water, are products of the life of plants which live only while they generate them. The same may be said of humus, for it can be formed in plants like the former substances. Smithson, Jameson, and Thomson, found that the black excretions of unhealthy elms, oaks, and horse chesnuts, consisted of humic acid in combination with alkalies. Berzelius detected similar products in the bark of most trees. Now, can it be supposed that the diseased organs of a plant possess the power of generating the matter to which its substance and vigour are ascribed?

How does it happen, it may be asked, that the absorption of carbon from the atmosphere by plants is doubted by all botanists and vegetable physiologists, and that by the greater number the purification of the air by means of them is wholly denied?

The action of plants on the air in the absence of light, that is during night, has been much misconceived by botanists, and from this we may trace most of the errors which abound in the greater part of their writings. The experiments of Ingenhouss

* Meyen, *Pflanzenphysiologie*, II. S. 141.

were in a great degree the cause of this uncertainty of opinion regarding the influence of plants in purifying the air. His observation that green plants emit carbonic acid in the dark, led De Saussure and Grischow to new investigations, by which they ascertained that under such conditions plants do really absorb oxygen and emit carbonic acid; but that the whole volume of air undergoes diminution at the same time. From the latter fact it follows, that the quantity of oxygen gas absorbed is greater than the volume of carbonic acid separated; for, if this were not the case, no diminution could occur. These facts cannot be doubted, but the views based on them have been so false, that nothing, except the total want of observation and the utmost ignorance of chemical relations of plants to the atmosphere, can account for their adoption.

It is known that nitrogen, hydrogen, and a number of other gases, exercise a peculiar, and in general, an injurious influence upon living plants. Is it, then, probable, that oxygen, one of the most energetic agents in nature, should remain without influence on plants when one of their peculiar processes of assimilation has ceased?

It is true that the decomposition of carbonic acid is arrested by absence of light. But then, namely, at night, a true chemical process commences, in consequence of the action of the oxygen in the air, upon organic substances composing the leaves, blossoms, and fruit. This process is not at all connected with the life of the vegetable organism, because it goes on in a dead plant exactly as in a living one.

The substances composing the leaves of different plants being known, it is a matter of the greatest ease and certainty to calculate which of them, during life, should absorb most oxygen by chemical action when the influence of light is withdrawn.

The leaves and green parts of all plants containing volatile oils or volatile constituents in general, which change into resin by the absorption of oxygen, should absorb more than other parts which are free from such substances. Those leaves, also, which contain either the constituents of nutgalls, or compounds in which nitrogen is present, ought to absorb more oxygen than those which do not contain such matters. The correctness of these inferences has been distinctly proved by the observations of De Saussure; for, while the tasteless leaves of the *Agave americana* absorb only 0·3 of their volume of oxygen in the dark, during 24 hours, the leaves of the *Pinus Abies*, which contain volatile and resinous oils, absorb 10 times, those of the *Quercus Robur* containing tannic acid 14 times, and the balmy leaves of the *Populus alba* 21 times that quantity. This chemical action is shown very plainly also, in the leaves of the *Cotyledon calycinum*, the *Cacalia ficoides*, and others; for they are sour like sorrel in the morning, tasteless at noon, and bitter in the evening. The formation of acids is effected during the night by a true process of oxidation : these are deprived of their acid properties during the day and evening, and are changed by separation of a part of their oxygen into compounds containing oxygen and hydrogen, either in the same proportions as in water, or even with an excess of hydrogen, which is the composition of all tasteless and bitter substances.

Indeed, the quantity of oxygen absorbed could be estimated pretty nearly by the different periods which the green leaves of plants require to undergo alteration in colour, by the influence of the atmosphere. Those which continue longest green will abstract less oxygen from the air in an equal space of time, than those the constituent parts of which suffer a more rapid change. It is found, for example, that the leaves of the *Ilex aquifolium*, distinguished by the durability of their colour, absorb only 0·86 of their volume of oxygen gas in the same time that the leaves of the poplar absorb 8, and those of the beech 9½ times their volume; both the beech and poplar being remarkable for the rapidity and ease with which the colour of their leaves changes.

When the green leaves of the poplar, the beech, the oak, or the holly, are dried under the air pump, with exclusion of light, then moistened with water, and placed under a glass globe filled with oxygen, they are found to absorb that gas in proportion as they change in colour. The chemical nature of this process is thus completely established. The diminution of the gas which occurs can only be owing to the union of a large proportion of oxygen with those substances which are already in the state of oxides, or to the oxidation of the hydrogen in those vegetable compounds which contain it in excess. The fallen brown or yellow leaves of the oak contain no longer tannin, and those of the poplar no balsamic constituents.

The property which green leaves possess of absorbing oxygen belongs also to fresh wood, whether taken from a twig or from the interior of the trunk of a tree. When fine chips of such wood are placed in a moist condition under a jar filled with oxygen, the gas is seen to diminish in volume. But wood, dried by exposure to the atmosphere and then moistened, converts the oxygen into carbonic acid, without change of volume ; fresh wood, therefore, absorbs most oxygen.

MM. Petersen and Schödler have shown, by the careful elementary analysis of 24 different kinds of wood, that they contain carbon and the elements of water, with the addition of a certain quantity of hydrogen. Oak wood, recently taken from the tree, and dried at 100° C. (212 F.,) contains 49,432 carbon, 6.069 hydrogen, and 44.499 oxygen.

The proportion of hydrogen which is necessary to combine with 44.498 oxygen in order to form water, is ⅛ of this quantity, namely, 5.56; it is evident, therefore, that oak wood contains $\frac{1}{12}$ more hydrogen than corresponds to this proportion. In *Pinus Larix, P. Abies,* and *P. picea,* the excess of hydrogen amounts to ⅐, and in *Tilia europæa* to ⅕. The quantity of hydrogen stands in some relation to the specific weight of the wood; the lighter kinds of wood contain more of it than the heavier. In ebony wood (*Diospyros Ebenum*) the oxygen and hydrogen are in exactly the same proportion as in water.

The difference between the composition of the varieties of wood, and that of simple woody fibre, depends, unquestionably, upon the presence of constituents, in part soluble, and in part insoluble, such as resin and other matters, which contain a large proportion of hydrogen : the hydrogen of such substances being in the analysis of the various woods superadded to that of the true woody fibre.

It has previously been mentioned that mouldering oak wood contains carbon and the elements of water, without any excess of hydrogen. But the proportions of its constituents must necessarily have been different, if the volume of the air had not changed during its decay, because the proportion of hydrogen in those component substances of the wood which contained it in excess is here diminished, and this diminution could only be effected by an absorption of oxygen, and consequent formation of water.

Most vegetable physiologists have connected the emission of carbonic acid during the night with the absorption of oxygen from the atmosphere, and have considered these actions as a true process of respiration in plants, similar to that of animals, and like it, having for its result the separation of carbon from some of their constitutents. This opinion has a very weak and unstable foundation.

The carbonic acid, which has been absorbed by the leaves and by the roots, together with water, ceases to be decomposed on the departure of daylight; it is dissolved in the juices which pervade all parts of the plant, and escapes every moment through the leaves in quantity corresponding to that of the water which evaporates.

A soil in which plants vegetate vigorously, contains a certain quantity of moisture which is indispensably necessary to their existence. Carbonic acid, likewise, is always present in such a soil, whether it has been abstracted from the air or has been generated by the decay of vegetable matter. Rain and wellwater, and also that from other sources, invariably contains carbonic acid. Plants during their life constantly possess the power of absorbing by their roots moisture, and, along with it, air and carbonic acid. Is it, therefore, surprising that the carbonic acid should be returned unchanged to the atmosphere, along with water, when light (the cause of the fixation of its carbon) is absent?

Neither this emission of carbonic acid nor the absorption of oxygen has any connection with the process of assimilation; nor have they the slightest relation to one another; the one is a purely mechanical, the other a purely chemical process. A cotton wick, inclosed in a lamp, which contains a liquid saturated with carbonic acid, acts exactly in the same manner as a living plant in the night. Water and carbonic acid are sucked up by capillary attraction, and both evaporate from the exterior part of the wick.

Plants which live in a soil containing humus exhale much more carbonic acid during the night than those which grow in dry situations; they also yield more in rainy than in dry weather. These facts point out to us the cause of the numerous contradictory observations, which have been made with respect to the change impressed upon the air by living plants, both in darkness and in common daylight, but which are unworthy of consideration, as they do not assist in the solution of the main question.

There are other facts which prove in a decisive manner that plants yield more oxygen to the atmosphere than they extract from it; these proofs, however, are to be drawn with certainty only from plants which live under water.

When pools and ditches, the bottoms of which are covered with growing plants, freeze upon their surface in winter, so that the water is completely excluded from the atmosphere by a clear stratum of ice, small bubbles of gas are observed to escape, continually, during the day, from the points of the leaves and twigs. These bubbles are seen most distinctly when the rays of the sun fall upon the ice; they are very small at first, but collect under the ice and form larger bubbles. They consist of pure oxygen gas. Neither during the night, nor during the day when the sun does not shine, are they observed to diminish in quantity. The source of this oxygen is the carbonic acid dissolved in the water, which is absorbed by the plants, but is again supplied to the water, by the decay of vegetable substances contained in the soil. If these plants absorb oxygen during the night, it can be in no greater quantity than that which the surrounding water holds in solution, for the gas, which has been exhaled, is not again absorbed. The action of water plants cannot be supposed to form an exception to a great law of nature, and the less so, as the different action of aerial plants upon the atmosphere is very easily explained.

The opinion is not new that the carbonic acid of the air serves for the nutriment of plants, and that its carbon is assimilated by them; it has been admitted, defended, and argued for, by the soundest and most intelligent natural philosophers, namely, by Priestley, Sennebier, De Saussure, and even by Ingenhouss himself. There scarcely exists a theory in natural science, in favour of which there are more clear and decisive arguments. How, then, are we to account for its not being received in its full extent by most other physiologists, for its being even disputed by many, and considered by a few as quite refuted?

All this is due to two causes, which we shall now consider.

One is, that in botany the talent and labour of inquirers has been wholly spent in the examination of form and structure: chemistry and physics have not been allowed to sit in council upon the explanation of the most simple processes; their experience and their laws have not been employed, though the most powerful means of help in the acquirement of true knowledge. They have not been used, because their study has been neglected.

All discoveries in physics and in chemistry, all explanations of chemists, must remain without fruit and useless, because, even to the great leaders in physiology, carbonic acid, ammonia, acids, and bases, are sounds without meaning, words without sense, terms of an unknown language, which awaken no thoughts and no associations. They treat these sciences like the vulgar, who despise a foreign literature in exact proportion to their ignorance of it; since even when they have had some acquintance with them, they have not understood their spirit and application.

Physiologists reject the aid of chemistry in their inquiry into the secrets of vitality, although it alone could guide them in the true path; they reject chemistry, because in its pursuit of knowledge it destroys the subjects of its investigation; but they forget that the knife of the anatomist must dismember the body, and destroy its organs, if an account is to be given of their form, structure, and functions.

When pure potato starch is dissolved in nitric acid, a ring of the finest wax remains. What can be opposed to the conclusion of the chemist, that each grain of starch consists of concentric layers of wax and amylin, which thus mutually protect each other against the action of water and ether? Can results of this kind, which illustrate so completely both the nature and properties of bodies, be attained by the microscope? Is it possible to make the gluten in a piece of bread visible in all its connections and ramifications? It is impossible by means of instruments; but if the piece of bread is placed in a lukewarm decoction of malt, the starch, and the substance called dextrine,* are seen to dissolve like sugar in water, and, at last, nothing remains except the gluten, in the

* According to Raspail, starch consists of vesicles inclosing within them a fluid resembling gum. Starch may be put in cold water without being dissolved: but, when placed in hot water, these spherules burst, and allow the escape of the liquid. This liquid is the *dextrine* of Biot, so called because it possesses the property of turning the plane of the polarization of light to the right hand. —ED.

form of a spongy mass, the minute pores of which can be seen only by a microscope.

Chemistry offers innumerable resources of this kind which are of the greatest use in an inquiry into the nature of the organs of plants; but they are not used, because the need of them is not felt. The most important organs of animals and their functions are known, although they may not be visible to the naked eye. But in vegetable physiology, a leaf is in every case regarded merely as a leaf, notwithstanding that leaves generating oil of turpentine or oil of lemons must possess a different nature from those in which oxalic acid is formed. Vitality, in its peculiar operations, makes use of a special apparatus for each function of an organ. A rose twig engrafted upon a lemon tree does not bring forth lemons, but roses. Vegetable physiologists in the study of their science have not directed their attention to that part of it which is most worthy of investigation.

The second cause of the incredulity with which physiologists view the theory of the nutrition of plants by the carbonic acid of the atmosphere is, that the art of experimenting is not known in physiology, it being an art which can be learned accurately only in the chemical laboratory. Nature speaks to us in a peculiar language, in the language of phenomena; she answers at all times the questions which are put to her; and such questions are experiments. An experiment is the expression of a thought: we are near the truth when the phenomena elicited by the experiment corresponds to the thought; while the opposite result shows that the question was falsely stated, and that the conception was erroneous.

The critical repetition of another's experiments must be viewed as a criticism of his opinions; if the result of the criticism be merely negative, if it do not suggest more correct ideas in the place of those which it is intended to refute, it should be disregarded; because the worse experimenter the critic is, the greater will be the discrepancy between the results he obtains and the views proposed by the other.

It is too much forgotten by physiologists, that their duty really is not to refute the experiments of others, nor to show that they are erroneous, but to discover truth, and that alone. It is startling, when we reflect that all the time and energy of a multitude of persons of genius, talent, and knowledge, are expended in endeavours to demonstrate each other's errors.

The question whether carbonic acid is the food of plants or not has been made the subject of experiments with perfect zeal and good faith; the results have been opposed to that view. But how was the inquiry instituted?

The seeds of balsamines, beans, cresses, and gourds, were sown in pure Carrara marble, and sprinkled with water containing carbonic acid. The seeds sprang, but the plants did not attain to the development of the third small leaf. In other cases, they allowed the water to penetrate the marble from below, yet, in spite of this, they died. It is worthy of observation, that they lived longer with pure distilled water than with that impregnated with carbonic acid; but still, in this case also, they eventually perished. Other experimenters sowed seeds of plants in flowers of sulphur and sulphate of barytes, and tried to nourish them with carbonic acid, but without success.

Such experiments have been considered as positive proofs, that carbonic acid will not nourish plants; but the manner in which they were instituted is opposed to all rules of philosophical inquiry, and to all the laws of chemistry.

Many conditions are necessary for the life of plants; those of each genus require special conditions; and should but one of these be wanting, although the rest be supplied, the plants will not be brought to maturity. The organs of a plant, as well as those of an animal, contain substances of the most different kinds; some are formed solely of carbon and the elements of water, others contain nitrogen, and in all plants we find metallic oxides in the state of salts. The food which can serve for the production of all the organs of a plant, must necessarily contain all its elements. These most essential of all the chemical qualities of nutriment may be united in one substance, or they may exist separately in several; in which case, the one contains what is wanting in the other. Dogs die although fed with jelly, a substance which contains nitrogen; they cannot live upon white bread, sugar or starch, if these are given as food, to the exclusion of all other substances. Can it be concluded from this, that these substances contain no elements suited for assimilation? Certainly not.

Vitality is the power which each organ possesses of constantly reproducing itself; for this it requires a supply of substances which contain the constituent elements of its own substance, and are capable of undergoing transformation. All the organs together cannot generate a single element, carbon, nitrogen, or a metallic oxide.

When the quantity of the food is too great, or is not capable of undergoing the necessary transformation, or exerts any peculiar chemical action, the organ itself is subjected to a change : all poisons act in this manner. The most nutritious substances may cause death. In experiments such as those described above, every condition of nutrition should be considered. Besides those matters which form their principal constituent parts, both animals and plants require others, the peculiar functions of which are unknown. These are inorganic substances, such as common salt, the total want of which is in animals inevitably productive of death. Plants, for the same reason, cannot live unless supplied with certain metallic compounds.

If we knew with certainty that there existed a substance capable alone of nourishing a plant and of bringing it to maturity, we might be led to a knowledge of the conditions necessary to the life of all plants, by studying its characters and composition. If humus were such a substance, it would have precisely the same value as the only single food which nature has produced for animal organization, namely, milk (Prout.) The constituents of milk are cheese or caseine, a compound containing nitrogen in large proportion; butter, in which hydrogen abounds; and sugar of milk, a substance with a large quantity of hydrogen and oxygen in the same proportion as in water. It also contains in solution, lactate of soda, phosphate of lime, and common salt; and a peculiar aromatic product exists in the butter, called butyric acid. The knowledge of the composition of milk is a key to the conditions necessary for the purposes of nutrition of all animals.

All substances which are adequate to the nourishment of animals contain those materials united, though not always in the same form; nor can any one be wanting for a certain space of time, without a marked effect on the health being produced. The employment of a substance as food presupposes a knowledge of its capacity of assimilation, and of the conditions under which this takes place.

A carnivorous animal dies in the vacuum of an air pump, even though supplied with a superabundance of food; it dies in the air, if the demands of its stomach are not satisfied; and it dies in pure oxygen gas, however lavishly nourishment be given to it. Is it hence to be concluded, that neither flesh, nor air, nor oxygen, is fitted to support life? Certainly not.

From the pedestal of the Trajan column at Rome we might chisel out each single piece of stone, if upon the extraction of the second we replaced the first. But could we conclude from this that the column was suspended in the air, and not supported by a single piece of its foundation? Assuredly not. Yet the strongest proof would have been given that each portion of the pedestal could be removed, without the downfall of the column.

Animal and vegetable physiologists, however, come to such conclusions with respect to the process of assimilation. They institute experiments, without being acquainted with the circumstances necessary for the continuance of life—with the qualities and proper nutriment of the animal or plant on which they operate—or with the nature and chemical constitution of its organs. These experiments are considered by them as convincing proofs, while they are fitted only to awaken pity.

Is it possible to bring a plant to maturity by means of carbonic acid and water, without out the aid of some substance containing nitrogen, which is an essential constituent of the sap, and indispensable for its production? Must the plant not die, however abundant the supply of carbonic acid may be, as soon as the first small leaves have exhausted the nitrogen contained in the seeds?

Can a plant be expected to grow in Carrara marble, even when an azotised substance is supplied to it, if the marble be sprinkled with an aqueous solution of carbonic acid, which dissolves the lime and forms bicarbonate of lime? A plant of the family of the *Plumbagineæ*, upon the leaves of which fine hornlike, or scaly processes of crystallised carbonate of lime are formed, might, perhaps, attain maturity under such circumstances; but these experiments are only sufficient to prove, that cresses, gourds, and balsamines, cannot be nourished by bicarbonate of lime, in the absence of matter containing nitrogen. We may, indeed, conclude, that the salt of lime acts as a poison, since the developement of plants will advance farther in pure water, when lime and carbonic acid are not used.

Moist flowers of sulphur attract oxygen from the atmosphere, and become acid. Is it possible that a plant can grow and flourish in presence of free sulphuric acid, with no other nourishment than carbonic acid? It is true, the quantity of sulphuric acid formed thus in hours, or in days, may be small, but the property of each particle of the sulphur to absorb oxygen and retain it, is present every moment.

When it is known that plants require moisture, carbonic acid, and air, should we choose as the soil for experiments on their growth, sulphate of barytes, which, from its nature and specific gravity, completely prevents the access of air?

All these experiments are valueless for the decision of any question. It is absurd to take for them any soil, at mere hazard, as long as we are ignorant of the functions performed in plants by those inorganic substances which are apparently foreign to them. It is quite impossible to mature a plant of the family of the *Gramineæ*, or of the *Equisetaceæ*, the solid framework of which contains silicate of potash, without silicic acid and potash, or a plant of the genus *Oxalis* without potash, or saline plants such as the saltworts (*Salsola* and *Salicornia*) without chloride of sodium, or at least some salt of similar properties. All seeds of the *Gramineæ* contain phosphate of magnesia; the solid parts of the roots of the *althæa* contain more phosphate of lime than woody fibre. Are these substances merely accidentally present? A plant should not be chosen for experiment, when the matter which it requires for its assimilation is not well known.

What value, now, can be attached to experiments in which all those matters which a plant requires in the process of assimilation, besides its mere nutriment, have been excluded with the greatest care? Can the laws of life be investigated in an organised being which is diseased or dying?

The mere observation of a wood or meadow is infinitely better adapted to decide so simple a question than all the trivial experiments under a glass globe; the only difference is that instead of one plant there are thousands. When we are acquainted with the nature of a single cubic inch of their soil, and know the composition of the air and rainwater, we are in possession of all the conditions necessary to their life. The source of the different elements entering into the composition of plants cannot possibly escape us, if we know in what form they take up their nourishment, and compare its composition with that of the vegetable substances which compose their structure.

All these questions will now be examined and discussed. It has been already shown that the carbon of plants is derived from the atmosphere: it still remains for us to inquire what power is exerted on vegetation by the humus of the soil and the inorganic constituents of plants and also to trace the sources of their nitrogen.

[*Editor's Note:* Chapters III, IV, and V have been omitted.]

of soda, and produces sulphate of soda From this fact follows the rule—that the quantity, which an acid requires of an alkali for its saturation, may be represented by a very simple number.

It is perfectly necessary to form a proper conception of what chemists denominate the "capacity for saturation of an acid," before we are able to form a correct idea of the functions performed in plants, by their inorganic constituents. The power of a base to neutralize an acid does not depend upon the quantity of radical which it contains, but altogether upon the quantity of its oxygen. Thus protoxide of iron contains 1 eq. of oxygen, and unites with 1 eq. of sulphuric acid in forming a neutral salt; but peroxide of iron contains 3 eq. of oxygen, and requires 3 eq. of the same acid for its neutralization. Hence when a given weight of an acid is neutralized by different bases, the quantity of oxygen contained in these bases must be the same as is exhibited by the following scale :—

501·17 parts of Sulphuric Acid neutralize
  258·35 Magnesia Oxygen= 100
  647·29 Strontia " = 100
 1451·61 Oxide of Silver " = 100
  956·8 Barytes " = 100

CHAPTER VI.

OF THE INORGANIC CONSTITUENTS OF PLANTS.

CARBONIC acid, water and ammonia, are necessary for the existence of plants, because they contain the elements from which their organs are formed; but other substances are likewise requisite for the formation of certain organs destined for special functions peculiar to each family of plants. Plants obtain these subtances from inorganic nature. In the ashes left after the incineration of plants, the same substances are found, although in a changed condition.

Although the vital principle exercises a great power over chemical forces, yet it does so only by directing the way in which they are to act, and not by changing the laws to which they are subject. Hence when the chemical forces are employed in the processes of vegetable nutrition, they must produce the same results which are observed in ordinary chemical phenomena. The inorganic matter contained in plants must, therefore, be subordinate to the laws which regulate its combinations in common chemical processes.

The most important division of inorganic substances is that of *acids* and *alkalies*. Both of these have a tendency to unite together, and form neutral compounds, which are termed salts. According to the doctrine of equivalents, these combinations are always effected in definite proportions, that is to say, one equivalent of an acid always unites with one or two equivalents of a base, whatever that base may be. Thus 501·17 parts by weight of sulphuric acid unite with 1 eq. of potash, and form one eq. of sulphate of potash; the same quantity unites with 1 eq.

It follows from the law of equivalents, that the quantity of oxygen in a base must stand in a simple relation to the quantity of oxygen in an acid which unites with it. By this is meant, that the quantities in both cases must either be equal or multiples of each other; for the doctrine of equivalents denies the possibility of their uniting in fractional parts. This will be rendered obvious by a consideration of the two following examples :

100 parts of Cyanic Acid contain 23·26 oxygen = 1.
100 parts of Cyanic Acid saturate 137·21 parts of potash, which contain 23·26 oxygen = 1.
100 parts of Nitric Acid contain 73·85 oxygen = 5.
100 parts of Nitric Acid saturate 214·40 parts of oxide of silver, which contain 14·77 oxygen = 1.

In the first of these cases, the relation of the oxygen of the base to that of the acid is as 1:1 ; in the second, as 1:5. The capacity for saturation of each acid, is, therefore, the constant quantity of oxygen necessary to neutralize 190 parts of it.

Many of the inorganic constituents vary according to the soil in which the plants grow, but a certain number of them are indispensable to their developement. All substances in solution in a soil are absorbed by the roots of plants, exactly as a sponge imbibes a liquid, and all that it contains, without selection. The substances thus conveyed to plants are retained in greater or less quantity, or are entirely separated when not suited for assimilation.

Phosphate of magnesia in combination with ammonia is an invariable constituent of the seeds of all kinds of grasses. It is contained in the outer horny husk, and is introduced into bread along with the flour, and also into beer. The bran of flour contains the greatest quantity of it. It is this salt which forms large crystalline concretions, often amounting to several pounds in weight, in the *cæcum* of horses belonging to millers; and when ammonia is mixed with beer, the same salt separates as a white precipitate.

Most plants, perhaps all of them, contain organic acids of very different composition and properties, all of which are in combination with bases, such as potash, soda, lime, or magnesia. These bases evidently regulate the formation of the acids, for the diminution of the one is followed by a decrease of the other: thus in the grape, for example, the quantity of potash contained in its juice is less when it is ripe than when unripe; and the acids, under the same circumstances, are found to vary in a similar manner. Such constituents exist in small quantity in those parts of a plant in which the process of assimilation is most active, as in the mass of woody fibre; and their quantity is greater in those organs whose office it is to prepare substances conveyed to them for assimilation by other parts. The leaves contain more inorganic matters than the branches, and the branches more than the stem. The potato plant contains more potash before blossoming than after it.

The acids found in the different families of plants are of various kinds; it cannot be supposed that their presence and peculiarities are the result of accident. The fumaric and oxalic acids in the liverwort, the kinovic acid in the *China nova*, the rocellic acid in the *Rocella tinctoria*, the tartaric acid in grapes, and the numerous other organic acids, must serve some end in vegetable life. But if these acids constantly exist in vegetables, and are necessary to their life, which is incontestable, it is equally certain that some alkaline base is also indispensable, in order to enter into combination with the acids which are always found in the state of salts. All plants yield by incineration ashes containing carbonic acid; all therefore must contain salts of an organic acid.*

Now, as we know the capacity of saturation of organic acids to be unchanging, it follows that the quantity of the bases united with them cannot vary, and for this reason the latter substances ought to be considered with the strictest attention both by the agriculturist and physiologist.

We have no reason to believe that a plant in a condition of free and unimpeded growth produces more of its peculiar acids than it requires for its own existence; hence, a plant, on whatever soil it grows, must contain an invariable quantity of alkaline bases. Culture alone will be able to cause a deviation.

* Salts of organic acids yield carbonates on incineration, if they contain either alkaline or earthy bases.

In order to understand this subject clearly, it will be necessary to bear in mind that any one of the alkaline bases may be substituted for another, the action of all being the same. Our conclusion is therefore by no means endangered by the existence of a particular alkali in one plant, which may be absent in others of the same species. If this inference be correct, the absent alkali or earth must be supplied by one similar in its mode of action, or in other words, by an equivalent of another base. The number of equivalents of these various bases which may be combined with a certain portion of acid must necessarily be the same, and therefore the amount of oxygen contained in them must remain unchanged under all circumstances and on whatever soil they grow.

Of course, this argument refers only to those alkaline bases which in the form of organic salts form constituents of the plants. Now, these salts are preserved in the ashes of plants as carbonates, the quantity of which can be easily ascertained.

It has been distinctly shown, by the analysis of De Saussure and Berthier, that the nature of a soil exercises a decided influence on the quantity of the different metallic oxides contained in the plants which grow on it; that magnesia, for example, was contained in the ashes of a pine-tree grown at Mont Breven, whilst it was absent from the ashes of a tree of the same species from Mont La Salle, and that even the proportion of lime and potash was very different.

Hence it has been concluded, (erroneously, I believe,) that the presence of bases exercises no particular influence upon the growth of plants: but even were this view correct, it must be considered as a most remarkable accident that these same analyses furnish proof for the very opposite opinion. For although the composition of the ashes of these pine-trees were so very different, they contained, according to the analyses of De Saussure, an equal number of equivalents of metallic oxides; or, what is the same thing, the quantity of oxygen contained in all the bases was in both cases the same.

100 parts of the ashes of the pine-tree from Mont Breven contained—

Carbonate of Potash	.	3·60
" Lime	.	46·34
" Magnesia	·	6·77
Sum of the carbonates		56·71
Quantity of oxygen in the Potash		0·41
" " " Lime		7·33
" " " Magnesia		1·27
Sum of the oxygen in the bases		9·01

100 parts of the ashes of the pine from Mont La Salle contained*—

* According to the experiments of Saussure, 1000 parts of the wood of the pine from Mont Brevon gave 11·87 parts of ashes; the same quantity of wood from Mont La Salle yielded 11·28 parts. From this we might conclude that the two pines, although brought up in different soils, yet contained the same quantity of inorganic elements.

Carbonate of Potash	•	7·36
" Lime	•	51·19
" Magnesia		00·00

Sum of the carbonates 58·55

Quantity of oxygen in the Potash	0·85
" " " Lime	8·10

Sum of the oxygen in the bases 8·95

The numbers 9·01 and 8·95 resemble each other as nearly as could be expected even in analyses made for the very purpose of ascertaining the fact above demonstrated which the analyst in this case had not in view.

Let us now compare Berthier's analyses of the ashes of two fir-trees, one of which grew in Norway, the other in Allevard (département de l'Isère). One contained 50, the other 25 per cent. of soluble salts. A greater difference in the proportion of the alkaline bases could scarcely exist between two totally different plants, and yet even here the quantity of oxygen in the bases of both was the same.

100 parts of the ashes of firwood from Allevard contained, according to Berthier, (Ann. de Chim. et de Phys. t. xxxii. p. 248,)

Potash & Soda	16·8	in which	3·42	must be oxygen.
Lime .	29·5	"	8·20	" "
Magnesia	3·2	"	1·20	" "
	49.5		12·82	

Only part of the potash and soda in these ashes was in combination with organic acids; the remainder was in the form of sulphates, phosphates, and chlorides. One hundred parts of the ashes contain 3·1 sulphuric acid, 4·2 phosphoric acid, and 0·3 hydrochloric acid, which together neutralize a quantity of base containing 1·20 oxygen. This number therefore must be substracted from 12·82. The remainder 11·62 indicates the quantity of oxygen in the alkaline bases, combined with organic acids in the firwood of Allevard.

The firwood of Norway contained in 100 parts,—*

Potash .	14·1	of which	2·4	would be oxygen.
Soda .	20·7	"	5·3	' "
Lime .	12·3	"	3·45	" "
Magnesia	4·35	"	1·69	" "
	51·45		12·84	

And if the quantity of oxygen of the bases in combination with sulphuric and phosphoric acid, viz. 1·37, be again substracted from 12·84, 11·47 parts remain as the amount of oxygen contained in the bases which were in combination with organic acids.

---

* This calculation is exact only in the case where the quantity of ashes is equal in weight for a given quantity of wood; the difference cannot, however, be admitted to be so great as to change sensibly the above proportions. Berthier has not mentioned the proportion of ashes contained in the wood.

These remarkable approximations cannot be accidental; and if further examinations confirm them in other kinds of plants, no other explanation than that already given can be adopted.

It is not known in what form silica, manganese, and oxide of iron, are contained in plants; but we are certain that potash, soda, and magnesia, can be extracted from all parts of their structure in the form of salts of organic acids. The same is the case with lime, when not present as insoluble oxalate of lime. It must here be remembered, that in plants yielding oxalic acid, the acid and potash never exist in the form of a neutral or quadruple salt, but always as a double acid salt, on whatever soil they may grow. The potash in grapes also is more frequently found as an acid salt, viz. cream of tartar, (bitartrate of potash,) than in the form of a neutral compound. As these acids and bases are never absent from plants, and as even the form in which they present themselves is not subject to change, it may be affirmed that they exercise an important influence on the developement of the fruits and seeds, and also on many other functions of the nature of which we are at present ignorant.

The quantity of alkaline bases existing in a plant also depends evidently on this circumstance of their existing only in the form of acid salts,—for the capacity of saturation of an acid is constant; and when we see oxalate of lime in the lichens occupying the place of woody fibre which is absent, we must regard it as certain that the soluble organic salts are destined to fulfil equally important though different functions, so much so that we could not conceive the complete developement of a plant without their presence, that is, without the presence of their acids, and consequently of their bases.

From these considerations we must perceive, that exact and trustworthy examinations of the ashes of plants of the same kind growing upon different soils would be of the greatest importance to vegetable physiology' and would decide whether the facts above mentioned are the results of an unchanging law for each family of plants, and whether an invariable number can be found to express the quantity of oxygen which each species of plant contains in the bases united with organic acids. In all probability such inquiries will lead to most important results; for it is clear that if the production of a certain unchanging quantity of an organic acid is required by the peculiar nature of the organs of a plant, and is necessary to its existence, then potash or lime must be taken up by it in order to form salts with this acid; that if these do not exist in sufficient quantity in the soil, other bases must supply their place; and that the progress of a plant must be wholly arrested when none are present.

Seeds of the *Salsola Kali*, when sown in common garden soil, produce a plant containing both potash and soda; while the plants grown from the seeds of this contain only salts of potash, with mere traces of muriate of soda. (Cadet.)

The examples cited above, in which the quantity of oxygen contained in the bases was shown to be the same, lead us to the legitimate conclusion that the developement of certain plants is not retarded by the substitution of the bases contained in them. But it was by no means inferred that any one base could replace all the others which are found in a plant in its normal condition. On the contrary, it is known that certain bases are indispensable for the growth of a plant, and these could not be substituted without injuring its developement. Our inference has been drawn from certain plants, which can bear without injury this substitution; and it can only be extended to those plants which are in the same condition. It will be shown afterwards that corn or vines can only thrive on soils containing potash, and that this alkali is perfectly indispensable to their growth. Experiments have not been sufficiently multiplied so as to enable us to point out in what plants potash or soda may be replaced by lime or magnesia; we are only warranted in affirming that such substitutions are in many cases common. The ashes of various kinds of plants contain very different quantities of alkaline bases, such as potash, soda, lime, or magnesia. When lime exists in the ashes in large proportion, the quantity of magnesia is diminished, and in like manner according as the latter increases the lime or potash decreases. In many kinds of ashes not a trace of magnesia can be detected.

The existence of vegetable alkalies in combination with organic acids gives great weight to the opinion that alkaline bases in general are connected with the developement of plants.

If potatoes are grown where they are not supplied with earth, the magazine of inorganic bases, (in cellars, for example,) a true alkali, called Solanin, of very poisonous nature, is formed in the sprouts which extend towards the light, while not the smallest trace of such a substance can be discovered in the roots, herbs, blossoms, or fruits of potatoes grown in fields. (Otto.) In all the species of the *Cinchona*, kinic acid is found; but the quantity of quinina, cinchonina, and lime, which they contain is most variable. From the fixed bases in the products of incineration, however, we may estimate pretty accurately the quantity of the peculiar organic bases. A maximum of the first corresponds to a minimum of the latter, as must necessarily be the case if they mutually replace one another according to their equivalents. We know that different kinds of opium contain meconic acid in combination with very different quantities of narcotina, morphia, codeia, &c., the quantity of one of these alkaloids diminishing on the increase of the others. Thus the smallest quantity of morphia is accompanied by a maximum

of narcotina. Not a trace of meconic acid* can be discovered in many kinds of opium, but there is not on this account an absence of acid, for the meconic is here replaced by sulphuric acid. Here, also, we have an example of what has been before stated, for in those kinds of opium where both these acids exist, they are always found to bear a certain relative proportion to one another. Attention to these facts must be very important in the selection of soils destined for the cultivation of plants which yield the vegetable alkaloids.

Now if it be found, as appears to be the case in the juice of poppies, that an organic acid may be replaced by an inorganic, without impeding the growth of a plant, we must admit the probability of this substitution taking place in a much higher degree in the case of the inorganic bases.

When roots find their more appropriate base in sufficient quantity, they will take up less of another.

These phenomena do not show themselves so frequently in cultivated plants, because they are subjected to special external conditions for the purpose of the production of particular constituents or particular organs.

When the soil, in which a white hyacinth is growing in a state of blossom, is sprinkled with the juice of the *Phytolacca decandra,* the white blossoms assume in one or two hours a red colour, which again disappears after a few days under the influence of sunshine, and they become white and colourless as before.† The juice in this case evidently enters into all parts of the plant, without being at all changed in its chemical nature, or without its presence being apparently either necessary or injurious. But this condition is not permanent, and when the blossoms have again become colourless, none of the colouring matter remains; and if it should occur that any of its elements were adapted for the purposes of nutrition of the plant, then these alone would be retained, whilst the rest would be excreted in an altered form by the roots.

Exactly the same thing must happen when we sprinkle a plant with a solution of chloride of potassium, nitre, or nitrate of strontia; they will enter into the different parts of the plant, just as the coloured juice mentioned above, and will be found in its ashes if it should be burnt at this period. Their presence is merely accidental; but no conclusion can be hence deduced against the necessity of the presence of other bases in plants. The experiments of Macaire-Princep have shown, that plants made to vegetate with their roots in a weak solution of acetate of lead, and then in rain water,

* Robiquet did not obtain a trace of meconate of lime from 300 lbs. of opium, whilst in other kinds the quantity was very considerable. Ann. de Chim. liii. p. 425.
† Biot, in the Comptes rendus des Séances de l'Académie des Sciences, à Paris, 1er Sémestre, 1837, p. 12.

yield to the latter all the salt of lead which they had previously absorbed. They return, therefore, to the soil all matters which are unnecessary to their existence. Again, when a plant, freely exposed to the atmosphere, rain, and sunshine, is sprinkled with a solution of nitrate of strontia, the salt is absorbed, but it is again separated by the roots and removed farther from them by every shower of rain, which moistens the soil, so that at last not a trace of it is to be found in the plant.

Let us consider the composition of the ashes of two fir-trees as analysed by an acute and most accurate chemist. One of these grew in Norway, on a soil the constituents of which never changed, but to which soluble salts, and particularly common salt, were conveyed in great quantity by rain-water. How did it happen that its ashes contained no appreciable trace of salt, although we are certain that its roots must have absorbed it after every shower?

We can explain the absence of salt in this case by means of the direct and positive observations referred to, which have shown that plants have the power of returning to the soil all substances unnecessary to their existence; and the conclusion to which all the foregoing facts lead us, when their real value and bearing are apprehended, is that the alkaline bases existing in the ashes of plants must be necessary to their growth, since if this were not the case they would not be retained.

The perfect developement of a plant, according to this view, is dependent on the presence of alkalies or alkaline earths; for when these substances are totally wanting its growth will be arrested, and when they are only deficient it must be impeded.

In order to apply these remarks, let us compare two kinds of trees, the wood of which contains unequal quantities of alkaline bases, and we shall find that one of these grows luxuriantly in several soils upon which the others are scarcely able to vegetate. For example, 10,000 parts of oak wood yield 250 parts of ashes, the same quantity of fir wood only 83, of linden wood 500, of rye 440, and of the herb of the potato plant 1500 parts.*

Firs and pines find a sufficient quantity of alkalies in granitic and barren sandy soils in which oaks will not grow; and wheat thrives in soils favourable for the linden tree, because the bases which are necessary to bring it to complete maturity, exist there in sufficient quantity. The accuracy of these conclusions, so highly important to agriculture and to the cultivation of forests, can be proved by the most evident facts.

All kinds of grasses, the *Equisetaceæ*, for example, contain in the outer parts of their leaves and stalk a large quantity of silicic acid and potash in the form of acid silicate

of potash. The proportion of this salt does not vary perceptibly in the soil of corn-fields, because it is again conveyed to them as manure in the form of putrefying straw. But this is not the case in a meadow, and hence we never find a luxuriant crop of grass* on sandy and calcareous soils, which contain little potash, evidently because one of the constituents indispensable to the growth of the plants is wanting. Soils formed from basalt, grauwacke, and porphyry are, *cæteris paribus*, the best for meadow land, on account of the quantity of potash which enters into their composition. The potash abstracted by the plants is restored during the annual irrigation. The potash contained in the soil itself is inexhaustible in comparison with the quantity removed by plants. But when we increase the crop of grass in a meadow by means of gypsum, we remove a greater quantity of potash with the hay than can under the same circumstances be restored. Hence it happens that, after the lapse of several years, the crops of grass on the meadows manured with gypsum diminish, owing to the deficiency of potash. But if the meadow be strewed from time to time with wood-ashes, even with the lixiviated ashes which have been used by soap-boilers, (in Germany much soap is made from the ashes of wood,) then the grass thrives as luxuriantly as before. The ashes are only a means of restoring the potash.

A harvest of grain is obtained every thirty or forty years from the soil of the Luneburg heath, by strewing it with the ashes of the heath plants (*Erica vulgaris*) which grow on it. These plants during the long period just mentioned collect the potash and soda, which are conveyed to them by rain-water; and it is by means of these alkalies that oats, barley, and rye, to which they are indispensable, are enabled to grow on this sandy heath.

The woodcutters in the vicinity of Heidelberg have the privilege of cultivating the soil for their own use, after felling the trees used for making tan. Before sowing the land thus obtained, the branches, roots, and leaves, are in every case burned, and the ashes used as a manure, which is found to be quite indispensable for the growth of the grain. The soil itself upon which the oats grow in this district consists of sandstone; and although the trees find in it a quantity of alkaline earths sufficient for their own sustenance, yet in its ordinary condition it is incapable of producing grain.

The most decisive proof of the use of strong manure was obtained at Bingen (a town on the Rhine,) where the produce and developement of vines were highly increased by

---

* It would be of importance to examine what alkalies are contained in the ashes of the sea-shore plants which grow in the humid hollows of downs, and especially in those of the millet-grass. If potash is not found in them, it must certainly be replaced by soda as in the *Salsola*, or by lime as in the *Plumbagineæ*.

* Berthier, Annales de Chimie et de Physique, xxx. p. 248.

manuring them with such substances as shavings of horn, &c.; but after some years the formation of the wood and leaves decreased to the great loss of the possessor, to such a degree that he has long had cause to regret his departure from the usual methods. By the manure employed by him, the vines had been too much hastened in their growth; in two or three years they had exhausted the potash in the formation of their fruit, leaves, and wood, so that none remained for the future crops, his manure not having contained any potash.

There are vineyards on the Rhine the plants of which are a hundred years old, and all of these have been cultivated by manuring them with a cow-dung, a manure containing a large proportion of potash, although very little nitrogen. All the potash, in fact, which is contained in the food consumed by a cow is again immediately discharged in its excrements.

The experience of a proprietor of land in the vicinity of Göttingen offers a most remarkable example of the incapability of a soil to produce wheat or grasses in general, when it fails in any one of the materials necessary to their growth. In order to obtain potash, he planted his whole land with wormwood, the ashes of which are well known to contain a large proportion of the carbonate of that alkali. The consequence was, that he rendered his land quite incapable of bearing grain for many years, in consequence of having entirely deprived the soil of its potash.

The leaves and small branches of trees contain the most potash; and the quantity of them which is annually taken from a wood for the purpose of being employed as litter,* contain more of that alkali than all the old wood which is cut down. The bark and foliage of oaks, for example, contain from 6 to 9 per cent. of this alkali; the needles of firs and pines, 8 per cent.

With every 2650 lbs. of firwood which are yearly removed from an acre of forest, only from 0·114 to 0·53 lbs. of alkalies are abstracted from the soil, calculating the ashes at 0·83 per cent. The moss, however, which covers the ground, and of which the ashes are known to contain so much alkali, continues uninterrupted in its growth, and retains that potash on the surface, which would otherwise so easily penetrate with the rain through the sandy soil. By its decay, an abundant provision of alkalies is supplied to the roots of the trees, and a fresh supply is rendered unnecessary.

The supposition of alkalies, metallic oxides, or inorganic matter in general, being produced by plants, is entirely refuted by these well-authenticated facts.

It is thought very remarkable, that those plants of the grass tribe, the seeds of which furnish food for man, follow him like the domestic animals. But saline plants seek the sea-shore or saline springs, and the Chenopodium the dunghill from similar causes. Saline plants require common salt, and the plants which grow only on dunghills need ammonia and nitrates, and they are attracted whither these can be found, just as the dung-fly is to animal excrements. So likewise none of our corn-plants can bear perfect seeds, that is, seeds yielding flour, without a large supply of phosphate of magnesia and ammonia, substances which they require for their maturity. And hence, these plants grow only in a soil where these three constituents are found combined, and no soil is richer in them than those where men and animals dwell together; where the urine and excrements of these are found corn-plants appear, because their seeds cannot attain maturity unless supplied with the constituents of those matters.

When we find sea-plants near our salt-works, several hundred miles distant from the sea, we know that their seeds have been carried there in a very natural manner, namely, by wind or birds, which have spread them over the whole surface of the earth, although they grow only in those places in which they find the conditions essential to their life.

Numerous small fish, of not more than two inches in length (*Gasterosteus aculeatus*,) are found in the salt-pans of the graduating house at Nidda (a village in Hesse Darmstadt.) No living animal is found in the salt-pans of Neuheim, situated about 18 miles from Nidda; but the water there contains so much carbonic acid and lime, that the walls of the graduating house are covered with stalactites. Hence the eggs conveyed to this place by birds do not find the conditions necessary for their developement, which they found in the former place.*

---

* [This refers to a custom some time since very prevalent in Germany although now discontinued. The leaves and small twigs of trees were gleaned from the forests by poor people, for the purpose of being used as litter for their cattle. The trees, however, were found to suffer so much in consequence, that their removal is strictly prohibited. The cause of the injury was that stated in the text.—ED.]

* The itch-insect (*Acarus Scabiei*) is considered by Burdach as the production of a morbid condition, so likewise lice in children; the original generation of the fresh-water muscle (*mytilus*) in fish-ponds, of sea-plants in the vicinity of salt-works, of nettles and grasses, of fish in pools of rain, of trout in mountain streams, &c., is according to the same natural philosopher not impossible. A soil consisting of crumbled rocks, decayed vegetables, rain and salt water, &c., is here supposed to possess the power of generating shell-fish, trout, and saltwort (*salicornia*.) All inquiry is arrested by such opinions, when propagated by a teacher who enjoys a merited reputation, obtained by knowledge and hard labour. These subjects, however, have hitherto met with the most superficial observation, although they well merit strict investigation. The dark, the secret, the mysterious, the enigmatic, is, in fact, too seducing for the youthful and philosophic

How much more wonderful and inexplicable does it appear, that bodies which remained fixed in the strong heat of a fire, have under certain conditions the property of volatilizing and, at ordinary temperatures, of passing into a state, of which we cannot say whether they have really assumed the form of a gas or are dissolved in one! Steam or vapours in general have a very singular influence in causing the volatilization of such bodies, that is, of causing them to assume the gaseous form. A liquid during evaporation communicates the power of assuming the same state in a greater or less degree to all substances dissolved in it, although they do not of themselves possess that property.

Boracic acid is a substance which is completely fixed in the fire; it suffers no change of weight appreciable by the most delicate balance, when exposed to a white heat, and, therefore, it is not volatile. Yet its solution in water cannot be evaporated by the gentlest heat, without the escape of a sensible quantity of the acid with the steam. Hence it is that a loss is always experienced in the analysis of minerals containing this acid, when liquids in which it is dissolved are evaporated. The quantity of boracic acid which escapes with a cubic foot of steam, at the temperature of boiling water, cannot be detected by our most sensible re-agents; and nevertheless the many hundred tons annually brought from Italy as an article of commerce, are procured by the uninterrupted accumulation of this apparently inappreciable quantity. The hot steam which issues from the interior of the earth is allowed to pass through cold water in the lagoons of Castel Nuova and Cherchiago; in this way the boracic acid is gradually accumulated, till at last it may be obtained in crystals by the evaporation of the water. It is evident, from the temperature of the steam, that it must have come out of depths in which human beings and animals never could have lived, and yet it is very remarkable and highly important that ammonia is never absent from it. In the large works in Liverpool, where natural boracic acid is converted into borax, many hundred pounds of sulphate of ammonia are obtained at the same time.

This ammonia has not been produced by the animal organism, it existed before the creation of human beings; it is a part, a primary constituent, of the globe itself.

The experiments instituted under Lavoisier's guidance by the *Direction des Poudres et Salpêtres*, have proved that during the evaporation of the saltpetre ley, the salt volatilizes with the water, and causes a loss which could not before be explained. It is known also, that in sea storms, leaves of

mind, which would penetrate the deepest depths of nature, without the assistance of the shaft or ladder of the miner. This is poetry, but not sober philosophical inquiry.

plants in the direction of the wind are covered with crystals of salt, even at the distance of from 20 to 30 miles from the sea. But it does not require a storm to cause the volatilization of the salt, for the air hanging over the sea always contains enough of this substance to make a solution of nitrate of silver turbid, and every breeze must carry this away. Now, as thousands of tons of sea water annually evaporate into the atmosphere, a corresponding quantity of the salts dissolved in it, viz. of common salt, chloride of potassium, magnesia, and the remaining constituents of the sea water, will be conveyed by wind to the land.

This volatilization is a source of considerable loss in salt works, especially where the proportion of salt in the water is not large. This has been completely proved at the salt works of Nauheim, by the very intelligent director of that establishment, M. Wilhelmi. He hung a plate of glass between two evaporating houses, which were about 1200 paces distant from each other, and found in the morning, after the drying of the dew, that the glass was covered with crystals of salt on one or the other side, according to the direction of the wind.

By the continual evaporation of the sea, its salts* are spread over the whole surface of the earth; and being subsequently carried down by the rain, furnish to the vegetation those salts necessary to its existence. This is the origin of the salts found in the ashes of plants, in those cases where the soil could not have yielded them.

In a comprehensive view of the phenomena of nature, we have no scale for that which we are accustomed to name, small or great; all our ideas are proportioned to what we see around us, but how insignificant are they in comparison with the whole mass of the globe! that which is scarcely observable in a confined district appears inconceivably large when regarded in its extension through unlimited space. The atmosphere contains only a thousandth part of its weight of carbonic acid; and yet small as this proportion appears, it is quite

* According to Marcet, sea-water contains in 1000 parts,
26·660 Chloride of Sodium.
4·660 Sulphate of Soda.
1·232 Chloride of Potassium.
5·152 Chloride of Magnesium.
0·153 Sulphate of Lime.
According to M'Clemm, the water of the North Sea contains in 1000 parts,
24·84 Chloride of Sodium.
2·42 Chloride of Magnesium.
2·06 Sulphate of Magnesia.
1·25 Chloride of Potassium.
1·20 Sulphate of Lime.
In addition to these constituents, it also contains inappreciable quantities of carbonate of lime, magnesia, iron, manganese, phosphate of lime, iodides and bromides, silica, sulphuretted hydrogen, and organic matter, together with ammonia and carbonic acid. (Liebig's *Annalen der Chemie*, Bd. xxxvii. s. 3.)

sufficient to supply the whole of the present generation of living beings with carbon for a thousand years, even if it were not renewed. Sea-water contains $\frac{1}{12400}$ of its weight of carbonate of lime; and this quantity, although scarcely appreciable in a pound, is the source from which myriads of marine mollusca and corals are supplied with materials for their habitations.

Whilst the air contains only from 4 to 6 ten-thousandth parts of its volume of carbonic acid, sea-water contains 100 times more, (10,000 volumes of sea-water contain 620 volumes of carbonic acid—Laurent, Bouillon, Lagrange.) Ammonia is also found in this water, so that the same conditions which sustain living beings on the land are combined in this medium, in which a whole world of other plants and animals exist.

The roots of plants are constantly engaged in collecting from the rain those alkalies which formed part of the sea-water, and also those of the water of springs, which penetrates the soil. Without alkalies and alkaline bases most plants could not exist, and without plants the alkalies would disappear gradually from the surface of the earth.

When it is considered, that sea-water contains less than one-millionth of its own weight of iodine, and that all combinations of iodine with the metallic bases of alkalies are highly soluble in water, some provision must necessarily be supposed to exist in the organization of sea-weed and the different kinds of Fući, by which they are enabled during their life to extract iodine in the form of a soluble salt from sea-water, and to assimilate it in such a manner, that it is not again restored to the surrounding medium. These plants are collectors of iodine, just as land plants are of alkalies; and they yield us this element, in quantities such as we could not otherwise obtain from the water without the evaporation of whole seas.

We take it for granted that the sea-plants require metallic iodides for their growth, and that their existence is dependent on the presence of those substances. With equal justice, then, we conclude, that the alkalies and alkaline earths, always found in the ashes of land-plants, are likewise necessary for their developement.

[*Editor's Note:* The remainder of this article has been omitted.]

11

CALORIC EQUIVALENTS FOR INVESTIGATIONS IN ECOLOGICAL ENERGETICS

K. W. Cummins and J. C. Wuycheck

Introduction

Since LINDEMAN (1942) exerted a synthesizing influence on trophic-dynamic theory, the tendency to utilize the calorie as a common denominator has increased steadily among ecologists. The progression has recently culminated in the, equilibration of ecosystem ecology and ecological energetics (e.g. PHILLIPSON. 1966).

This equivalence intended by ecological investigators has created a demand for calorific values. Several summaries of caloric data have appeared previously (SLOBODKIN & RICHMAN, 1961; GOLLEY, 1961; STRAŠKRABA, 1968) including former unpublished versions of the present tabulation (CUMMINS, 1966, 1967). The current presentation is an attempt to expand former summaries and to indicate ecologically and taxonomically defined areas where caloric data are lacking.

Discussion of calorimetry methods

A number of problems associated with the measurement and interpretation of calorific equivalents need to be stressed in order to place the present paper in proper context.

The total theoretical range of organismic caloric values is 5400 cal/ash-free gm, from an average for pure carbohydrate of 4100 to 9500 for fats. However, actual organisms would be expected to exhibit only a portion of this range with an average closer to that for protein (5100 cal/ash-free gm). Assuming that most field sampling programs in ecosystem analysis do well to operate at a variance of less than 10%, only differences of 500 to 1000 calories per gram would be considered significant in most ecosystem studies at present. Calorific variations of lesser magnitude probably would represent reliable differences in detailed energy budget studies at the population level. Therefore, given the large variance with which ecosystem ecologists are often forced to deal, it may be more realistic at present to use a median caloric value, or a grand mean, or producer and consumer means.

Within the range of calorific values determined for whole individuals of a given species, variables such as season of collection, life history stage, sex, reproductive condition and nutritional history assume considerable importance. Of particular interest, with regard to seasonal differences, are the extremely high values of premigratory birds resulting from the storage of large fat reserves. Life history stage is also of importance relative to lipid reserves, for example the elaboration of fat bodies in prepupal insects.

Fruit-bearing producers, spore-bearing microconsumers and egg-bearing animal consumers usually yield the highest caloric content for a given species. Male organisms frequently show lower values than females, again related to differential fat reserves.

The diet of animals prior to collection for calorie determinations is of significance not only because of its relation to the condition of fat stores, but also since most literature values for whole individuals include the gut contents of specimens. Special considerations are necessary in this case. For example, an investigator interested in the caloric content of a prey organism as part of a food chain study undoubtedly should use values which include gut contents-at least until assimilation experiments demonstrate the material not to be utilized by the next trophic level. In other words, in calorimetry, just as in many other facets of ecosystem studies, it is important to distinguish between tissue and gut phenomena.

In addition to variability associated with the nature of the material analyzed, there is the problem of methodology; standard methods have yet to be achieved. The four most commonly employed procedures are: 1. various models of oxygen bomb calorimeters manufactured by the Parr Instrument Company (Moline, Illinois, U.S.A. 61265); 2. the Phillipson-type microbomb calorimeter (PHILLIPSON, 1964) or its modified form supplied by Gentry-Wiegert Instruments (313 Silver Bluff Road, Aiken, South Carolina, U.S.A. 29801); 3. wet dichromate oxidation as described by MACIOLEK (1962); and, 4. calculation from protein (usually estimated from nitrogen determinations), fat (usually measured by soxhlet ether extraction) and carbohydrate (usually obtained by difference) content, using the caloric equivalents given above (e.g. BIRGE & JUDAY, 1922; SPOEHR & MILNER, 1949; KETCHUM & REDFIELD, 1949; VINOGRADOV, 1953). A miniature adiabatic oxygen bomb of the "Parr-type" developed by McEWAN & ANDERSON (1955) has been used by a few investigators (see Table 3).

Although considerable disagreement is to be expected, the authors' recommendations are summarized in Fig. 1. Pretreatment techniques need to be standardized. Since a number of suitable instruments are available for making the actual caloric determinations, the selection of an instrument is dependent primarily on the range of sample sizes to be combusted. Because of the assumptions that must be made when caloric values are calculated from protein, fat and carbohydrate determinations, an equally fruitful approach may be to assign an average caloric value, from Table 2, for the ecological or taxonomic group in question.

Concerning the problem of expressing calorific values on a per gram wet, dry or ash-free dry weight basis, it seems that all three are required, at least for certain kinds of studies. Since most ecologists are presently concerned with converting biomass data to calorific equivalents, the most useful form of the caloric data is that which conforms to the most frequently employed form of the biomass data. This is clearly dry weight, until ash-free (i.e. organic) weight values are more frequently utilized. However, there is no doubt that comparisons of the caloric content of organisms, for example along phylogenetic lines, should be made on an ash-free dry weight basis.

Dry weights are most frequently determined after oven drying at between 80 and 105° C and storage in a desiccator, both treatments being of a least 24-hour duration. As shown in Fig. 1, we recommend standardization of oven drying at 105° C. but if the material contains more than 1 % organics volatilized

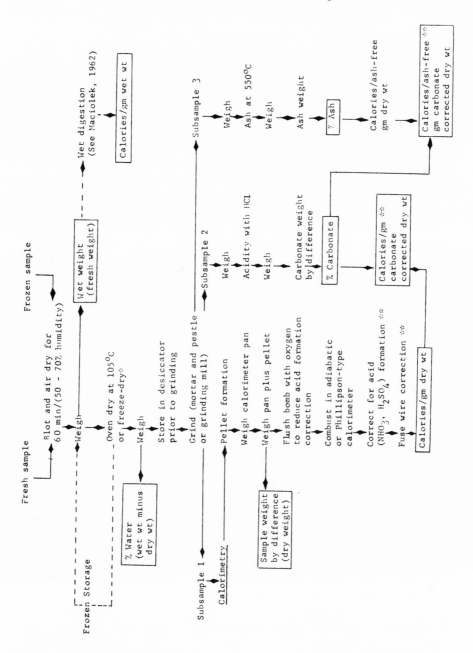

Fig. 1. Methods for determining caloric values (per gram wet, dry and ash-free dry weight) with recommended corrections. °Freeze drying, oven drying at 50° C (or air drying for very small samples) followed by desiccation over P_2O_5 is recommended especially if the material is known to have a high lipid content or to contain greater than 5% organics volatilized on drying. °°In many cases these corrections are negligible. Calcium carbonate content can be determined by combusting subsamples at 950° C.

on drying, freeze drying is recommended. Best results are obtained if the material to be combusted is compressed into a pellet. Such pellets, stored in a desiccator until use, should be weighed just prior to combustion (balance chamber should be supplied with desiccant). To avoid loss of material in transferring the weighed pellet to the bomb chamber, the most satisfactory procedure is to weigh the platinum combustion cup without the pellet and then with the pellet. Then the cup plus pellet is placed in the holder within the chamber and the fuse wire positioned.

The large variations associated with wet weight (i.e. fresh weight or as used by some investigators, air dry weight) determinations and the lack of information on % of water content for most organisms render this the least desirable form for calorific data. Nevertheless, in certain growth (production) studies in which the same individual must be weighed at intervals, wet weight caloric conversion values are necessary.

Since fuse wire contamination of the residual sample is always a problem in bomb calorimetry, ash values should be determined on separate samples by combustion in a muffle furnace at 550° C for three hours. As PAINE (1964, 1966) has pointed out, high inorganic salt and hydrated skeletal material can be sources of errors in both calorimetry and ashing. Aside from the problem of furnace accuracy (PAINE, 1964), temperatures between 500 and 900° C can cause weight loss due to the breakdown of carbonates ($MgCO_3$ at 350° C, $CaCO_3$ at 898° C). If more than 25% of the dry weight of the organism is carbonate a correction is necessary for endothermic reactions (PAINE, 1966) occurring during combustion in the calorimeter. The empirically determined correction was 0.14 cal/mg $CaCO_3$ (PAINE, 1966). Although most organisms can be expected to contain less than 25% carbonates, for greatest accuracy the correction should be made based on independent determinations of carbonate content by combustion of subsamples at 925° C.

A correction for acid formation (HNO_3 and H_2SO_4) during combustion in the calorimeter is also desirable (Parr Instrument Co.; GOLLEY, 1961) although in many cases it is negligible.

Of course, it is always desirable to avoid utilization of benzoic acid, membrane filter or some other material of known or previously determined caloric content as a "carrier". However, if the sample size is of necessity below the accurate range for the Phillipson-type calorimeter (< 5 mg), there may be no alternative. For this reason, caloric values for some miscellaneous materials have been included in Table 3. If the material is greater than 30% ash it will be difficult to obtain complete combustion. In such cases a high energy carrier such as mineral oil is necessary in order to obtain reliable values.

Regardless of the problems associated with calorimetry, it seems desirable to continue to take stock of the caloric values obtained thus far. In most instances, these values carry with them specific data as to the nature of the material burned and some estimate, such as standard deviation, of the variation encountered within a given set of samples. Comparison of caloric values obtained for the same species by different investigators should allow conclusions to be drawn concerning seasonal, habitat, dietary and other differences that might be expected. Thus, if either extreme prevails, that is, very narrow ranges cutting across vastly different taxonomic and ecological groupings or wide ranges of variation even within the same species, we will have the data from which to select an approach that will yield maximum benefit to ecosystem research.

Discussion of tabular material

The data certainly indicate the desirability of using separate values for producers (mean from Table 1 = 4685), microconsumers (mean = 4958) and macroconsumers (mean = 5821). The values for detritus, which include both the organic substrates and the microflora (and, undoubtedly, certain microfaunal elements also), do not differ very .much from the primary producer values (500 calorie/ash-free gm difference). Although there is only approximately a 200 calorie difference between the aquatic and terrestrial invertebrate means, over 1200 calories separate the aquatic and terrestrial vertebrate means. This results from the heavy dominance of low fish values in the former and of high bird values in the latter.

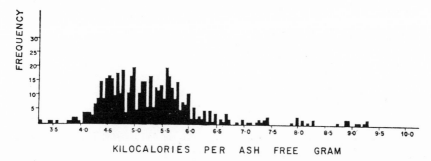

Fig. 2. Frequency distribution per ash-free gram dry weight caloric values for organisms (from Table 1).

A frequency distribution of the species caloric values, on a per ash-free gram basis, has been plotted in Fig. 2. A similar presentation of calorific data by SLOBODKIN & RICHMAN (1961) showed a distribution skewed in the direction of lower caloric values. The authors related such a distribution to maximization of progeny output but only sporadic selection for high energy content per unit weight. Fig. 2 shows a skewed distribution between the range of 3300 — 9400 calories/ash-free gram. This sort of distribution would be expected given the predominance of plant values in the data that were plotted.

Obviously the tabulation is not complete, particularly since many data undoubtedly exist in theses and manuscripts not yet brought to our attention. We wish to gratefully acknowledge the assistance of a great many ecologists throughout the world who supplied data for inclusion in Table 3 and brought published and unpublished values to our attention. We are particularly grateful for the helpful suggestions made by Drs. R. T. PAINE, R. G. WIEGERT, R. G. WETZEL, W. OHLE and V. SLÁDEČEK.

The data in Tables 1—3 have been organized according to categories intended to be of maximum use to ecosystem ecologists. The primary organization is according to trophic levels, the secondary organization according to habitat. Within secondary categories, the data are organized by taxonomic grouping down to the family level. When generic and specific designations were available, they have been included along with common names. Mean values have been summarized in Tables 1 and 2. These are not "true means" since entries from Table 3 were averaged rather than individual determinations. The outline of the tabular presentations is as follows:

Primary Producers
Aquatic
Terrestrial

Microconsumers
Aquatic
Terrestrial
Detritus (microconsumers plus substrate)
Aquatic
Terrestrial

Macroconsumers
Aquatic
Terrestrial

Literature cited in text

BIRGE, E. A. & JUDAY, C., 1922: The inland lakes of Wisconsin. The Plankton. I. Its quantity and chemical composition. — *Bull. Wisconsin Geol. Nat. Hist. Sur.* 64 (Sci. Ser. 13), 1—222.

CUMMINS, K. W., 1966: Calorific equivalents for studies in ecological energetics. — *Unpubl. Mimeo. Rept., University of Pittsburgh.* 26 pp.

— 1967: Calorific equivalents for studies in ecological energetics. — *Unpubl. Mimeo. Rept., University of Pittsburgh.* 52 pp.

GOLLEY, F. B., 1961: Energy values of ecological materials. — *Ecology* 42, 581—584.

KETCHUM, B. H. & REDFIELD, A. C., 1949: Some physical and chemical characteristics of algae grown in mass culture. — *J. Cell. Comp. Physiol.* 33, 281—300.

LINDEMAN, R. L., 1942: The trophic-dynamic aspect of ecology. — *Ecology* 23, 399—418.

MACIOLEK, J. A., 1962: Limnological organic analyses by quantitative dichromate oxidation. — *Rept. Bur. Sport Fish. Wildlife* 20, 1—61.

McEWAN, W. S. & ANDERSON, C. M., 1955: Miniature bomb calorimeter for the determination of heats of combustion samples of the order of 50 mg mass. — *Rev. Sci. Instruments* 26, 280—284.

PAINE, R. T., 1964: Ash and calorie determinations of sponge and opisthobranch tissues. — *Ecology* 45, 384—387.

— 1966: Endothermy in bomb calorimetry. — *Limnol. Oceanogr.* 11, 126—129.

PHILLIPSON, J., 1964: A miniature bomb calorimeter for small biological samples. — *Oikos* **15**, 130—139.

— 1966: *Ecological energetics.* — St. Martins Press, N. Y. 57 pp.

SPOEHR, H. A. & MILNER, H. W., 1949: The chemical composition of *Chlorella;* effect of environmental conditions. — *Plant Physiol.* **24**, 120—149.

SLOBODKIN, L. B. & RICHMAN, S., 1961: Calories/gm in species of animals. — *Nature* **191**, 299.

STRAŠKRABA, M., 1968: Der Anteil der höheren Pflanzen an der Produktion der stehenden Gewässer. — *Mitt. int. Ver. Limnol.* **14**, 212—230.

VINOGRADOV, A. P., 1953: The elementary chemical composition of marine organisms. — *Mem. Sears Found. Mar. Res.* **2**, 1—647.

Address of the authors:

Dr. KENNETH W. CUMMINS and JOHN C. WUYCHECK, W. K. Kellogg Biological Station, Michigan State University, Hickory Corners, Michigan 49060, U.S.A.

[*Editor's Note:* Tables 2 and 3 are not reproduced here.]

Table 1. Grand mean caloric values for organisms, arranged by trophic level, habitat and taxonimic category. Notations are those used in all 3 tables; see notes following Table 3. (Total sample number minimal since for those cases in which sample number was not given unity was used.)

Ecological and Systematic Position	Cal/gm dry wt.	Total samples	Number averaged	Cal/gm ash-free dry wt.	Total samples	Number averaged	Cal/gm wet wt.	Total samples	Number averaged	Notations: (life stage, sex and parts of organisms used)
Primary Producers	4135	1070	342	4681	769	255				
Aquatic (Grand Mean)	3482	502	126	4639	4I0	126				
Algae	3277	396	93	4628	359	89	611	171	40	1
Chlorophyta	3850	89	22	4780	55	15	847	22	5	1
Chrysophyta	3814	9	3	5310	4	2				1
Phaeophyta	3056	127	25	4496	120	25	459	81	15	1
Rhodophyta	3170	120	39	4582	120	39	666	68	20	1
Cyanophyta	1367	46	3	4882	53	6				1
Mixed Algae	4477	5	1	4669	5	1				1
Periphyton				4520	2	1				1
Bryophyta				4303	7	4				3a
Pteridophyta				4440	1	1				
Spermatophyta	4062	106	33	4716	43	32				
Angiospermae										5b, 6a, 5a, 5c, 5d, 8d, 8b, 5f, 8a, 5p, 5m, 51, 5k
Monocotyledoneae	4099	101	30	4770	32	24				
Dicotyledoneae	3695	5	3	4555	11	8				
Terrestrial (Grand Mean)	4516	568	216	4758	359	115				5a, 5f, 6a
Eumycetes	3856	2	2							2b

Table 1   11

Ecological and Systematic Position	Cal/gm dry wt.	Total samples	Number averaged	Cal/gm ash-free dry wt.	Total samples	Number averaged	Cal/gm wet wt.	Total samples	Number averaged	Notations: (life stage, sex and parts of organisms used)
Bryophyta				4458	12	6				3a, 3d
Lichenes				4324		9				3b
Pteridophyta				4609		4				3a
Spermatophyta	4539	566	204	4824	164	96				
Gymnospermae	6005	29	8	5729	11	4				4f, 4b, 4h, 4i
Angiospermae	4479	537	196	4785	153	92				
Monocotyledoneae	4365	267	61	4580	16	8				5i, 5c, 8c, 8b, 5a, 5h, 5k, 5b, 6a, 5m, 5g
Dicotyledoneae	4558	258	125	4879	137	70				5m, 5i, 5c, 6a, 8b, 5a, 5b, 5d, 5k, 8c, 5f, 5o, 5j, 5i, 5g, 7a, 7d, 7e, 8a, 8e, 6b, 6d, 6c
Alpine Vegetation (Grand $\bar{X}$)				4383	173	12				5b, 5g, 5c, 5n
Mixed Woodland		22		4719	10	2				5b
Old field Vegetation	4189		10							5b, 5g
Microconsumers										
Aquatic	4713	8	2	4958	8	2				
Detritus	4414	141	44	4885	264	52				1
Aquatic	4422	111	38	5168	111	38				5g, 6a, 5o
Terrestrial	4371	30	6	4117	153	14	6243	81	27	5b, 22d, 22e, 22g, 22c

Ecological and Systematic Position	Cal/gm dry wt.	Total samples	Number averaged	Cal/gm ash-free dry wt.	Total samples	Number averaged	Cal/gm wet wt.	Total samples	Number averaged	Notations: (life stage, sex and parts of organisms used)
Macroconsumers (Grand Mean)	4953	1061	357	5821	765	298	1093	232	72	
Aquatic (Grand Mean)	4301	631	155	5465	524	139	672	127	47	
Invertebrates (Grand Mean)	4229	600	142	5470	514	135	635	125	45	
Protozoa				5938	1	1				1
Porifera	1295	6	1	6475	4	1				1, 9a
Platyhelminthes				6332	3	2	1330	1	1	9a
Coelenterata	2886	2	1	5882	2	2	494	1	1	9b
Mollusca	3120	54	16	5492	163	20	480	17	6	9b, 9a, 16a, 10a
Annelida	3910	77	22	4700	1	1	645	32	13	9a, 10a
Echinodermata	2020	20	8				351	25	8	9a
Arthropoda	4726	441	94	5445	339	107	792	49	16	9a, 10c, 10k, 10l, 10m, 9i, 10f, 10g, 16f, 16g, 10o, 10p, 10h, 10e, 10d, 10b, 12g, 11a, 16b, 14a, 10a, 9f, 11b, 9c, 9d, 9e, 13d, 13e, 13f, 13l, 13i, 13u, 10u, 10s, 11d, 16a, 16h, 13a, 10i, 13c, 13x, 10v, 13g, 12a
Brachiopoda				4397	1	1				
Vertebrates				5296	10	4				
Chondrichthyes (R. orinacea) eggs				5600	1	1				16a
Osteichthyes				5296	10	4	1493	2	2	
Terrestrial (Grand Mean)	5453	430	202	6099	243	167	1884	105	25	9a, 16a, 16e, 10t, 10d
Invertebrates	5274	224	93	5673	157	85	2008	2	5	

Table 1 — page 13

Ecological and Systematic Position	Cal/gm dry wt.	Total samples	Number averaged	Cal/gm ash-free dry wt.	Total samples	Number averaged	Cal/gm wet wt.	Total samples	Number averaged	Notations: (life stage, sex and parts of organisms used)
Platyhelminthes	4569	3	2	5684	1	1				9a
Annelida				5628	1	1	782	1	1	9a
Arthropoda	5289	221	91	5673	155	83	2319	1	4	10a, 9a, 16a, 13d, 13a, 13f, 13g, 13h, 11a, 10c, 12f, 13e, 13k, 10n, 10r, 16l, 15a, 15b, 15c, 15d, 15e, 15f, 15g, 15h, 17b, 17c, 17d, 12a, 16a, 13m, 13n, 13o, 13p, 13q, 13r, 13s, 13t, 22a, 12d, 12e, 10v
Vertebrates	5606	206	109	6542	86	82	1853	103	20	16d, 13w, 31a, 16a, 3b, 20a, 20d, 20e, 20g, 20b, 20c, 20f, 20h, 10a, 20i, 20j, 20l, 20m, 20k, 18e, 18f, 22b, 11a, 9g, 16k, 10c

[*Editor's Note:* Material has been omitted at this point.]

III. Life stage, sex and parts of organisms used

A. Bacteria, algae, protozoa. sponges
    1. Entire cells or colonies of cells or loosely grouped cells (marine algae)

B. Fungi
    2a. hyphae and sporulating portions
    2b. sporulating portions
    2c. spores
    2d. hyphae

C. Mosses, liverworts, lichens, ferns
    3a. entire plant
    3b. thallus
    3c. non-spore bearing portions
    3d. capsules with spores

D. Gymnosperms
    4a. entire plant
    4b. needles
    4c. stem
    4d. flowers
    4e. cones, with seeds
    4f. seeds
    4g. roots
    4h. twigs
    4i. pollen

E. Angiosperms
    5a. entire plant
    5b. stems, leaves, fruits, flowers (all above ground parts)
    5c. stems, leaves (above ground parts, except fruits and flowers)
    5d. stems, leaves and flowers
    5e. stems, leaves and fruits
    5f. stems
    5g. roots
    5h. stems, leaves and seeds
    5i. twigs
    5j. xylem
    5k. roots, stems, leaves, flowers
    5l. roots, stems, leaves, fruits
    5m. roots, stems, leaves
    5n. roots, stems, leaves, flowers
    5o. bark
    5p. entire young plants
    6a. leaves
    6b.—6d. sized green leaves: b = small; c = medium; d = large
    7a. flowers
    7b. male flowers
    7c. female flowers
    7d. bracts
    7e. receptacle plus peduncle
    7f. nectar
    7g. pollen
    8a. fruits
    8b. seeds

      8c. hulled seeds
      8d. seeds and flowering heads
      8e. pericarps

F. Invertebrates
      9a. entire animals of varying stages (age classes or sizes) and both sexes (or age and sex not specified)
      9b. entire animals without shells (Mollusca)
      9c.—9e. entire animals of both sexes (or sex not specified) grouped into general size categories: c = small; d = medium; e = large
      9f. immature (i.e. not sexually mature) stages (non-arthropod groups)
      9g. immature males
      9h. immature females
      9i. entire male animals, all stages

      10a. adults, both sexes (or sex not specified)
      10b. adult females, reproductive
      10c. adult females, non-reproductive
      10d. adult females with eggs (eggs and embryos at various stages of development, or stage not specified)
      10e. adult females without eggs
      10f. females with summer eggs
      10g. females with winter eggs
      10h. females with and without eggs, mixed
      10i. subimagos, both sexes (Ephemeroptera)
      10j. female subimagos
      10k.—10m. various size classes, females without eggs: k = 2—5 mm; l = 6—7 mm; m = 10—12 mm
      10n. V instar females
      10o. non-reproductive adults < 1.5 mm (sex not determined)
      10p. females > 1.5 mm with eggs (or young)
      10q. females > 1.5 mm without eggs (or young)
      10r. VI instar females
      10s. adults without chelipeds, both sexes (or sex not specified)
      10t. adults, cephalothorax (and appendages) only
      10u. adults, cephalothorax, without chelipeds
      10v. newly emerged adults

      11a. adult males, reproductive
      11b. adult males, non-reproductive
      11c. male subimagos (Ephemeroptera)
      11d. 7 cm males

      12a. pupae, all stages and sexes (or sex and stage not specified)
      12b. male pupae
      12c. female pupae
      12d. prepupae (same designations as for 12a.)
      12e. 11 day pupae
      12f. molted exoskeleton
      12g. chitin

      13a.—13u. larvae (or nymphs) of both sexes (or sex not specified) grouped according to various size (stage, age or instar) categories:

          a. = various sizes (or size and age not specified)
          b. = early stages
          c. = late stages
          d. = instar I
          e. = instar II
          f. = instar III

          g. = instar IV
          h. = instar V
          i. = stage VA
          j. = stage VB
          k. = instar VI
          l. = instar VII
          m. = instar VIII

n. = instar IX   r. = instar XIII
o. = instar X   s. = instar XIV
p. = instar XI   t. = instar XV
q. = instar XII   u. = post larval (nymphal) stages

13v. just prior to emergence (nymphs) or pupation (larvae)
13w.—13.y. reared in an aquarium from eggs
     w. = one month   y. = three months
     x. = two months

14a. naupliar stages (specific instars not specified)
14b.—14g. copepodite stages
     b. = I   e. = IV
     c. = II   f. = V
     d. = III   g. = V, fat only

15a.—15d. male larval (nymphal) stages (or instars)
     a. = I   c. = III
     b. = II   d. = IV
15e.—15h. female larval (nymphal) stages (or instar)
     e. = I   g. = III
     f. = II   h. = IV

16a. eggs (and embryos, various stages of development, or stage not given)
16b. egg sacs with eggs
16c. ovaries with eggs
16d. newly fertilized eggs
16e. yolk only
16f. summer eggs
16g. winter eggs
16h. eggs, no development
16i. oöthecae, newly laid
16j. oöthecae, fully developed
16k. embryos various stages (or stage not given), but some development completed

17a.—17e. social insects
     a. = all castes   d. = reproductive males (kings)
     b. = workers   e. = reproductive females (queens)
     c. = soldiers

G. Vertebrates
18a. adults, all sizes (or ages) and both sexes (or sex not specified)
18b.—18c. sexually mature adults
     b. = males   c. = females
18d. fat extracted adult females
18e. fat extracted adults (size and sex as in 18a.)
18f. adult body fat (size and sex in 18a.)
18g.—18h. breeding adults
     g. = males   h. = females
18i. adult females with eggs (eggs and embryos at various stages of development or stage not specified)
18j.—18k. non-reproductive adults
     j. = males   k. = females

19a. immatures (or juveniles), all sizes (or ages) and both sexes (or sex not specified)
19b. immature males
19c. immature females
19d. immature (not sexually mature)

20a.—20m. ages in days
     a. = 1   c. = 18   e. = 26
     b. = 8   d. = 19   f. = 28

g. = 37	j. = 60	m. = 180
h. = 38	k. = 65	
i. = 46	l. = 92	

21a. ages in months
    a. = 7

22a. eggs (and embryos, various stages of development, or stage not specified)
22b. newly fertilized eggs
22c. yolk only
22d. embryos, various stages (or stages not specified) but some development completed
23a. 144 hr tadpoles
23b. tadpole larvae (hours or days of development not specified)

H. Detritus (+ microconsumers)
    24a. feces, larval (nymphal) or juvenile
    24b. feces, adult
    24c. particulate organic matter
    24d. dead leaves
    24e. dead twigs (or stems)
    24f.—24h. sized dead leaves
        f. = small   h. = large
        g. = medium

IV. Methods of analysis

1. not given (left blank in the table)
2. wet digestion (specific technique not given)
    2a. dichromate digestion method (MACIOLEK, 1962)
3a.—3d. Parr oxygen bomb (Parr Instrument Co., Moline, Illinois)
    a. macrobomb non-adiabatic
    b. macrobomb adiabatic
    c. semi-microbomb non-adiabatic
    d. semi-microbomb adiabatic
    e. peroxide bomb
4. Phillipson microbomb
5. Gentry-Wiegert bomb (Modified Phillipson bomb) Gentry-Wiegert Instruments, Aiken, South Carolina
6. BERTKELAT bomb
7. Values calculated from organic analysis (protein, i.e. nitrogen, and/or lipid, and/or carbohydrate — the latter usually by difference)
8. McEWAN & ANDERSON miniature bomb
9. GALLENHAMP ballistic bomb calorimeter
10. Unknown oxygen bomb calorimeter

V. Authors and sources

1. COMITA, W. G. & SCHINDLER, D. W., 1963: Calorific values of microcrustacea. — *Science* 140, 1394—1395.
2. SMIRNOV, N. N., 1962: On nutrition of caddis worms *Phryganea grandis* L. — *Hydrobiologia* 19, 252—261.
3. IVLEV, I. S., 1934: Eine Micromethode zur Bestimmung des Kaloriengehalts von Nährstoffen. — *Biochem. Z.* 275, 49—55.
4. JOHSI, BHARAT S., 1965: Original data (under direction of R. T. HARTMAN). — Biology Department, University of Pittsburgh, Pittsburgh, Pa.
5. COFFMAN, W. P., WUYCHECK, J. C. & CUMMINS, K. W., 1966: Original data. — Pymatuning Laboratory of Ecology, University of Pittsburg, Pittsburg, Pa.

6. RICHMAN, S., 1958: The transformation of energy by *Daphnia pulex*. — *Ecol. Monogr.* **28**, 275—291.

7. KETCHUM, B. H. & REDFIELD, A. C., 1949: Some physical and chemical characteristics of algal growth in mass cultures. — *J. Cell. Comp. Physiol.* **33**, 281—299.

8. TRAMA, F. B., 1957: The transformation of energy by an aquatic herbivore, *Stenonema pulchellum*. — Unpubl. Ph.D. Dissertation, Univ. Mich., Ann Arbor, Mich.

9. KUENZLER, E. J., 1961: Unpublished data. — Dept. Zool., Univ. Georgia, Athens, Ga.

10. BLISS, L. C., 1962: Caloric and lipid content in alpine tundra plants. — *Ecology* **43**, 753—754.

11. LONG, F., 1934: Application of calorimetric methods to ecological research. — *Plant Physiol.* **9**, 323—337.

12. GOLLEY, F. B., 1958: Energy dynamics of a food chain of the old field community. — Unpubl. Ph.D. Dissertation. Mich. State Univ., East Lansing, Michigan, and unpubl. data, Dept. Zool., Univ. of Georgia, Athens, Ga.

13. KENDEIGH, C. S. & WEST, G. C., 1965: Caloric values of plant seeds eaten by birds. — *Ecology* **46**, 553—555.

14. CONNELL, C., 1961: Unpubl. data. — Dept. Zool., Univ. Georgia, Athens, Ga.

15. SCHMID, W. C., 1965: Energy intake of the mourning dove *Zenaidura macronra marginella*. — *Science* **150**, 1171—1172.

16. TRYON, C. A., 1962—1966: Original data. — Pymatuning Laboratory of Ecology, Univ. of Pittsburgh, Pittsburgh, Pa.

17. BLISS, L. C., 1961: Unpubl. data. — Dept. Botany, Univ. Illinois, Urbana, Ill.

18. SLOBODKIN, L. B. & RICHMAN, S., 1961: Calories/gm in species of animals. — *Nature* **191**, 299.

19. McEWAN, W. S. & ANDERSON, C. M., 1955: Miniature bomb calorimeter for the determination of heats of combustion samples of the order of 50 mg mass. — *Rev. Sci. Instrum.* **26**, 280.

20. PAINE, R. T., 1965: Natural history, limiting factors and energetics of the opisthobranch *Navanex inermis*. — *Ecology* **46**, 603—619.

21. PAINE, R. T., 1964: Ash and calorie determinations of sponge and opisthobranch tissues. — *Ecology* **45**, 384—387.

22. TEAL, J. M., 1957: Community metabolism in a temperate cold spring. — *Ecol. Monogr.* **27**, 283—302.

23. ENGLEMANN, M. D., 1961: The role of soil arthropods in the energetics of an old field community. — *Ecol. Monogr.* **31**, 221—238.

24. TUBB, R. A. & DORRIS, T. C., 1965: Herbivorous insect populations in oil refinery effluent holding pond series. — *Limnol. Oceanogr.* **10**, 121—134.

25. GIBBS, J., 1957: Food requirements and other observations on captive tits. — *Bird Study* **4**, 207—215.

26. GOLLEY, F. B., 1961: Energy values of ecological materials. — *Ecology* **42**, 581—584.

27. Parr Instrument Co., Moline, Illinois, Standard benzoic acid pellets or powder supplied having the calorie value given in the table.

28. ODUM, E. P., MARSHALL, S. G. & MARPLES, T. G., 1965: The calorie content of migrating birds. — *Ecology* **46**, 901—904.

29. WIEGERT, R. G., 1965: Intraspecific variation in calories/g of meadow spittlebugs (*Philaenus spumarius* L.). — *Bioscience* **15**, 543—545. And: SLOBODKIN, L. B., 1962: Energy in animal ecology. — pp. 69—101. In: CRAGG, J. B. (ed.), *Advances in ecological research*. Vol. I. Academic Press, New York. 203 pp.

30. GORHAM, E. & SANGER, J., 1967: Calorie values of organic matter in woodland, swamp and lake soils. — *Ecology* 48, 492—494.

31. DAVIS, G. E. & WARREN, C. E., 1965: Trophic relations of a sculpin in laboratory stream communities. — *J. Wildlife Management* 29, 846—871.

32. SMALLY, A. E., 1960: Energy flow of a salt marsh grasshopper population. — *Ecology* 41, 672—677.

33. COMITA, G. W., MARSHALL, S. M. & ORR, A. P., 1966: On the biology of *Calanus finmarchicus*. XIII. Seasonal change in weight, calorific value and organic matter. — *J. Mar. Biol. Assoc. U. K.* 46, 1—17.

34. CONOVER, R. J., 1964: Food relations and nutrition of zooplankton. — *Occ. Publs., Narragansett Mar. Lab.* 2, 81—91.

35. KENDEIGH, S. C., 1967: Unpubl. data. — Dept. Zool., Univ. Illinois, Urbana, Ill.

36. JENKINS, R. C., 1959: Monthly variations in the population of stream invertebrates in the vicinity of Roberts, Illinois. — Unpubl. M.S. Thesis, Dept. Zool., Univ. Illinois, Urbana, Ill.

37. MINSHALL, G. W., 1969: Unpubl. data. — Dept. Biology, Idaho State Univ., Pocatello, Idaho.

38. TOETZ, D. W., 1966: The change from endogenous to exogenous sources of energy in bluegill sunfish larvae. — *Invest. Indiana Lakes Streams* 7, 115—146.

39. MALONE, C. R., 1968: Variation in caloric equivalents for herbs as a possible response to environment. — *Bull. Torrey Bot. Club* 95, 23—34.

40. BELICHICK, R. L., 1968: The effect of supplemental food on bioenergetics and population processes of *Microtus pennsylvanicus* in an old-field area. — Unpubl. M.S. thesis. Johns Hopkins Univ., Baltimore, Maryland.

41. THOMAS, W. A., 1968: Energy content of dogwood trees. *Oak Ridge Nat '1. Lab., Health Physics Div. Annual Prog. Rept.* (ORNL) 4316, 94—97.

42. BOAG, D. A. & KICENIAK, J. W., 1968: Protein and caloric content of lodge pole pine needles. — *Forestry Chronicle* 4, 1—4.

43. LEAR, S. I., 1969: Unpubl. data. — Dept. Botany, Ohio Univ., Athens, Ohio.

44. REICHLE, D. E. & CROSSLEY, D. A., JR., 1967: Investigation on heterotrophic productivity in forest insect communities. Pp. 563—587. In: PETRUSEWICZ, K. (ed.). *Secondary productivity of terrestrial ecosystems.* Panstwowe Wydawnictwo Naukowe, Warsaw-Krakow.

45. WILLIAMS, E. C., JR., & REICHLE, D. E., 1968: Radioactive tracers in the study of energy turnover by a grazing insect (*Chrysochus auratus* FAB.; Coleoptera Chrysomelidae). — *Oikos* 19, 10—18.

46. PAINE, R. T. & VADAS, R., 1969: Unpubl. data. — Dept. Zool., Univ. Washington, Seattle, Wash.

47. WIEGERT, R. G. & EVANS, F. C., 1964: Primary production and the disappearance of dead vegetation on an old field in southeastern Michigan. — *Ecology* 45, 49—63.

48. MALONE, C. R. & SWARTOUT, M. B., 1969: Size, biomass and caloric content of particulate organic matter in old field and forest soils. — In preparation.

49. WISSING, T. E. & HASLER, A. D., 1968: Calorific values of some invertebrates in Lake Mendota, Wisconsin. — *J. Fish. Res. Bd. Can.* 25, 2515—2518.

50. SITARAMAIAH, P., 1967: Water, nitrogen and calorific values of freshwater organisms. — *J. Cons. perm. int. Explor. Mer.* 31, 27—30.

51. LAWTON, J. H., 1969: Energy flow through *Pyrrhosoma nymphula* populations in a river. — Unpubl. data. Dept. Zool., Oxford Univ., Oxford, England.

52. BRAWN, V. M., PEER, D. L. & BENTLEY, R. J., 1968: Caloric content of the standing crop of benthic and epibenthic invertebrates of St. Margaret's Bay, Nova Scotia. — *J. Fish. Res. Bd. Can.* 25, 1803—1811.

53. EDWARDS, C. A., REICHLE, D. E. & CROSSLEY, D. A., JR., 1969: The role of soil invertebrates in organic matter and nutrient turnover. In: *Analysis of an ecosystem: the temperate forest.* REICHLE, D. E. (ed.). Springer-Verlag, Berlin. In press.

54. BLEM, C. R., 1969: Unpubl. data. — Dept. Zool., Univ. Illinois, Urbana, Ill.

55. KITCHELL, J. F. & NORRIS, J. S., 1969: Unpubl. data. — Fish Physiology Laboratory, University of Colorado, Boulder, Colo. And: KITCHELL, J. F. & WINDELL, J. T., 1969: Nutritional value of algae to bluegills. — In preparation.

56. WILHM, J. L. & GRAHAM, V., 1969: Unpubl. data. — Dept. Zool., Oklahoma State Univ., Stillwater, Oklahoma.

57. MOSHIRI, G. A. & CUMMINS, K. W., 1969: Calorific values for *Leptodora kindtii* (FOCKE) (Crustacea Cladocera) and selected food organisms. — *Arch. Hydrobiol.* 66, 91—99.

58. SNOW, N. B., 1969: Unpubl. data, Ph.D. thesis in preparation. — University of Southampton, Southampton, England.

59. MOSHIRI, G. A., 1969: Unpubl. data. — Institute of Ecology, Univ. California, Davis, California.

60. HARGRAVE, B., 1969: Unpubl. data, Ph.D. thesis in preparation. — Dept. Zool., Univ. British Columbia, Vancouver, Canada.

61. WOODLAND, D. J., 1969: Unpubl. data. — Dept. Zool., Univ. New England, Armidale, N.S.W., Australia.

62. MATHIAS, J. A., 1966: Unpubl. data. — Univ. British Columbia, Vancouver, Canada.

63. WOODLAND, D. J., HALL, B. K. & CALDER, J., 1968: Gross bioenergetics of *Blattella germanica.* — *Physiol. Zool.* 41, 424—431.

64. REICHLE, D. E., 1967: Radioisotope turnover and energy flow in terrestrial isopod populations. — *Ecology* 48, 351—366.

65. WIEGERT, R. G., 1965: Energy dynamics of the grasshopper populations in old field and alfalfa field ecosystems. — *Oikos* 16, 161—176.

66. HINTON, J. M., 1968: A study of the energy flow in a natural population of the spittlebug *Neophilaenus lineatus* L. (Homoptera, Cercopidae). — Unpubl. Ph.D. thesis, Univ. Exeter, England.

67. QASRAWI, H., 1966: A study of the energy flow in a natural population of the grasshopper *Chorthippus parallelus* ZETT. (Acrididae). — Unpubl. Ph.D. thesis. Univ. Exeter, England.

68. O'NEILL, R. V., 1969: Unpubl. data. — Radiation Ecology Section, Health Physics Division, Oak Ridge National Laboratory, Oak Ridge, Tennessee.

69. DUTTON, R., 1968: Unpubl. data. — Univ. of Durham, England.

70. BRISBIN, I. L., 1969: Unpubl. data. — Savannah River Ecology Laboratory, AEC Savannah River Operations Office, Aiken, South Carolina.

71. BRISBIN, I. L., 1966: Energy utilization in a captive hoary bat. — *J. Mammology* 47, 719—720.

72. HUGHES, R. N., 1968: The population ecology and energetics of *Scrobiculata plana* da Costa. — Unpubl. Ph.D. thesis. Univ. College of North Wales, Bangor, Wales.

73. BRISBIN, I. L., 1968: A determination of the caloric density and major body components of large birds. — *Ecology* 49, 792—794.

74. KEVERN, N. R. & BALL, R. C., 1965: Primary productivity and energy relationships in artificial streams. — *Limnol. Oceanogr.* 10, 74—87.

75. BRAY, J. R., 1962: Estimates of energy budgets for a *Typha* (cattail) marsh. — *Science* 136, 119—120.

76. STRAŠKRABA, M., 1968: Der Anteil der höheren Pflanzen an der Produktion der stehenden Gewässer. — *Mitt. int. Ver. Limnol.* 14, 212—230.

77. GENG, H., 1925: Der Futterwert der natürlichen Fischnahrung. — Z. *Fischerei* **23**, 137—165. (See Note No. 128.)

78. JABLONSKAJA, E. A., 1935: Zur Kenntnis der Fischproduktivität der Gewässer. Mitt. V. Die Ausnutzung der natürlichen Futterarten seitens der Spiegelkarpfen und die Wertung des Futterreichtums der Wasserbecken von diesem Standpunkt. — (in Russian). *Arb. Limnol. Stat. Kossino* **20**, 99—127. (See Note No. 128.)

79. KAUSHIK, N. K., 1969: Autumn-shed leaves in relation to stream ecology. — Unpubl. Ph.D. thesis. University of Waterloo. Canada.

80. RYBAK, J. I., 1969: Bottom deposits of lakes of different trophic types. — *Ecol. Pol. s.A.* (English). (Reported by W. LAWACZ, Polish Academy of Science, Institute of Ecology, Dept. of Hydrobiology. Warszawa, Nowy Swiat 72, Poland.) (Personal Communication.)

81. CHILARECKI, R., 1968: Seasonal changes in calorific value of bottom fauna in Mikolajskie Lake. — Unpubl. M.Sc. thesis. (See reference 80.)

82. CHILARECKA, M., 1968: Calorific values of benthic organisms in different trophy lakes. — Unpubl. M.Sc. thesis. (See reference 80.)

83. SMIRNOV, A. I., KAMYSHNAYA, M. S. & KALASHNIKOVA, Z. M., 1968: Dimension, biochemical characteristics and caloric values of mature eggs of members of the genera *Oncorhynchus* and *Salmo*. — Dept. of Ichthyology, M. V. Lomonosov State University, Moscow, U.S.S.R. *Problems of Ichthyology* **8**, 524—529. (American Fisheries Society, Translated and Produced by Scripta Technica, Inc.)

84. STOCKNER, J. G., 1968: A comparative study of the ecological energetics of two thermal spring communities. — Unpubl. Ph.D. thesis. Dept. Zool., Univ. Washington, Seattle, Washington.

[*Editor's Note:* Material has been omitted at this point.]

Part II
MEASURING PRIMARY PRODUCTIVITY

Editor's Comments
on Papers 12 Through 24

METHODS OF MEASURING PRIMARY PRODUCTIVITY

The papers in Part I suggested that primary production is measured as gas exchange, dry matter, and energy. Two further dimensions need to be added—time and space. Based on these we can define *primary productivity* in three analogous ways:

1. The amount of CO_2 (or carbon) uptake (assimilation) by plants per unit of horizontal earth surface and per unit of time.
2. The amount of dry matter produced by plants per unit of horizontal earth surface and per unit of time.
3. The amount of caloric energy produced by plants per unit of horizontal earth surface and per unit of time.

These definitions are straightforward and easy to understand. However, measuring these values accurately is quite cumbersome. The three definitions are not the same, although they approach similarity on a gross scale. Carbon fixation is not the same as apparent CO_2 uptake. Carbon (or CO_2) fixation is not synonymous with dry matter production and energy accumulation, neither is dry matter formation the same as energy production. The logic can easily be deducted from Papers 11 by Cummins and Wuycheck in Part I, Transeau (1926), and Paper 18 by Golley in this section. For example, if ash content in plant material is variable, the caloric content and the amount of carbon needed to produce one weight unit of dry matter must be different. The first common product of CO_2 assimilation is sugar, the final product of the plant matter formation varies along the list of compounds mentioned by Morowitz (1968). Depending on the state of oxidation of the carbon molecules produced, the energy content varies from about 3 K cal/g to over 10 K cal/g. Therefore, using the three definitions independently is necessary since all three stress important aspects of primary plant productivity.

Agriculture and forestry deals with productivity in different ways. Economic considerations require us to consider the salable or usable product as productivity. This usage of the word is different from biological terminology inasmuch as the productivity in the economic sense constitutes only a portion of the total dry matter produced and the product contains variable amounts of water. We can call this kind of productivity "yield" and can convert yield data into primary productivity values with proper constants. Such conversions require a knowledge of the rules of allometric growth in the manner of Huxley and Bertallanfy (Bertallanfy 1951). This approach is one of the most important in assessing and comparing the productivity of large regions and must be discussed, therefore, as an important method (see Papers 16 and 20).

The papers presented in Part II demonstrate the development of production measurement techniques in the following sequence: gas analysis, dry matter–productivity assessment of terrestrial plants, aquatic productivity measurements, and caloric value measurement. These papers again are presented in an historical perspective to show when and by whom the present methods were initiated.

Gas Analysis

Priestley, Ingenhousz, De Saussure, and the others mentioned in Part I, had established that there is an exchange of components between plants and air. The following generation therefore had the task of establishing the levels and rates, as well as the basic properties of the exchange process. At the same time that Liebig, Wolff, Sachs, and others were analyzing the mineral nutrients needed for plant growth, Boussingault analyzed gas exchange and dry matter accumulation over time. He was apparently the first one to design a cuvette with controlled stream flow (see Figure 1). His work had an enormous impact on the plant physiologists of his time and his books have been translated into many languages. Paper 12 is an excerpt from the German translation, chosen because it contains an important updating footnote by the translator. The method was used by the two generations following Boussingault to analyze production processes in detail especially with regard to environmental constraints.

Around the turn of the century, Blackman and Lundegårdh made exceptional contributions to our present knowledge of a plant's gas exchange. Lundegårdh's work on gas circulation within ecosystems

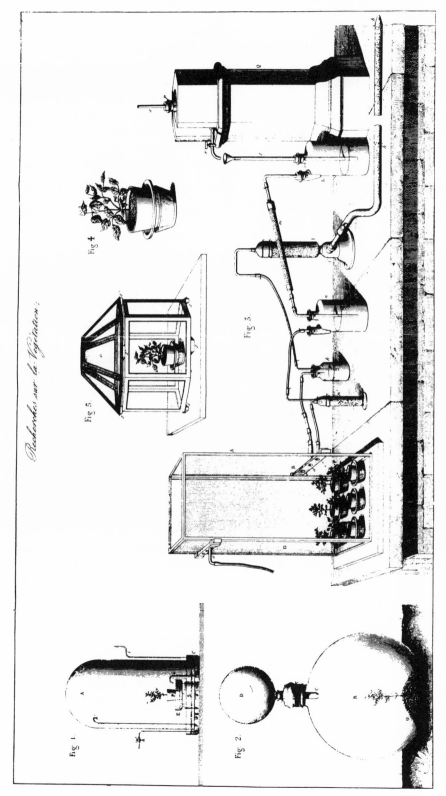

Figure 1. Boussingault's CO_2 cuvette with controlled stream flow. (Source: Boussingault 1886)

and on a global scale was so substantial that his books published mostly in Swedish and German were translated into English long after his retirement. During the 1920s and 1930s, a common practice for the world's physiological ecologists was to visit Lundegårdh's laboratory, just as it was common in earlier times for them to visit the laboratories of Sachs and Liebig. Lundegårdh's most significant contribution to the topic discussed in this volume is his book, *Kreislauf der Kohlensäure in der Natur* (Lundegårdh 1924). A chapter from the English translation of one of his textbooks is presented here as Paper 13. It gives an interesting account of the people working in this field during the early part of this century as well as the way results were presented at that time. All results were obtained with the Barium chloride titration method. This method does not allow the measurement of instantaneous CO_2 uptake of plants but it allows for the measurement of average uptake rates. Lundgårdh reports the first breakthroughs in measuring instantaneous CO_2 exchange when he mentions the heat absorption method.

All methods in use in the 1930s were superseded, however, by the introduction of the infrared-gas analyzer. This instrument was developed by two engineers of the Badische Analin and Soda Fabriken (BASF) for the coal mines in the Ruhr district of Germany that were heavily permeated with monoxide gas. Seybold (1942) first established its usefulness in physiological work. After further experimentation, Egle and Schenk published their basic paper about the successful use of the IRGAS (URAS) in physiology. Paper 14 contains a translated summary of this work, which made the method instantly famous among plant ecologists. The method is now almost the only one in use, and hundreds, if not thousands, of papers are available that demonstrate the successful use of the IRGAS all over the world. A paper by Billings, Clebsch, and Mooney demonstrating the typical application of the method is included here as Paper 15. Dr. Billings was probably the most prominent among the American ecologists to introduce and refine this method for ecological purposes, and according to him, the alpine environment described in Paper 15 was chosen for the first experiments because the temperature control of the cuvettes was easiest to achieve there.

Dry Matter Assessment

Air dry weight measurements of crops were probably made during Roman times. Whether they were actually made as weight unit or by volume is difficult to judge now. However, in no case can we see in the older literature, even up to van Helmont's famous experiment, that total dry weight per unit area and unit time were assessed. This assessment probably starts with Boussingault and Liebig although these two had not yet considered the underground portion to the full extent. However, the following figures from Boussingault's (1840) famous treatise have survived for almost 75 years as the standard for agricultural productivity. They are taken from his third experiment on cropping wheat as reported in Graeger's translation of Boussingault's book (1851):

Forty-six wheat grains were sowed at the beginning of August in burnt sand. The plants grew to 36–38 cm in height by the end of September. Most lower leaves were yellow. The roots had substantially expanded and were intermingled such that washing off the soil became very difficult.

Result of the analysis:

	Grain		Yield
Carbon	46.60		48.1
Hydrogen	5.80		5.8
Nitrogen	3.45		2.0
Oxygen	44.15		44.0
	100.00		100.0

			Net Gain
The dry weight in g	1.644	3.022	1.378
Carbon	0.767	1.457	0.690
Hydrogen	0.095	0.175	0.080
Oxygen	0.725	1.330	0.605
Nitrogen	0.057	0.060	0.003

Later, Boussingault (1846) submitted to the Akademie in Paris new experiences about the gradual increase of plant matter in wheat, which merit being recorded at this point.

Date of investigation	Dried plant weight per Ha in Kg	C in Kg	H in Kg	O in Kg	N in Kg	Minerals (Ash) in Kg
19 May 1844– 9 June 1844	689 2631	257.0 1007.7	40.0 163.1	354.1 1370.7	12.4 23.7	25.5 65.8
Increase over period yield 15 August	1942 4666	750.7 1735.8	123.1 317.3	1016.6 2324.3	11.3 42.0	40.3 186.6
Increase 9 June– 15 August	2035	728.1	154.2	953.6	18.3	120.8

J. B. Boussingault (1802–1887) described the first photosynthetic cuvette and measured annual productivity rates for crops in France. (Source: Chardon 1953)

E. Ebermayer (1829–1908) measured the primary productivity and nutrient budgets of Bavarian forests (Source: courtesy of the Institute für Bodenkunde und Standortslehre, Forstl. Forschungsanstalt München)

A bit later in time came another scientist, Ebermayer, a man whose lifetime objective was to measure the dry matter productivity and nutrient uptake of forests. He has been cited in almost every summary and discussion on primary productivity work. The basis of these citations is surprising: about 1,000 pages on wood production, litter fall, and nutrient content analyses in the style we now use for the Coweeta watershed (for example, see Johnson and Swank 1973) and the Hubbard brook watershed analysis by Likens, Bormann et al. (1970). Ebermayer's figures are just as valid today as they were a hundred years ago, as can be seen from the following excerpt:

> In order to know the total dry matter production of a forest, one must determine first the weight of leaves per ha.
>
> In my book "Die gesammte Lehre der Waldstreu etc." (a complete textbook of forest litter etc.)[1] the fundamental figures are reported for such a calculation, gathered in large numbers in the forests of the State of Bavaria. Accordingly the annual litter per ha contains in dry matter:
>
> | | |
> |---|---|
> | In harvestable beech stands | 3270–3331 kg |
> | In harvestable spruce stands | 2783–3007 kg |
> | In harvestable pine stands | 3002–3186 kg |
>
> Adding to these figures the annual wood increment, we get the total annual dry matter production from good habitats:
>
> a) in beech forests
>
> | | | |
> |---|---|---|
> | wood | 3952 kg/ha without ash[2] | 3912 kg organic substance |
> | leaves | 3331 kg/ha without ash | 3145 kg organic substance |
> | total | 7283 kg dry matter | 7057 kg organic substance |
>
> b) in spruce forests
>
> | | | |
> |---|---|---|
> | wood | 4065 kg | 4024 |
> | leaves | 3007 kg | 2896 |
> | total | 7072 kg dry matter | 6896 kg organic substance |
>
> c) in pine forests
>
> | | | |
> |---|---|---|
> | wood | 3000 kg | 2970 |
> | leaves | 3186 kg | 3139 |
> | total | 6186 kg dry matter | 6109 kg organic substance |
>
> Similar results were obtained in direct measurements in the Bavarian State Forests by the King's forest officers in which the annual wood increment was estimated by mass estimates and on the same areas leaf litter was determined by weighing for several continuing years. As average for several stand classes they found:
>
> | | *in wood* *kg/ha* | *dry matter* *in leaves* *kg/ha* | *total* *kg/ha* | *organic* *substance* *of total* *kg/ha* |
> |---|---|---|---|---|
> | mature beech stands | 3474 | 3270 | 6744 | 6526 |
> | mature spruce stands | 3480 | 2783 | 6263 | 6103 |
> | mature pine stands | 2897 | 3186 | 6083 | 6009 |

[1] Ebermayer 1876, p. 67.
[2] As ash content was substructed for wood 1%, beech leaves 5.6% spruce needles 4.5%, and pine needles 1.46%.

> This means that the dry matter production of different forest species
> of similar stand classes are almost alike . . .
> [Ebermayer 1882, pp. 43–44].

By the early part of this century, partial productivity measurements
had become routine in agriculture and forestry. The scientists of the
day therefore started to use the concept for further theoretical work.
With regard to methods and production concepts, the work of Boysen
Jensen of Copenhagen became important. His treatise (1932) on the
primary production of plants was a landmark in the field.

Boysen Jensen died in 1960 at the time when a symposium on the
dry matter production of plants was being organized (Lieth 1962).
At this event Dr. Mueller, his student and subsequent holder of the
chair for plant physiology at the University of Copenhagen presented
a paper in homage to Dr. Boysen Jensen. The following extract from
this paper (Mueller 1962) demonstrates the concepts and methods used
during the early part of this century.

> . . .The first dry matter production paper was published by Boysen
> Jensen in 1910 (Tideskrift for Skovvaesen, Copenhagen 22, 1–116) . . . .
> In his paper, "Production of matter in light- and shadow-plants,"
> *Botanisk Tidsskr.* **36**:219–262 (1918), he analyzed the dry matter
> production of a light plant, *Sinapis alba,* and a shade plant, *Oxalis
> acetosella.* He found that the percent daily dry matter production in
> July amounted for *Sinapis* to about 16% and for *Oxalis acetosella,*
> 2% . . . .
> Boysen Jensen tried to explain the fact that the dry matter produc-
> tion of "light trees" is depressed in the shade more than that of "shade
> trees" with the following statement:
>
> Dry matter production or net production equals gross production
> minus matter loss by respiration in roots, trunk, branches, and
> leaves and minus matter losses by root, branches and leaves.
>
> In his time, Boysen Jensen could not put numbers into this equation.
> In the book of P. Boysen Jensen: "Die Stoffproduction der Pflanzen"
> (The Dry Matter Production of Plants) published by Gustav Fischer in
> 1932 is anticipated the outline for this first symposium about the dry
> matter production of green plants . . . . In his own words I want to
> honor Boysen Jensen and open this symposium:
>
> The dry matter production of the green plant may not only be esti-
> mated but analyzed as well . . . [pp. 6–13].
>
> A compilation of all Boysen Jensen's papers appeared under the title
> *Bibliografi P. Boysen Jensen* (Kφbenhavns Universitets plantefysilogiske
> Laboratorium, Kφbenhaven K Ö Farimagsgade 2A).

Jensen's objective was to analyze the laws and principles of the production process. This philosophy was expressed unusually well by R. H. Whittaker's studies (1961, 1962) of the primary productivity of plant communities in the great Smoky Mountains. Paper 16 presents one of his first papers, which became a model for many similar studies all over the world.

A slightly different approach to productivity studies was taken by a research group in Minnesota in the early 1950s. Drs. Lawrence and Bray, and several of their students and associates, began a series of analyses of the productivity of forests and grasslands, litter fall, chlorophyll content, and other parameters. Their papers had great influence on the concepts studied by the present generation of field workers. The most influential paper was probably their joint paper with Pearson (Lawrence, Bray, and Pearson 1959) and Ovington (Lawrence, Bray, and Ovington 1962).

The methods developed in recent times are more or less the same as those used in the studies described above. Only the degree of accuracy and the scope have changed. Typically, however, many different approaches to the productivity measurement are used in the same study. One of the earliest examples of such a complex study is contained in Paper 17 by Kira et al.

Energy Measurements

The measurements of energy accumulation during the primary production process requires calorimetric measurements of samples previously collected in dry matter productivity studies. The choice method for this analysis is the bomb calorimeter. The first ecological approach to plant energy production is probably Transeau's study reported in 1926. Quite possibly, a number of similar studies on animal feeding were made at an earlier time. Transeau's (1926) quotations of older research (before 1800) make clear that animal physiologists dealt with gas exchange problems much earlier than plant physiologists. This is in line with the medieval philosophy that one of the material differences between plants and animals is the fact that the latter breathe air. However, such early work went unnoticed by ecologists, whereas Transeau's paper is quoted by almost all early ecologists working on ecological efficiency. His study is also interesting with regard to the early history of gas exchange and nutrient demand of plants and animals.

The calorimetric analysis has been a standard procedure in food and feed analysis for several decades now. Terrestrial ecologists, however, have only taken the energy approach seriously since Lindemann (1942). One of the first detailed analyses, which was carried out by

Golley, is presented here as Paper 18. In the last decade, calorimetric analyses have become customary. Many of the papers in this collection on analyses of dry matter types and chemical components simultaneously produced details about the energy content as well.

Productivity Measurements in Aquatic Environments

Aquatic systems have different requirements for productivity measurements. Gases exchanged are dissolved mostly in the surrounding water. The dry matter produced does not accumulate per vegetation period as conveniently as in terrestrial systems, and the production occurs in some cases at the bottom of the water body itself. For the vast majority of the open ocean, the plankton production that occurs in the euphotic zone is most important. Large-scale productivity studies in aquatic environments started much later than terrestrial productivity studies. An early method—oxygen titration—reported by Winkler (1883) made plankton productivity studies possible and is, therefore, referred to as Winkler titration.[1] Early measurements using this method were reported by Boysen Jensen (1928) and others in the early part of this century. Large-scale studies of ocean productivity were done with a combination of oxygen titration, chlorophyll determination, ^{14}C-measurements, and the determination of the light-depth profile. Recent summaries of these methods and numerous applications are available in Goldman (1966) and Hall and Moll (1975).

Paper 19 by Ryther is included here as an example of the papers that preceded the wholesale measurements of net primary productivity of the oceans. More than any other, this paper stimulated the hope that the oceans of the world may become the protein basket for mankind. Although this hope proved to be erroneous, Ryther's work as well as that of Steeman Nielsen (1952), Sverdrup (1955), Sverdrup, Johnson, and Fleming (1942), Fleming (1957/1966), Rodhe (1958), Gessner (1949), and many others laid the foundation for our sound knowledge about the marine productivity pattern. For many years the ocean productivity pattern was better understood than the productivity patterns on land.

Other Methods

Direct assessment of primary productivity is not always feasible. Many authors have, therefore, attempted to predict the productive power of a given plant community on the basis of sufficiently cor-

[1] The Winkler titration method is today largely replaced by the oxygen probe.

related properties that are easier to assess. This approach is especially valuable in difficult areas or for inconveniently large vegetation covers. In recognition of the importance of this type of approach to developing countries with climates adverse to extensive field work, a summary of the experience collected during the first productivity symposium (Lieth 1962) was prepared, by request, for a UNESCO symposium in Montpellier (Lieth 1965). One of the methods involves the use of agricultural yield data that can be converted with appropriate allometric functions into productivity averages for large areas. The pattern analysis of agricultural yield was first attempted by Filzer (1951) in central Europe. The large-scale assessment of regional productivity patterns using allometric functions was first attempted by the US-IBP Eastern Deciduous Forest Biome. The first of these studies, a productivity profile of North Carolina by Whigham, Lieth, Noggle, and Gross (1971), was soon followed by similar assessments in Wisconsin, Tennessee, New York, and Massachusetts. The summary of these assessments was recently published by D. Sharpe and appears here as Paper 20.

Recent Productivity Measurements

Around 1960 three independent searches for existing terrestrial productivity data were undertaken by Lieth (1962), Westlake (1963), and Rodin and Bazilevitch (1965). These three collections served a dual purpose: first demonstrating the extreme scarcity of reliable productivity data, especially from natural vegetation, and second, initiating global and regional pattern analysis. These collections were quickly absorbed by the early organizing committees of the International Biological Program, and together with the papers discussed previously, they served as the first reading material for many young people who began working in productivity studies. In 1962 when the first IBP organizing meeting was held in Brussels, about twenty ecologists with actual experience in obtaining terrestrial productivity values were present. About five years later a second meeting was held at the same place at which time an overwhelming corps of over two hundred young field workers were present to report or discuss their recent projects or plans. A flood of papers on primary productivity has appeared since. Obviously, the International Biological Program has been mainly responsible for the sound knowledge we now have on primary productivity patterns.

Papers 21 through 24 are representative of the contributions to the Program by various nations. All authors have been in one way or another connected with IBP, and most field work was done before 1970. The selection of these papers aims to present information on the main vegetation types and from as many continents as possible. Paper 21 by Art and Marks tabulates a large amount of primary productivity data for the ten-year period following the first attempt to collect such data (Lieth 1962) and includes all data reported by Westlake (1963) and Rodin and Bazilevitch (1965). Paper 22 contains the tundra productivity data recently summarized by Wielgolaski. The most recent global summaries of the forests and grasslands are not yet available, but extracts from a prepublication by Van Dyne et al. are presented in Paper 23 to provide some of the tabulated productivity data for grasslands and the correlation to gross environmental parameters. Various other tabulations of productivity data appeared recently in Cooper (1975).

Since IBP summaries are not available at this time and the reprint of large amounts of original IBP studies would be too voluminous, Paper 24 contains a sampling of photographs of several IBP cooperators' sites accompanied by some key data on their environmental regimes and net primary productivity. The photographs are sorted into forest and nonforest vegetation. Only terrestrial vegetation types are represented.

REFERENCES

Bertalanffy, L. von. 1951. *Theoretische Biologie*, Vol. 2, 2nd ed. Bern: A. Francke.

Boussingault, J. B. 1840/1851. *Die Landwirtschaft in ihren Beziehungen zur Chemie, Physik, und Meteorologie*, Vol. 1, translated into German by R. Graeger, 2nd ed. Halle: Ch. Graeger.

——. 1846. Recherches sur le développement successif de la matière végétale dans la culture du froment extract. *Comptes Rendus Acad. Sci. Paris* **22**(15):617–618.

——. 1886. *Agronomie, Chimie Agricole et Physiologie*, Vol. 1. Paris: Gauthier-Villars.

Boysen Jensen, P. 1932. *Die Stoffproduktion der Pflanzen*. Jena: Gustav Fischer Verlag.

Chardon, C. E. 1953. *Boussingault, Juicio critico del eminente agronomo del siglo XIX su viaje a la gran Columbia . . . sus relaciones con el libertador y Manuelita Saenz*. Ciudad Trujillo, R. D. Editora Montalvo.

Cooper, J. P. (ed.) 1975. *Photosynthesis and Productivity in Different Environments*. Cambridge: Cambridge University Press.

Doty, M. S. 1959. Phytoplankton photosynthetic periodicity as a function of latitude. *J Mar. Biol. Assoc. India* 1:66–68.

Ebermayer, E. 1876. *Die gesamte Lehre der Waldstreu mit Rücksicht auf die chemische Statik des Waldbaues*. Berlin-Heidelberg: Springer-Verlag.

——. 1882. *Physiologische Chemie der Pflanzen*. Berlin: Springer.

Filzer, P. 1951. *Die Natürlichen Grundlagen des Pflanzenertrages in Mitteleuropa*. Stuttgart: Gustav Fischer Verlag.

Fleming, R. H. 1957. General features of the oceans, pp. 87–107 in *Treatise on Marine Ecology and Paleoecology*, Vol. 1: Ecology, Memoir 67, Geological Society of America (reprinted in 1966), J. W. Hedgpeth, ed.

Gessner, F. 1949. Der Chlorophyllgehalt im See und seine photosynthetische Valenz als geophysikalisches Problem. *Schweiz. Z. Hydrob.* 11:378–410.

Goldman, C. R. (ed.) 1966. *Primary Productivity in Aquatic Environments,* Proc. IBP PF Symp., Pallanza, Italy, 1965. Berkeley and Los Angeles: University of California Press.

Hall, C. A. S., and R. Moll. 1975. Methods of assessing aquatic primary productivity, pp. 11–53 in *Primary Productivity in the Biosphere,* H. Lieth and R. H. Whittaker, eds. New York: Springer-Verlag.

Johnson, P. L., and W. T. Swank. 1973. Studies of cation budgets in the Southern Appalachian on four experimental watersheds with contrasting vegetation. *Ecology* 54:70–80.

Lawrence, D. B., J. R. Bray, and J. D. Ovington, 1962. Quantitative ecology and woodland ecosystem. *Adv. Ecol. Res.* 1:151–175.

Lawrence, D. B., J. R. Bray, and L. C. Pearson. 1959. Primary production in some Minnesota terrestrial communities for 1957. *Oikos* 10:38–49.

Lieth, H. (ed.) 1962. *Die Stoffproduktion der Pflanzendecke.* Stuttgart: Gustav Fischer Verlag.

——. 1965. Indirect methods of measurement of dry matter production, pp. 513–518 in *Methodology of Plant Eco-physiology,* Proc. Montpellier Symp., 1962, F. E. Eckardt, ed. Paris: UNESCO.

Likens, G. E., F. H. Bormann, N. M. Johnson, D. W. Fisher, and R. S. Pierce. 1970. Effects of forest cutting and herbicide treatment on nutrient budgets in the Hubbard Brook watershed-ecosystem. *Ecol. Mono.* 40:23–47.

Lindemann, R. L. 1942. The trophic-dynamic aspect of ecology. *Ecology* 23: 399–418.

Lundegårdh, H. 1924. *Der Kreislauf der Kohlensäure in der Natur.* Jena: Gustav Fischer Verlag.

Morowitz, H. J. 1968. *Energy Flow in Biology.* New York: Academic Press.

Mueller, D. 1962. Boysen Jensen and the dry matter production of plants, pp. 6–10 in *Die Stoffproduktion der Pflanzendecke,* H. Lieth, ed. Stuttgart: Gustav Fischer Verlag.

Rodhe, W. 1958. Primärproduktion und Seentypen. *Verhandl. Internat. Ver. Limnol.* 13:121–141.

Rodin, L. E., and N. I. Bazilevich. *Production and Mineral Cycling in Terrestrial Vegetation.* Edinburgh: Oliver and Boyd.

Seybold, A. 1942. Pflanzenpigmente und Lichtfeld als physiologisches, geologisches, und landwirtschaftliches Problem. *Ber. Deut. Bot. Ges.* 60:46–85.

Steeman Nielsen, E. 1952. The use of radioactive carbon (C^{14}) for measuring organic production in the sea. *J. Cons. Internat. Explor. Mer* 18:117–140.

Sverdrup, H. U. 1955. The place of physical geology in oceanographic research. *J. Mar. Res.* 14:287–294.

Sverdrup, H. U., M. W. Johnson, and R. H. Fleming. 1942. *The Oceans, Their Physics, Chemistry, and General Biology.* Englewood Cliffs, N.J.: Prentice-Hall.

Transeau, E. N. 1926. The accumulation of energy by plants. *Ohio J. Sci.* 26:1–10.

Westlake, D. F. 1963. Comparison of plant productivity. *Biol. Rev.* 28:385–425.

Whigham, D., H. Lieth, R. Noggle, and D. Gross. 1971. *The North Carolina Productivity Profile 1971,* US-IBP Eastern Deciduous Forest Biome Memo Rep. 71:23 (mimeo.).

Winkler, C. 1883. *Die Massanalyse nach neuem titrimetrischem System.* Freiberg, Leipzig.

ADDITIONAL READINGS

Egle, K., and A. Ernst. 1940. Die Verwendung des URAS fur die vollautomatische und fortlaufende CO_2—Analyse bei Assimilations und Atmungsmessungen an Pflanzen. *Zeit. f. Naturf.* **46**:351.

Huber, B. 1950. Registrierung des CO_2 Gefalles und Berechnung des CO_2 Stromes uber Pflanzengeselleschaften mittels URAS. *Ber. Deut. Bot. Ges.* **63**:52.

Lundegårdh, H. 1960/1966. *Plant Physiology*, F. M. Irvine, trans., W. O. James, ed., New York: American Elsevier.

Strugger, S., and W. Baumeister. 1951. Zur Anwendung des URAS für CO_2 Messungen im Laboratorium. *Ber. Deut. Bot. Ges.* **64**:5.

12

GAS EXCHANGE CUVETTE

J. B. Boussingault

The following excerpt was translated expressly for this Benchmark volume by H. H. Lieth from pages 40–41 of Die Landwirtschaft in ihren Beziehungen zur Chemie, Physik, und Meteorologie von J.B. Boussingault, *R. Graeger, transl. Halle: Verlag von Ch. Graeger, 1851, 399pp.*

. . . This experiment (Sennebier's) documents, probably not conclusively, the splitting of CO_2, if this is diluted in a large amount of air. It appears, however, that leaves while hanging in the air have the purpose to split the CO_2 therein, and do that with a surprising velocity.

I built, therefore, in the summer of 1840, a special apparatus to conduct accurate investigations about this matter. A (glass)-balloon of 15 1 volume with three openings was made. Through the lower opening a fresh grape shoot was entered and fastened there with a rubber tube. The shoot carried 20 leaves. Through the top hole entered a small tube which connected the balloon with the outside air. The openings on one side connected (the balloon) with a special apparatus which allowed us to analyze the CO_2 content of the air with high accuracy.

The air passed first along the branch in the balloon before entering the apparatus. The air flow velocity amounted to 15 1/hr. determined by the water outflow from an aspirator. The leaves were exposed to the light and the experiment lasted from 11:00 A.M. until 3:00 P.M.

The first experiment showed that the air after having passed through the balloon contained 0.0002 CO_2 (after considering all the corrections required). The air analyzed at the same time near the apparatus contained 0.00045.

A second experiment yielded values of 0.0001 CO_2 for air passing the balloon vs 0.0004 of the air outside. Thus the air lost in the first experiment five-ninths, and in the second three-quarters, of its CO_2.

The experiment during the night yielded opposite results . . .

13

THE INFLUENCE OF EXTERNAL FACTORS ON THE INTENSITY OF PHOTOSYNTHESIS

Henrik Lundegårdh

[*Editor's Note:* In the original, material precedes this excerpt and
Figure 20 is not reproduced here.]

Because of its central position in the metabolism of green plants photosynthesis
dominates growth and development. For these reasons light and temperature are the
dominant ecological factors.

1 THE LIGHT FACTOR

The absorptions pectrum of chlorophyll (Fig. 20, p. 36) affords an indication of the
value of different wavelengths; but it cannot be assumed that there is a complete
parallel between light absorption and photosynthetic ability, even if there is a striking
agreement, especially in the red and yellow regions (NODDAK and EICHHOFF 1939).
Considerable photosynthesis also takes place in green light, that is in the region of
minimum light absorption (HOOVER 1937).

Different plants give rather variable results. WARBURG found with *Chlorella* a light
utilisation of 59% in the red-orange (610-690 mμ), 54% in the yellow (578 mμ),
44% in the green (546 mμ) and 34% in the blue (439 mμ). GABRIELSEN (1935) found
with *Sinapis alba* a utilisation of 26% in the red (600-705 mμ), 19% in the whole
region from 480 to 640 mμ and 13% in the region 400-510 mμ. Later GABRIELSEN
found a photosynthesis in the green which amounted to two-thirds of that in the red,
whilst blue light was used only to the extent of one-third (see also EMERSON and LEWIS
1943).

The maximum of light utilisation falls generally at the characteristic red absorption
band (655 mμ; HOOVER 1937), but a second maximum occurs in the blue (440 mμ),
even if green light has about the same effect (BURNS 1942). Infra-red appears not to be
used by higher plants. This applies also for those Cyanophyceae which live in the
surface layer of the soil (BAATZ 1939). As mentioned on p. 115, purple bacteria,
however, assimilate in the infra-red. For the ultra-violet BURNS (1942) states that higher
plants assimilate at 365 mμ. The failure of certain leaves, e.g. the needles of pines, to
utilise the ultra-violet is probably due to the presence of pigments that shield the
chlorophyll. Short-wave rays, 320-340 mμ, are however, in general, photosynthetically
ineffective.

Detailed investigations on the spectral sensitivity of photosynthesis assume, of
course, exact methods for determining the incident light energy. The absorption of a
black surface is best measured bolometrically or thermoelectrically. The results are,
of course, strongly affected by the extent of absorption and reflection of the object.
The varying values for energy yields which are found in nature often depend on losses
of light by reflection or conversion to heat in the photosynthetic organs or cells (see
on this also SCHANDERL 1931; SEYBOLD 1932, 1933). GABRIELSEN states e.g. that the
occurrence of anthocyanin in the epidermis can reduce the total light yield by 43%
(cf. Fig. 69; discordant data about anthocyanin effects are given by KUILMAN 1930
and KOSAKA 1933). The cell sap, according to METZNER (1930), absorbs the ultra-
violet strongly in the region 350-400 mμ. Tannins also exercise a similar protective
effect (PÉNZES 1938).

In investigations of the light yield the technical difficulties of getting monochromatic
light of a sufficient intensity are considerable. Light sources such as the mercury

lamp, which give a spectrum with few sharp lines, are naturally to be preferred and are frequently used. Of monochromatic filters, which, however, only separate out narrow regions, those of CHRISTIANSEN-WEIGERT (MCALISTER 1935) may be mentioned and particularly the interference filters prepared by Schott. With monochromators the quartz prism apparatus is to be preferred because of the purity of the spectrum, unless one uses double monochromators, the only really reliable ones.

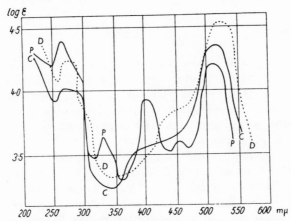

Fig. 69. Absorption spectra of three anthocyanidin pigments:
D-Delphinidin, C-Cyanidin, P-Pelargonidin (after GILMAN 1944).

Red and brown algae possess pigments which by means of light-transformations (p. 103) make possible a fuller utilisation of infra-red and green-blue-violet rays. *Phycoerythrin* absorbs more green light than chlorophyll; *phycoxanthin*, and *phycocyanin* even more, absorb blue-violet rays in particular (MONTFORT 1933, 1934).

Ecologically these colour variations indicate an adaptation to the screening off of certain regions of the spectrum in water. Ontogenetic adaptations also occur (HARDER *et al.* 1936). SEYBOLD describes changes of concentration between chlorophyll a and b in plants which live in 'green shade' or in 'blue shade'. Similar differences also occur among green water plants.

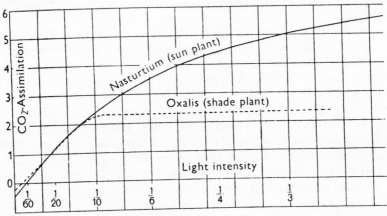

Fig. 70. Apparent (ecological) assimilation curves of a sun plant and a shade plant (after LUNDEGÅRDH 1921).

The relation between intensity of photosynthesis and light-intensity follows a 'logarithmic curve', i.e. it rises initially almost in a straight line and gradually turns parallel to the abscissa (Fig. 70). WASSINK and KERSTEN (1944) found in Diatomaceae a direct proportionality between photosynthesis and light-intensity up to a value 1.5×10^1 erg per cm^2 per second.

The curve of apparent CO_2-assimilation begins at a point below the abscissa, because at the light-intensity *nil* only respiration would take place. At a definite low light-intensity as much CO_2 is used in photosynthesis as is formed in respiration. This is the so-called *light-compensation point*, whose position depends also on temperature. In general, the light value of the compensation point drops with temperature (MÜLLER 1928, RUSSEL 1940). The difference between the courses of the light-assimilation curves usually observed for shade leaves and sun leaves (Fig. 70), which is treated in ecological handbooks, is allied to the general problems of limiting and minimum factors.

The minimum law introduced by LIEBIG in the first half of last century was applied by F. F. BLACKMAN (1905) to photosynthesis. BLACKMAN spoke of limiting factors.

Light and carbon dioxide are both absolutely necessary for photosynthesis. Both therefore control the speed of the process: with weak light only a little assimilate is formed, even when the CO_2-supply is at a maximum; with a low CO_2-content photosynthesis proceeds slowly, even when the light-factor is at a maximum. The factor occurring at the minimum thus limits the process; if it is increased the photosynthetic yield rises until one or other essential factor becomes minimal, and so on. These easily understood relationships can also be expressed as a collaboration or balance between the production factors. The ecological application of these facts is expressed in the *biological relativity law* (LUNDEGÅRDH 1957), which is referred to here.

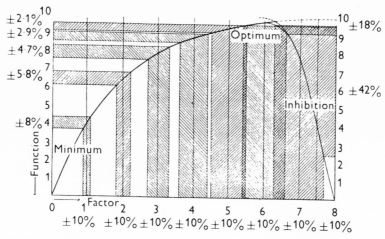

Fig. 71. Graphic representation of the biological relativity law. The shaded vertical columns indicate a ±10% variation of the factor. The horizontal fields show the corresponding variation of the biological function dependent on the factor.

According to the ideas of LIEBIG and BLACKMAN the retarding effect of the minimum factor should set in suddenly, so that the curves would show a sharp bend between the rising and the asymptotic branches. Shade leaves approach this type (cf. Fig. 70). but ordinarily the curves bend down very gradually. This fact is explained by interaction of the factors. Consider for example carbon dioxide: it is supplied from the air, but must be dissolved in the protoplasm and taken up by its buffers (see p. 107) before it passes over into the chloroplasts. An adequate surplus of CO_2 thus means that all these intermediate processes can keep pace with the CO_2-consumption regulated by the light. If now the supply of CO_2 to the chloroplasts follows a strictly constant speed, its retarding effect must of course set in suddenly at a definite light value. But the speed is dependent on the diffusion gradients between the air and the chloroplasts, which can reach the peak value only if the internal concentration is nil. The limit set by the retarding effect depends on a number of circumstances, e.g. the length of the transport path, the character of the intermediate processes, protoplasmic streaming, etc. It is pointless, under these conditions, to speak of a greater or lesser correctness of the various types of curve. The chemical reactions which co-operate in photosynthesis naturally follow curves of the first or higher orders only approximately, but they all show curves of convex form.

[*Editor's Note:* Material has been omitted at this point.]

ERRATUM

Page 120, line 4 should read "the absorption spectrum . . ."

2 THE CARBON DIOXIDE FACTOR

On account of its low concentration, carbon dioxide readily affects photosynthesis as a limiting factor. This has been shown by numerous physiological investigations (LUNDEGÅRDH 1921, 1924, 1957, STOCKER 1938) in which the increase in the intensity of photosynthesis on raising the CO_2-content of the air above 0·03% was determined. The effect of the CO_2-factor is of course dependent on light; an increase in the region where light is the limiting factor has only an unimportant, though measurable, effect which increases many-fold with optimal light (Fig. 72).

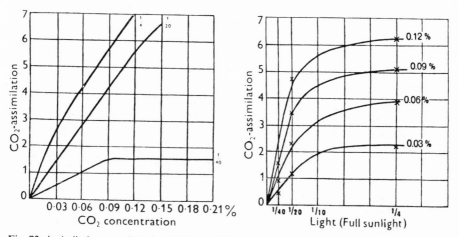

Fig. 72. Assimilation experiments with leaves of *Oxalis acetosella*. *A*-Variation of the CO_2-concentration with three light intensities (1/40, 1/20 and 1/4 of maximum daylight).
B-Variation of the light intensity with four CO_2-concentrations (0·03%, 0·06%, 0·09% and 0·12% by volume) (after LUNDEGÅRDH 1924).

The interaction between the CO_2-factor and the light-factor is influenced in its turn by temperature, as shown below. At 18°C-20°C and with strong light (1/4—1/1 of full daylight) a three-fold increase of the CO_2-content (from 0·03% to 0·09%) caused an increase of apparent photosynthesis by 2·1 times in *Oxalis*, 2·8 times in *Phaseolus* and 2·6 times in *Solanum*. Some experiments with *Chlorella* (WHITTINGHAM 1952) illustrate the effect of the CO_2-factor on photosynthesis (Table 3).

Table 3

Photosynthesis of *Chlorella* (after WHITTINGHAM 1952)

CO_2 (%)	0·0013	0·0035	0·007	0·0089	0·0108	0·030
Photosynthesis	6·8	11·5	14·6	16·3	16·8	18·0

A further increase in the CO_2-concentration raises photosynthesis still higher, but individual plants react in different ways. It was mentioned earlier (p. 123) that CO_2

has to travel a long way before it arrives at the interior of the chloroplasts. According to CLENDENNING and GORHAM (1952) the CO_2 taken up by the cells distributes itself quickly in the cytoplasm and cell sap. The abundance of buffers already mentioned of course facilitates the uptake. In this way a local concentration far above the external concentration is attained; which is undoubtedly very important for the result produced by the chlorophyl! apparatus. The pervasive CO_2-enrichment in the cytoplasm and cell sap probably explains isolated statements that the CO_2 concentration of the environment is, on the whole, unimportant. With experiments of long duration, however, the external concentration must finally be decisive.

With very high contents the effect of the CO_2 on the protoplasm, respiration, etc., becomes more important on account of a CO_2-inhibition or poisoning. STEEMANN NIELSEN (1955) reports for *Chlorella* that at CO_2-concentrations over 1% an unknown enzymatic process is blocked, so that photosynthesis decreases. Investigations on the respiration system of wheat roots have shown that CO_2 inhibits the oxidation of cytochrome b. This must undoubtedly cause an inhibition of respiration which, if green cells behave similarly, could secondarily influence the chloroplast mechanism.

The CO_2-inhibition can work out in various ways, and even the root-systems of different plants can show differences in this respect..For short periods leaves seem to tolerate 20%-50% CO_2 or even higher concentrations. BALLARD (1941), however, states that, with low temperatures and strong light, leaves of *Ligustrum* already suffered an inhibition of photosynthesis at 2%-$2\cdot5\%$ CO_2. A complicating factor in the leaves is the reaction of the stomata to CO_2. The investigations of BALLARD already mentioned also show that CO_2-inhibition increases at low temperatures.

As the CO_2-content of free air only exceptionally rises above about $0\cdot1\%$ (see LUNDEGÅRDH 1924, 1957), the upper limit of the positive effect of an increase in concentration is of practical interest, particularly for plant culture under glass. Comprehensive investigations on 'CO_2-fertilisation' (see LUNDEGÅRDH 1924, 1928, 1957) have shown, however, that optimal effects are attained with a moderate increase to about $0\cdot1\%$. This is probably connected, among other things, with the fact that the conducting tissues of the leaf stalk and of the stem in general are not large enough for the transport of a great excess of assimilates. The same morphological-anatomical obstacles may also explain the different reactions of different species of plants to CO_2-fertilisation.

[*Editor's Note:* Material has been omitted at this point.]

REFERENCES

[*Editor's Note:* The following references appear exactly as they are in the original publication, including any inconsistencies.]

Baatz, I. 1939. Arch. Mikrobiol., 10, 508.
Ballard, L. A. T. 1941. New Phytologist, 40, 276.
Blackman, F. F. 1905. Ann. Bot., 19, 281.
Burns, G. R. 1942. Amer. J. Bot., 29, 381.
Clendenning, K. A. and Gorham, P. R. 1952. Arch. Biochim. Biophys., 37, 56.
Emerson, R. L., and Lewis, C. M. 1943. Amer. J. Bot., 30, 165.
Gabrielsen, E. K. 1935. Planta, 23, 474.
Gabrielsen, E. K. 1948. Nature, 161, 139.
Gilman, H. 1944. *Organic Chemistry*, Vol. 2, New York: Wiley.
Harder, R., *et al.* 1936, 1938. Nachr. Ges. Wiss. Göttingen, 2, 1136; 3, 135.
Hoover, W. H. 1937. Smithsonian Miscell. Coll., 95, no. 21.
Kosaka, H. 1931, 1933. J. Dept. Agric. Kyushu Imp. Univ., 3, 29, 251; 4, 95.
Kuilman, L. W. 1930. Rec. Trav. Bot. Néerl., 27, 287.
Lundegårdh, H. 1921. Ber. deutsch. bot. Ges., 39, 195, 223.
——. 1921. Svensk Bot. Tidskr., 15, 46.
——. 1924. Biochem. Z., 146, 564; 149, 207; 151, 296; 154, 195.
——. 1924. *Der Kreislauf der Kohlensäure in der Natur.* Jena.
——. 1928. Medd. Centralanstalten, no. 331.
——. 1957. *Klima und Boden in ihrer Wirkung auf das Pflanzenleben.* 5th Edit. Jena. (Earlier Eng. trans. *Environment and Plant Development* trans. E. Ashby, 1931).
McAlister, E. D. 1935. Smithsonian Misc. Coll., 93, no. 7.
Metzner, P. 1930. Planta, 10, 281.
Montfort, C. 1934 (1933). Jahrb. wiss. Bot. 79, 493.
Müller, D. 1928. Planta, 6, 23.
Noddack, W. 1939. Z. phys. Chem., A, 185, 207, 241.
Penzes, A. 1938. Bot. Közlemények, Budapest, 35, 22.
Russell, R. S. 1940. J. Ecol., 28, 289.
Schanderl, H. and Kaempfert, W. 1931. Planta, 18, 700.
Seybold, A. 1932. Planta, 18, 479.
——. 1933. Planta, 20, 577; 21, 251.
Steemann Nielsen, E. 1955. Physiol. Plantarum, 8, 317, 945.
Stocker, O. 1935. Planta, 24, 40; Jahrb. wiss. Bot., 81, 464.
Stocker, O., *et al.* 1938. Jahrb. wiss. Bot., 86, 556.
Wassink, E. C. and Kersten, J. A. H. 1944. Enzymologia, 11, 282.
Wittingham, C. P. 1956 (1952). J. Exp. Bot., 7, 273.

14

THE APPLICATION OF THE INFRARED GAS ANALYZER IN PHOTOSYNTHESIS RESEARCH

Karl Egle and Walter Schenk

These excerpts were translated expressly for this Benchmark volume by H. H. Lieth from "Die Anwendung des Ultra-rotabsorptionsschreibers in der Photosyntheseforschung," in Ber. Deut. Bot. Ges. **64**:180–196 (1951), *by permission of the publisher, Gustav Fischer Verlag.*

The infrared gas analyzer (abbreviated URAS) of the Badische Anilin and Soda Fabrik Ludwigshafen represents a continuously and automatically working instrument for gas analysis that has recently gained wider use in plant physiology to measure CO_2 exchange. The measuring principle . . . has been described in biological literature by several authors during the last two years, Egle and Ernst (1949), Huber (1950), Strugger and Baumeister (1951). . . .

The URAS (IRGAS in English) was designed as a fully automatic gas analysis device for control and survey in the chemical industry. Its high sensitivity combined with constant and exact reproducible values soon made this new instrument a universal laboratory instrument. It was director Dr. H. Mehner, of Heidelberg, who discovered the potential of the URAS for plant physiological research and who introduced the instrument into botanical use. The first measurements of CO_2 exchange were conducted by Prof. Dr. Seybold, also of Heidelberg (Seybold 1942). . . .

The physiological apparatus is depicted in figure 1. It consists of the moistening pipe (Af), assimilation cuvette (Ak), drying device (Tr), and safety valve (Si). These parts are connected with glass and rubber tubing and placed in a water thermostat (W. Th.). The cuvette may contain several glass plates (GP) to decrease the volume of the chamber. Above the cuvette is the light source (Li) mounted at variable distances. Pressured air enters at the point indicated as Pressluft and leaves to the flowmeter (Strömungsmesser) and URAS. . . .[free translation]

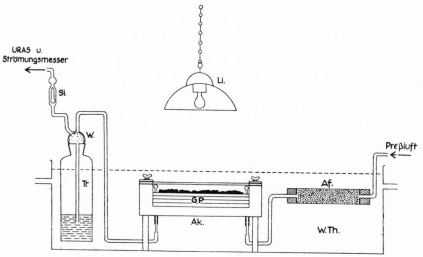

Figure 2 shows how large CO_2 concentrations can change after a short time in metropolitan air. (Maximum variation from 0.033–0.045 volume percent CO_2 within one minute.) The figure demonstrates also the images produced by the recording instrument. . . .

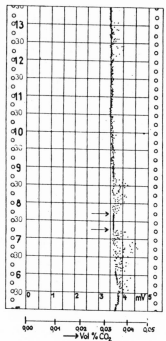

Figure 3 shows an assimilation and respiration experiment with Fegatella conica at different air speeds (pressured air source), at constant temperature ($20 \pm 0.1°C$), at constant light intensity of 3,000 lux from a 200w bulb at a distance of 40cm with the light passing through 3cm of water and 5mm of glass. Table 1 contains the values calculated from the chart shown in figure 3a and demonstrates that CO_2 assimilation drops at low flow rates (5 1/h) as well as dark respiration (Dunkelatmung) due to a shortage or excess of CO_2 in the chamber. The CO_2 content never drops below a certain minimum value (CO_2 compensation point). . . .[free translation]

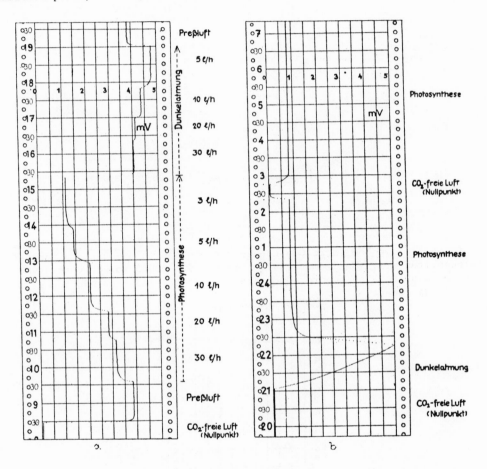

Fegatella conica 3000 Lux	mV	Korr. mV	Vol.% CO$_2$	Differenz		mg CO$_2$/h	Relativwerte	
				Vol.% CO$_2$	mg.CO$_2$/ℓ		App Assmil bzw Dunkelatmung	CO$_2$-Gehalt der austretenden Luft
Nullpunkt	0,04	–	–	–	–	–		
Preßluftwert	3,84	3,80	0,03510	–	–	–	–	100
App-Assimilation 30ℓ/h	3,16	3,12	0,02791	0,00719	0,132	3,96	100	80
" " 20ℓ/h	2,86	2,82	0,02494	0,01016	0,187	3,74	94	71
" " 10ℓ/h	2,12	2,08	0,01782	0,01728	0,318	3,18	80	51
" " 5ℓ/h	1,40	1,36	0,01144	0,02366	0,435	2,18	55	45
" " 3ℓ/h	1,10	1,06	0,00884	0,02626	0,484	1,45	37	25
Dunkelatmung 30ℓ/h	4,02	3,98	0,03718	0,00208	0,0383	1,149	100	106
" 20ℓ/h	4,12	4,08	0,03828	0,00318	0,0585	1,170	102	109
" 10ℓ/h	4,37	4,33	0,04126	0,00616	0,117	1,170	102	118
" 5ℓ/h	4,78	4,74	0,04636	0,01126	0,207	1,035	90	132

TABLE 1.

REFERENCES

Egle, K., und W. Schenk: Untersuchungen über die Reassimilation der Atmungs-
    kohlensäure bei der Photosynthese der Pflanzen. *Beitr. Biol. Pflanzen*
    (1951), (im Druck).
Huber, B.: Registrierung des CO$_2$-Gefälles und Berechung des CO$_2$-Stromes über
    Pflanzengesellschaften mittels Ultrarotabsorptionsschreiber. *Ber. Dtsch.*
    Bot. Ges. 63 (1950) 52.
Seybold, A.: Pflanzenpigmente und Lichtfeld als physiologisches, geologisches
    und landwirtschaftliches Problem. *Ber. Dtsch. Bot. Ges.* 60 (1942) 46–85.
Strugger, S., und W. Baumeister: Zur Anwendung des Ultrarotabsorptionsschreibers
    für CO$_2$-Messungen im Laboratorium. *Ber. Dtsch. Bot. Ges.* 64 (1951) 5.

15

Reprinted from *Amer. Midl. Natur.* 75(1):34–44 (1966)

Photosynthesis and Respiration Rates of Rocky Mountain Alpine Plants under Field Conditions[1]

W. D. BILLINGS, E. E. C. CLEBSCH,[2] and H. A. MOONEY[3]
Department of Botany, Duke University, Durham, North Carolina 27706

ABSTRACT: Field measurements of photosynthesis and respiration rates of whole alpine plants *in situ* were made at an elevation of 3,307 m in the Medicine Bow Mountains, Wyoming, using an infrared gas analyzer. Most measurements were for periods of 30 min to 2 hr but two "whole-day" measurements lasted from before sunrise to after sunset. Maximum net photosynthesis rates ranged from 4.15 to 13.4 mg CO_2/dm^2 two surfaces/hr with an average rate of 9.32. An all-day measurement on *Geum turbinatum* during somewhat cloudy weather showed net photosynthesis paralleling and responding to changes in light and temperature. Similar measurements on *Polygonum bistortoides* in clear, sunny weather showed a midday depression in net photosynthesis; this may be due to high leaf temperatures under intense solar radiation. Net photosynthesis of whole sod blocks averaged 6.23 mg CO_2/g dry wt/hr. Soil CO_2 output averaged 1,846 mg CO_2/m^2 hr which is higher than published results from lower altitudes. Rough estimates of dry weight gain as calculated from sod-block metabolic rates averaged 2.79 g dry wt/m^2/day. Measurements of CO_2 concentration in ambient air were made. Morning observations averaged 0.430 mg/l while afternoon measurements averaged 0.387 mg/l. The lower afternoon values probably reflect photosynthetic activity of the surrounding forest vegetation on the lower slopes.

INTRODUCTION

Alpine plants must produce enough food in a short, cold growing season to provide energy and substance for a rapid burst of growth at even lower temperatures during the first few days of the next growing season. Although photosynthesis and respiration rates of alpine plants measured under laboratory conditions provide some information on the speed of food manufacture at low temperatures, measurement under the impact of the natural environment provides a more realistic answer. Some field determinations have been made on alpine plants by Henrici (1918), Mönch (1937), and Cartellieri (1940) in the Alps, Blagowestschenski (1935) in the Pamirs, Mooney and Billings (1961) and Scott and Billings (1964) in the Rocky Mountains, Hadley and Bliss (1964) in the northern Appalachians, and Mooney and his co-workers (1964) in the White Mountains of California. In addition, Tranquillini (1955, 1959) has measured photosynthesis and respiration rates of timberline trees in the Tyrolean Alps. Tranquillini and the American workers have used infrared gas

---

[1] This research is part of that supported by National Science Foundation Grant G3832.

[2] Present address: Department of Botany, University of Tennessee, Knoxville 37916.

[3] Present address: Department of Botany, University of California, Los Angeles 90024.

analyzers for these measurements, a more precise technique for the measurement of rate than was available to the earlier investigators. The methods and results reported in most of these papers are included in the excellent reviews by Pisek (1960), Larcher (1963), and Tranquillini (1964).

The present investigation was concerned with four aspects of high altitude ecology: (1) maximum photosynthesis and respiration rates, (2) the course of net photosynthesis in individual plants throughout an entire day, (3) the carbon budget of undisturbed alpine turf, and (4) measurements of ambient CO_2 at different times of the day at high altitudes. The results reported here were obtained in the field during the summer of 1958 in the native alpine tundra near timberline in the Medicine Bow Mountains of Wyoming at an elevation of 10,850 ft (3,307m), except where otherwise noted. All nomenclature is that of Harrington (1954).

METHODS

A portable, d-c operated, infrared gas analysis system was utilized for all photosynthesis and respiration measurements (Fig. 1). Routinely, a portion of, or the entire shoot of an intact normally rooted plant was sealed within a one liter cubical Plexiglas chamber. The CO_2 concentrations in ambient air and in an air stream flowing through the chamber were analyzed alternately. Airflow rate was maintained at 80.6 1/hr (corrected for pressure). Rate calculations are based on

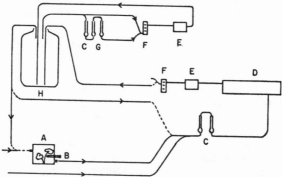

Fig. 1.—Schematic diagram of field CO_2 analysis system. Recorder (Esterline-Angus Model AW d-c milliammeter) and 6 volt d-c power supply are not figured. Arrows indicate direction of sample flow through Tygon tubing. Items A through F connected by solid lines comprise the plant sample system. All items except A and B are utilized in the calibration system. Dashed lines represent switchable pathways of tubing. A. Plexiglas plant sample chamber, B. Shielded mercury-in-glass thermometer, C. Drying agent, D. Beckman Instrument Co., Model 29 Liston-Becker Infrared Gas Analyzer, E. Six-volt Gast Mfg. Corp. Air Pump, Model AO44O, F. Dwyer Mfg. Co. Flowmeter, No. 490-10, 0-10 cu ft/hr, G. "Ascarite" CO_2 absorbent, H. Twenty-liter mixing bottle for the preparation of standard CO_2 samples for calibration.

the difference in CO_2 content between the two air streams, knowing the flow rate and the amount of plant tissue. Prior to each sample period, the gas analyzer was calibrated with freshly prepared gas mixtures. The bulk of the measurements involved continuous photosynthetic rate determinations on a given plant for a period of one to two hours. The chamber was then covered with black cloth and dark-respiration rates were monitored for 15 minutes.

In addition to the above short-duration analyses, photosynthetic rates were measured on two different plants throughout the entire daylight period. Carbon dioxide exchange rates were also determined for several intact plant communities. In the latter cases, the Plexiglas chamber was fitted over a portion of sod having a surface area of 100 cm². Soil and plants were removed from around this area so that the chamber fitted tightly over the sod block. CO_2 concentration of ambient air, of air passing through the chamber when illuminated and when darkened, and of air passing through the chamber after the above-ground portions of plants had been removed were all measured. These measurements provided necessary data for CO_2 exchange calculations on the entire microcommunity.

During the course of all CO_2 exchange determinations, air temperature was measured at five-minute intervals both inside and outside of the analysis chamber. A Corning infrared glass filter (CS1-69 no. 4600 M-2040) was placed over the chamber in an attempt to prevent excessive overheating. Snow packed at the base of the chamber also aided in temperature control. Light intensity at the top surface of the chamber, but under the infrared filter, was also measured at five-minute intervals with a Weston sunlight illumination meter.

RESULTS AND THEIR INTERPRETATION

MAXIMUM OBSERVED PHOTOSYNTHESIS AND RESPIRATION RATES

Table 1 lists the maximum rates of net photosynthesis and dark-respiration as measured in the field on each of two or more gas analyzer runs per species. Each run lasted from 1 to 2 hours. Net photosynthesis was continuously monitored for the first 80-90% of the run, after which a dark cloth was put over the chamber and dark-respiration measured for 10-15 minutes. Maximum gross photosynthesis is estimated in the table by adding the maximum figures for net photosynthesis and dark-respiration. However, since net photosynthesis and respiration could not be measured simultaneously and chamber temperatures for the two measurements were usually different, maximum gross photosynthesis as reported here is only approximate.

While most of our maximum net photosynthesis figures are based on fresh weight of total shoot sample, we do have some data in Table 1 based on "two-surface" leaf area for the following species: *Geum turbinatum*, *Oxyria digyna*, *Polygonum bistortoides*, and *Saxifraga rhomboidea*. The average maximum net photosynthesis for plants of

TABLE 1.—Maximum observed photosynthesis and respiration rates of alpine plants of several species measured under field conditions during the summer of 1958 in the Medicine Bow Mts., Wyoming, 3307 m elevation

Species	Date	Chamber temp. (°C) Ps	Chamber temp. (°C) Rs	Light (ft-c)	Net Ps (mg CO_2/g fresh wt)	Dark Rs (mg CO_2/g fresh wt)	Approx. gross Ps (mg CO_2/g fresh wt)	Net Ps (mg CO_2/dm2)	Dark Rs (mg CO_2/dm2)	Approx. gross Ps (mg CO_2/dm2)
Calamagrostis purpurascens	28 Jun	23	16	5,800	8.45	1.49	9.94			
Calamagrostis purpurascens	10 Jul	25	21	8,100	6.60	3.85	10.45			
Geum turbinatum	26 Jun	26	.....	7,500	2.10					
Geum turbinatum	14 Jul	26	4	7,800	2.90	1.35	4.25	8.98	4.42	13.40
Oxyria digyna	9 Jul	27	21	1,600	3.65	0.85	4.50	5.79	1.35	7.14
Oxyria digyna	22 Jul	25	14	9,000	3.74	0.18	3.92	7.39	0.35	7.74
Pedicularis parryi	16 Jul	21	12	4,200	1.64	2.01	3.65			
Pedicularis parryi	17 Jul	26	15	2,700	1.52	3.36	4.88			
Poa alpina	30 Jun	28	27	8,000	0.78	2.03	2.81			
Poa alpina	16 Jul	24	20	6,800	4.77	4.17	9.34			
Polygonum bistortoides	24 Jun	24	20	5,000	5.89	1.38	7.27	13.4	2.5	15.9
Polygonum bistortoides	2 Jul	22	16	1,500	3.17	3.93	7.10	8.96	11.11	20.07
Polygonum bistortoides	11 Jul	18	11	4,200	5.75	15.77	21.52	11.15	30.5	41.65
Polygonum bistortoides*	8 Jul	26	16	10,000	4.14	0.84	4.98	8.15	1.66	9.81
Polygonum bistortoides**	7 Jul	33	24	8,400	3.62	7.57	11.19	5.87	12.29	18.6
Saxifraga rhomboidea	27 Jun	21	18	8,300	2.87	0.55	3.42	7.52	1.46	8.98
Saxifraga rhomboidea*	8 Jul	25	20	3,000	3.08	1.75	4.83	4.15	2.36	6.51
Sibbaldia procumbens	30 Jun	26	18	6,300	0.88	0.88	1.76			
Sibbaldia procumbens	21 Jul	24	18	3,600	3.74	3.42	7.16			
Silene acaulis	28 Jun	27	20	8,700	1.32	1.66	2.98			
Silene acaulis	15 Jul	26	21	9,200	1.47	3.99	5.46			
Trifolium parryi	3 Jul	27	14	9,700	1.60	2.51	3.11			
Trifolium parryi	16 Jul	26	18	7,600	1.34	3.76	5.10			
Trisetum spicatum	30 Jun	25	22	8,700	2.80	3.80	6.60			
Trisetum spicatum	12 Jul	24	19	5,500	4.60	7.50	12.10			

* 2972 m elevation.   ** 2576 m elevation.

these four species is 9.32 mg CO_2/dm^2 two surfaces/hr with a range of 4.15 to 13.4. These figures compare very well with the 9 to 13.5 mg CO_2/dm^2 two surfaces/hr reported by Cartellieri (1940) at 2600 m in the Tyrolean Alps with *Doronicum clusii, Ranunculus glacialis, Sieversia reptans, Primula glutinosa,* and *Salix herbacea.* Both of these sets of figures are higher than maximum net photosynthesis rates of low elevation wild herbaceous plants reported by Larcher (1963) as 7 to 12 mg CO_2/dm^2 two surfaces/hr for sun plants and 2 to 8 mg/hr for shade plants. Crop plants, however, according to Larcher have rates of from 10 to 15 mg/dm² two surfaces/hr, not very much higher than those of alpine plants. Indeed, Scott and Billings (1964) in Wyoming found that on moist, fertile alpine soils net productivity per m² per day was 11.1 gm, a figure quite comparable with the rate of 11.0 gm per m² per day for maize in Minnesota reported by Ovington *et al.* (1963).

WHOLE-DAY CURVES OF NET PHOTOSYNTHESIS

All-day photosynthesis runs from sunrise to sunset were made twice during the summer with concomitant measurements of light and chamber temperatures. Dark-respiration was measured by covering the chamber with a black cloth shortly before sunrise and shortly after sunset. Figures 2 and 3 present curves of light and temperature associated with whole-day curves of net photosynthesis for plants of *Geum turbinatum* and *Polygonum bistortoides,* respectively.

The diurnal course of net photosynthesis of *Geum turbinatum* (Fig. 2) rather closely followed the fluctuations in light on this partly cloudy day. Light apparently was the principal limiting factor for

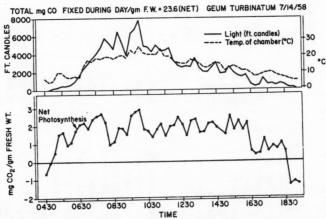

Fig. 2.—Net photosynthesis of a plant of *Geum turbinatum* from before sunrise to after sunset on a partly cloudy day at 3,307 m and concomitant chamber temperatures and light.

Geum during most of the day. Temperature and water supply were probably adequate for *Geum* at this site through most of the daylight hours.

Quite a different situation can be seen in the curve for *Polygonum bistortoides* (Fig. 3). Correlations of net photosynthesis with light and chamber temperature were good up to ca. 4,000 ft-c and 20 C which were reached about 0730. After that time, further increases in light and temperature on this clear morning were accompanied by rapid fluctuations in rate and no increase in net photosynthesis. At several times, particularly in the middle of the day, photosynthesis dropped below compensation in spite of more than 8,000 ft-c of light and chamber temperatures which ranged between 25 and 30 C. Similar behavior has been observed in other plants, especially those of dry regions (Stocker, 1954, 1960). Zalenskij (1941) (cited by Pisek, 1960) noticed that increased altitude accentuated the midday photosynthesis depression in potato plants grown in the Pamirs.

We can only speculate about the causes for the midday depression and irregular behavior of the photosynthesis rate in *Polygonum*. It is probably related in some way to the heat balance of the leaf under intense solar radiation. Leaves of mountain plants under intense solar radiation may have temperatures 5 to 20 C above ambient air temperature, and this is more likely to happen at higher temperatures (Tranquillini and Turner, 1961; Salisbury and Spomer, 1964). When our photosynthesis measurements were made, we had no means of measuring leaf temperatures. In the case of the *Polygonum* whole-day curve in Figure 3, the most likely explanation is that leaf temperatures, in bright sunlight and at air temperatures

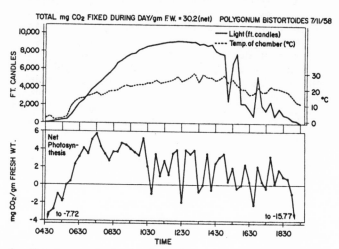

Fig. 3.—Net photosynthesis of a plant of *Polygonum bistortoides* from before sunrise to after sunset on clear day (until late afternoon) at 3,307 m and concomitant chamber temperatures and light.

of ca. 25 C, may have been in the vicinity of 40 C. This could have allowed respiration rate to exceed photosynthesis rate for short periods of time and resulted in net photosynthesis dropping below compensation. Certainly, higher leaf temperatures at midday would have the effect of lowering net photosynthesis despite adequate illumination (Pisek, 1960; Scott and Billings, 1964: 253, Fig. 2). In the case of *Geum* (Fig. 2), illumination and ambient temperatures were changeable and relatively low thus allowing net photosynthesis to respond quickly in a positive way to these fluctuations. Our data for *Geum* (Fig. 2) correspond rather well to Pisek's (1960) idealized curves for partly cloudy weather (his Fig. 14) and our curves for *Polygonum* are similar to those he postulates to show the midday depression in net photosynthesis in clear weather (his Fig. 13).

Many Rocky Mountain alpine plants have either very small scale-like leaves (cushion plants) or finely divided pinnatifid or dissected leaves. The leaves of *Geum* are of this latter type. On the other hand, *Polygonum* leaves are entire and relatively large. Theoretically, the small divisions of the *Geum* leaves should be relatively more efficient in giving up heat to the environment than the larger simple leaves of *Polygonum*. This morphological situation may also be involved in the maintenance of a steadier photosynthesis rate in *Geum*, and, if true, could be of adaptive value.

While high leaf temperatures seem the most reasonable explanation for the depression and irregularity in the *Polygonum* curve, other factors may be involved. Mooney, Hillier, and Billings (1965), working with the same species in the Sierra Nevada of California, found a midday depression of transpiration rate in bright, clear weather. This would indicate a partial closing of stomates during periods of high evaporation stress. However, Bierhuizen and Slatyer (1964) have shown that stomatal resistance appears to have a greater influence on transpiration rate than it does on photosynthesis rate. Therefore, partial stomatal closure in *Polygonum* may not be the best explanation for the midday depression in photosynthesis rate.

Another possible contributor to the depressed and irregular photosynthesis rate could be temporary carbohydrate accumulation in the leaves. This type of inhibition has been reported by Kursanov (1933) and others, and may be involved here.

CARBON BUDGETS OF UNDISTURBED ALPINE TURF

Net productivity of alpine turf was estimated by measuring net photosynthesis, dark-respiration, and output of CO_2 from the soil surface after clipping. Runs were made on three undisturbed sods of 100 cm^2 each. The following parameters were measured at a uniform flow rate:

(1) mg CO_2/hr in plant and soil (unclipped sod) air stream in light
(2) mg CO_2/hr in ambient air stream
(3) mg CO_2/hr in soil (clipped sod) air stream in light
(4) mg CO_2/hr in plant and soil (unclipped sod) air stream in dark

Using the above parameters (by number), calculations were made as follows: net photosynthesis $(Ps_n) = (4) — (1)$ in mg CO_2/hr, plant dark-respiration $(Rs) = (4) — (3)$ in mg CO_2/hr, soil CO_2 output $(S) = (3) — (2)$ in mg CO_2/hr.

Gross photosynthesis (Ps_g) of the turf block per hour was computed as follows: $Ps_n + Rs + S = Ps_g$ (in mg CO_2/hr).

Net photosynthesis per day was then: $15 Ps_g — 24 Rs — 24 S = 24 Ps_n$, where $15 =$ number of hours of daylight/day, and $24 =$ number of hours/day.

Net carbon productivity (as mg C fixed per m² per day) was computed as 27.27% of $24 Ps_n \times 100$. This makes possible a rough comparison of productivity as measured by net photosynthesis with harvest data productivity measurement by multiplying these latter figures by 44.59%, the average carbon content of dry matter of the whole corn plant (*Zea mays*) as reported by Transeau (1926).

The results of these computations of carbon budgets of alpine sod are presented in Table 2. Samples No. 1 and No. 3 were mixtures of several species (*Polygonum bistortoides, Trifolium parryi, Arenaria obtusiloba, Sedum stenopetalum*, et al.) while Sample No. 2 was principally (85% cover) *Carex elynoides*.

Since these measurements of *in situ* sod-block metabolism were few and short term, they are only roughly indicative of photosynthesis and productivity rates of alpine vegetation. Still when the net photosynthesis results (as mg CO_2/g dry wt/hr) are plotted against light intensity and chamber temperature, they fall into place on the curvilinear multiple regression diagram (Scott and Billings 1964: Fig. 2,

TABLE 2.—Carbon budgets of undisturbed alpine turf

Sample no.	1	2	3	Average
Date	1 Jul. 1958	4 Jul. 1958	9 Jul. 1958	
Mean chamber temp, light (°C)	30.2	20.0	17.4	
Mean chamber temp, dark (°C)	23.9	13.9	16.1	
Mean light intensity (ft-c)	5,000	3,800	2,230	
Net photosynthesis (mg CO_2/g d wt/hr)	5.33	7.47	5.89	6.23
Net photosynthesis (mg CO_2/m²/hr)	2,137	1,610	2,511	2,086
Dark respiration of shoots (mg CO_2/g d wt/hr)	1.92	2.36	4.93	3.07
Dark respiration of shoots (mg CO_2/m²/hr)	771	504	2,095	1,123
Soil CO_2 output (mg CO_2/m²/hr)	1,701	1,777	2,061	1,846
Gross photosynthesis (mg CO_2/m²/hr)	4,609	3,891	6,667	5,056
Net carbon fixed (g C/m²/day)	2.67	0.99	0.07	1.24
Estimated dry wt gain (g d wt/m²/day)	5.99	2.22	0.16	2.79

p. 253) which was derived from other measurements in the same region. So the net photosynthesis figures appear to be reliable. The dark respiration results in Sample No. 3 are unusually high and must not be due to the effects of temperature alone. We have no ready explanation since by its late date and relatively advanced phenology it could scarcely be due to the "early season" increase in respiration rate reported by Hadley and Bliss (1964).

The values for soil CO_2 output are somewhat higher than those reported by Walter (1960). Actually, soil CO_2 output may be higher in alpine sites since Walter does not include values for alpine tundra vegetation (in fact, none for over 1,000 m elevation) and it is well established that *in situ* plant respiration increases with altitude. On the other hand, our figures are much lower than those observed by Wallis and Wilde (1957) in forest soils of Wisconsin using a vacuum pump extraction technique. Airflow through our chamber approached or was even less than normal airflow in the site, so it is improbable that suction affected the results to any extent. Scott and Billings (1964) noticed that there was a sharp increase in the CO_2 output from the surface of closely clipped sod immediately following rapid clipping. This surge lasted for only a minute or two and probably comes from the cut stem bases. Since, in the present study, clipping took about 15 minutes and no CO_2 surge was noted, it is not likely that the relatively high CO_2 output figures are due to this phenomenon.

Using the CO_2 exchange data from the sod blocks, estimated dry weight gain (productivity) per m^2 per day appears to correlate fairly well with published data obtained by the harvest method. Since our estimate is based on total carbon metabolism, the present results can best be compared with only those productivity studies in which both above-ground and below-ground parts were harvested. The average estimated dry weight gain per m^2 per day of 2.79 g for our sod blocks falls between the 11.08 g on a mesic site and the 1.17 g for a xeric station reported by Scott and Billings (1964). This appears about right since our stations were intermediate to those of Scott and Billings in species composition and soil moisture characteristics. Sample No. 1, our most mesic site, actually showed a fairly high rate of 5.99 g/m^2 per day. Because of a possible seasonal increase in the photosynthesis/respiration ratio (Hadley and Bliss 1964), these estimated dry weight gain figures are probably slightly too high considering the entire growing season of about sixty days.

CARBON DIOXIDE CONCENTRATION IN AMBIENT AIR

Data on carbon dioxide concentration of ambient air at high altitudes in terms of mg CO_2 per liter are not abundant and often are not readily available. The data shown in Table 3 are derived from ambient air stream analyses immediately following system calibration and are probably fairly accurate. Early afternoon values are notably lower than those of the rest of the day. While this may partially

Table 3.—CO_2 concentration of ambient air stream
at 5 dm above soil surface (in mg/liter)

Date	Before noon Time	Concentration	Date	After noon Time	Concentration
10 Jul.	0855	0.432	9 Jul.	1215	0.435
16 Jul.	0900	0.400	15 Jul.	1225	0.392
17 Jul.	0950	0.385	24 Jun.	1240	0.370
27 Jun.	1000	0.470	30 Jun.	1330	0.320
3 Jul.	1010	0.412	1 Jul.	1355	0.360
2 Jul.	1015	0.420	4 Jul.	1400	0.360
26 Jun.	1030	0.462	9 Jul.	1440	0.470
28 Jun.	1100	0.466			
22 Jul.	1120	0.425			
	Avg.	0.430		Avg.	0.387

reflect the photosynthetic activity of the tundra vegetation, it probably is influenced more by photosynthesis in the forest and subalpine meadow vegetation which surrounds this tundra "island" on the lower slopes.

References

Bierhuizen, J. F. and R. O. Slatyer. 1964. Photosynthesis of cotton leaves under a range of environmental conditions in relation to internal and external diffusive resistances. *Austral. J. Biol. Sci.,* **17**:348-359.

Blagowestschenski, W. A. 1935. Über den Verlauf der Photosynthese im Hochgebirge des Pamir. *Planta,* **24**:276-287.

Cartellieri, E. 1940. Über Transpiration und Kohlensäureassimilation an einem hochalpinen Standort. *Sitzungsber. Akad. Wiss. Wien, Abt. I,* **149**:95-143.

Hadley, E. B. and L. C. Bliss. 1964. Energy relationships of alpine plants on Mt. Washington, New Hampshire. *Ecol. Monogr.,* **34**:331-357.

Harrington, H. D. 1954. Manual of the plants of Colorado. Sage Books, Denver. 666 p.

Henrici, M. 1918. Chlorophyllgehalt und Kohlensäureassimilation bei Alpen- und Ebenenpflanzen. *Ver. naturforsch. Ges. Basel,* **30**:43-136.

Kursanov, A. L. 1933. Über den Einfluss der Kohlenhydrate auf den Tagesverlauf der Photosynthese. *Planta,* **20**:535-548.

Larcher, W. 1963. Die Leistungsfähigkeit der CO_2-Assimilation höherer Pflanzen unter Laboratoriumsbedingungen und am natürlichen Standort. *Mitt. Floristischsoziol. Arbeitsgemeinsch., N.F.,* **10**:20-33.

Mooney, H. A. and W. D. Billings. 1961. Comparative physiological ecology of arctic and alpine populations of *Oxyria digyna. Ecol. Monogr.,* **31**: 1-29.

———, R. D. Hillier and ———. 1965. Transpiration rates of alpine plants in the Sierra Nevada of California. *Amer. Midl. Natur.,* **74**:374-386.

———, R. D. Wright and B. R. Strain. 1964. The gas exchange capacity of plants in relation to vegetation zonation in the White Mountains of California. *Ibid.,* **72**:281-297.

Mönch, I. 1937. Untersuchungen über die Kohlensäurebilanz von Alpenpflanzen am natürlichen Standort. *Jahrb. wiss. Bot.,* **85**:506-553.

OVINGTON, J. D., D. HEITKAMP AND D. B. LAWRENCE. 1963. Plant biomass and productivity of prairie, savanna, oakwood, and maize field eco-systems in central Minnesota. *Ecology,* **44**:52-63.

PISEK, A. 1960. Pflanzen der Arktis und des Hochgebirges. *Handb. d. Pflanzenphysiol.,* Band 5, Teil 2:376-414.

SALISBURY, F. B. AND G. G. SPOMER. 1964. Leaf temperatures of alpine plants in the field. *Planta,* **60**:497-505.

SCOTT, D. AND W. D. BILLINGS. 1964. Effects of environmental factors on standing crop and productivity of an alpine tundra. *Ecol. Monogr.,* **34**:243-270.

STOCKER, O. 1954. Der Wasser- und Assimilationshaushalt südalgerischer Wüstenpflanzen. *Ber. deutsch. bot. Ges.,* **67**:289-299.

———. 1960. Die photosynthetischen Leistungen der Steppen- und Wüstenpflanzen. *Handb. d. Pflanzenphysiol.,* Band 5, Teil 2:460-491.

TRANQUILLINI, W. 1955. Die Bedeutung des Lichtes und der Temperatur für die Kohlensäureassimilation von *Pinus cembra-* Jungswuchs an einem hochalpinen Standort. *Planta,* **46**:154-178.

———. 1959. Die Stoffproduktion der Zirbe (*Pinus cembra* L.) an der Waldgrenze während eines Jahres. I. Standortsklima. *Ibid.,* **54**:107-129.

———. 1964. The physiology of plants at high altitudes. *Ann Rev. Plant Physiol.,* **15**:345-362.

——— AND H. TURNER. 1961. Untersuchungen über die Pflanzentemperaturen in der subalpinen Stufe mit besonderer Berücksichtigung der Nadeltemperaturen der Zirbe. *Mitt. d. forstl. Bundes-Versuchsanstalt Mariabrunn.,* **59**:127-151.

TRANSEAU, E. N. 1926. The accumulation of energy by plants. *Ohio J. Sci.,* **26**:1-10.

WALLIS, G. W. AND S. A. WILDE. 1957. Rapid method for the determination of carbon dioxide evolved from forest soils. *Ecology,* **38**:359-361.

WALTER, H. 1960. Einführung in die Phytologie. Band 3, Grundlagen der Pflanzenverbreitung. Teil 1, Standortslehre. 2 Aufl. Eugen Ulmer, Stuttgart. 566 p.

16

Reprinted from *Ecology* 42(1):177–180 (1961)

ESTIMATION OF NET PRIMARY PRODUCTION OF FOREST AND SHRUB COMMUNITIES

R. H. WHITTAKER

Despite the current interest in productivity, one major group of natural communities—those dominated by woody plants—has largely resisted effective productivity measurement. Complexity of structure and size of plants in forests render such research difficult, though not impossible as indicated by work of Burger (1953) and Ovington (1957, Ovington & Madgwick 1959). Some woody communities, in climates producing a marked annual rhythm of growth, offer valuable aids to such research in the forms of wood rings and bud-scale scars by which this year's growth and past years' accumulated biomass may be distinguished. It is the purpose of this paper to discuss an approach to net primary production based on these marks of annual rhythm.

COMMUNITY SAMPLING

Two principal approaches to measurement of current growth in the field are possible (excluding for the purposes of this paper gaseous exchange and other molecular techniques)—weighing or other measurement of foliage and current twigs, and determination of wood radial increments. In sampling techniques used in the Great Smoky Mountains, a 0.1 hectare quadrat extending 10 m on each side of a 50-m steel tape is laid out and marked. All trees and arborescent shrubs in the quadrat are tallied by diameter and species from the one-inch diameter class up. Increment borings are taken for major species from a sufficient number of trees within or close by the quadrat to represent the range of size-classes, and individual trees of rarer species and largest sizes are bored. For each tree bored, radial wood increment for the past decade, bark thickness, age, height, and diameter at breast height are determined. Current foliage of undergrowth plants is clipped in 20 small (0.5 × 2 m) quadrats scattered (random numbers of meters out on each side of 5-m intervals along the tape) over the area of the larger quadrat. In each of these the herbs are clipped at ground level, shrubs at the last bud-scale scar. Per cent foliage removal by animals is recorded when this can be observed and estimated. Live weights of each species in each quadrat are determined in the field for statistical purposes; live and dry weights for each species from all 20 quadrats combined are determined in the laboratory. Since shrubs of 0.5 in. or more dbh are treated as tree stems, only those below this limit are clipped. A quadrat area greater than 0.1 ha is desirable for forests containing large trees. Shrub community samples are based on clippings from 10, 0.5 × 2.0 m quadrats scattered in the stand.

[1] This is the first publication from a project on "Productivities of plant communities in the Great Smoky Mountains," supported by funds from the National Science Foundation. Field data from the hemlock-beech cove forest were taken, and the methods described under "Community Sampling" and Production Analysis of Shrubs" developed, by the author in collaboration with R. W. Becking. The project was made possible by permission to take vegetation samples without disturbance in the Great Smoky Mountains National Park, and permission to cut trees and large shrubs for production analysis on National Forest lands outside the Park. The author is indebted to Stephen H. Spurr for comments on the manuscript.

TREE-STRATUM COMPUTATIONS

Data taken from the stand permit computation of basal areas of wood plus bark and of wood only, and bole volume if suitable volume tables are available. When they are not, a preliminary approximation of volume can be obtained from Spurr's (1952) volume equations without species corrections, or by the assumption that the bole approaches a paraboloid of rotation for which the volume

is represented by $V_p = \frac{\pi}{2} r^2 h$, in which r is the wood

radius at breast height and h is tree height. The limitations of this "parabolic volume" may be granted, but it permits comparison of stands; and on it one approach toward estimation of wood growth may be based. Basal area increment and height growth are separately estimated, according to the two-way method of Spurr (1952). Basal area increments are computed from the wood basal areas and average or individual radial increments. Height growth estimation is based on curves relating height to age for major species in the stand, and the age and height data, in relation to height curves of normal shape, for individual trees of rarer species. Given estimates of basal area increment and height growth, one may compute volume growth for each size class of major species and each individual of rarer species and largest sizes by the difference in parabolic (or otherwise determined) volumes before and after a year's growth ("parabolic volume increment"). This estimate weights heavily the height growth, as this reflects elongation of the stem or of both stem and branches. One-half basal area increment times height provides an alternative estimate of volume growth by the stem or bole itself; this value is termed here "estimated stem volume increment."

CONVERSION FACTORS

There now result two independent, part-measurements of production—estimated volume increment for trees and larger shrubs, and clipping production for smaller shrubs. These must be related to each other and to the desired community-wide measurement—net annual production, or the primary productivity, exclusive of that expended in respiration by the plants, measured as dry weight of current growth per year. For the smaller shrubs ratios are needed relating total shoot growth, and total shoot and root growth, to clipping production. For larger shrubs and trees, estimated volume increment must be related to actual volume and weight growth of bole, and to total shoot and total shoot and root growth. To determine these relations, the distribution of current production in the various major fractions of woody plants and the relation between estimated volume increment and actual growth must be studied. Ovington's (1957) approach through age-sequences of average individuals in plantations is thought inappropriate to uneven-age climax stands; methods determining production distribution for each individual studied are needed. Such methods will be described as they apply to shrubs.

PRODUCTION ANALYSIS OF SHRUBS

A shrub to be studied is felled, and stem sections or "logs" (usually 0.5 m in length) are marked off. Branches are removed from the base upward; for each branch are recorded height above stem base, branch basal diameter, length and age, and weight and number of current twigs and leaves, of older leaves, if any, by years, and of fruits or fruit clusters. For each log are recorded basal and mid diameter, basal age, bark thickness, radial wood increment, and live and dry weight of the log or a sample disc. For some sample branches live/dry weight ratios are determined for branch wood of different ages, current twigs, leaves, separated by ages if the plant is evergreen, and fruits. For some sample logs bark is stripped off for live and dry weighing, and log surface and average bark thickness are determined. For as many individuals as possible, root systems are dug up with as much care as possible, and root live and dry weights determined. Diameters of broken roots are recorded, and basal diameter and weight are determined for enough intact roots to permit construction of curves relating average weight to basal diameter.

BRANCH AGES

Age determination of branches proves to be a special difficulty, subject to several types of errors: (a) Ages determined only from wood rings at the bases of branches are often in error by several years (cf. Reukema 1959). (b) Age determination by bud-scale rings and branching nodes is often more reliable, but these cannot be counted with confidence in older branch wood. (c) Some twigs of older, lower branches fail to form a current twig in a given year. (d) Secondary current twigs or Lammas shoots are formed during the season by vigorous upper branches in some species. (e) Although branch age will usually accord with stem age at the point of branching, development from dormant or adventitious buds occurs in a few branches of some species, and in some other species is common. Branch ages should be determined not by any single means, but by careful checking against one another of both bud-scale and wood-ring ages of both branches and logs. Reasonable confidence in branch ages is possible when the shrub is thus studied as a whole.

PRODUCTION DISTRIBUTION COMPUTATIONS

Given the preceding measurements one may determine or estimate the dry weight production of the different fractions of the shrub: (a) Clipping (current twig and leaf) production, and fruit production if any, are directly measured. (b) Older leaf growth, in evergreen species, may be estimated from the increase of average dry weight per leaf from current to older leaves, together with the total numbers and weights of older leaves by years. If the greater average weight of older leaves results in part from loss of smaller leaves, such estimates will be in error; but counts of leaves of different ages may give some assurance that apparent growth does not result from loss of smaller leaves. (c) Stem wood growth may be determined from the ratio of the cross-section area increase for the current year to the cross-section area of the wood, times the wood dry weight for the log. It has not been thought feasible to separate sap- and heartwood in shrubs. To allow for curved taper, cross-section areas should be based on the mid point of the log, or the average of cross-section areas, not of diameters, for the ends of the log (Huber and Smalian formulas). (d) Relative bark thickness increase at different ages can be estimated from curves of bark thickness in relation to age, although sloughing from the surface of older bark cannot be adequately allowed for. Bark growth for a log can be estimated from relative bark thickness increase as related to log age, wood growth of the log as it increases bark surface area, and dry weight of bark present, whether

measured or estimated from bark thickness and surface area. (e) Growth of branches (wood and bark together, excluding current twigs) is very difficult to measure directly. Estimation has been based on construction of curves of branch weights in relation to age, with the branch population stratified by relative vigor of growth, as indicated by weight in relation to age or current twig and foliage weight. Average rates of branch weight increase/branch weight at different ages may thus be determined, and branch dry weights multiplied by these rates. Rates of growth differ widely in different branches of the same shrub, and a production estimate for a single branch may be substantially in error. Combining such estimates for all branches may give a reasonable value for branch growth, though a value which is necessarily an estimate, not a measurement. (f) Root growth also is inaccessible to direct measurement. Estimation of root production has been based on curves of total root weights in relation to total woody shoot weights at different ages, and on the assumption that the ratio of root and woody shoot growth approximates the ratio of root and woody shoot biomass. Roots, like twigs and branches, die and are lost during the life of the plant; and the magnitude of this loss cannot be accurately determined.

RESULTS—PRODUCTION RELATIONS OF *Rhododendron maximum*

Table I summarizes biomass and production distribution, with and without roots, for *R. maximum* in a hemlock-beech ravine cove forest, 1400 ft. elev., Laurel Hollow, Cherokee National Forest, Great Smoky Mountains, Tennessee. Data are based on sets of 10 older (20-50 years) and more vigorously growing "dominant" individuals, 10 smaller (12-25 years), less vigorously growing "intermediate" or "suppressed" individuals, and 10 younger (9-13 years) small and mostly unbranched plants. Other ratios needed for production estimation are given for "dominant" individuals in Table II. Regression lines, to which correspond high coefficients of correlation, have been computed for some of these values; but presentation in the form of ratios with standard errors of means for sets of 10 individuals has been preferred here. Since the values given are independently computed averages of sets of 10 ratios, they will check against one another approximately, not exactly. Standard errors of means for 10 measurements of total production/clipping production are of the order of 4 to 5% for *R. maximum* and other shrub species studied. Because of differences in branch pattern and relations of radial increment to height, and for other reasons, estimates based on volume increment are subject to wider errors; standard errors of means for ratios of total production to estimated volume increment are of the order of 8 to 10%. Root/shoot ratios are widely variable with age and from individual to individual for a given age. Ratios for total root dry weights to total stem and branch wood and bark dry weights, with estimates of dead root and branch weights from basal diameters, were 2.51 for seedlings of 3-7 years, 1.78 and 1.24 for small shrubs, 10-13 and 13-17 years, and 0.75 for medium-sized shrubs, 14-25 years. Root/shoot ratios of individuals among the latter set varied from 0.36 to 1.12, with a standard deviation of ±0.23.

APPLICATION—SPECIES AND STAND PRODUCTION

From these ratios the net annual production by *R. maximum* in this stand may be estimated. Clipping dry weight for smaller individuals was 2.6 g/m² indicating, with the ratios for intermediate shrubs, 6.5 g/m²/yr of shoot and 9.0 g/m²/yr of shoot and root production. Estimated stem volume increment for a population of

TABLE I. Biomass and production distribution for *Rhododendron maximum* in a hemlock-beech cove forest. All values are based on averages of 10 shrubs except root data for dominant shrubs, based on 5 shrubs.

| | \multicolumn{7}{c}{PERCENTAGES IN—} | | | |
	Stem wood	Stem bark	Branches	Old leaves	Fruit	Current twigs	Roots	Biomass Accumulation Ratio	Total production Clipping dry weight
BIOMASS									
Dominant									
Shoot only............	44.1	3.9	31.1	16.7	.2	4.0	—	8.9	—
Shoot & root..........	28.9	2.6	20.4	11.0	.1	2.6	34.4	—	—
Intermediate									
Shoot only............	48.3	4.0	15.1	26.4	—	6.2	—	5.6	—
Shoot & root..........	32.8	2.7	10.2	17.9	—	4.2	32.2	—	—
Small									
Shoot only............	16.4	6.4	3.8	59.2	—	14.2	—	4.1	—
Shoot & root..........	11.3	4.4	2.6	41.0	—	9.8	30.9	—	—
PRODUCTION									
Dominant									
Shoot only............	24.2	2.7	22.1	15.7	1.6	33.7	—	—	3.00 ± .11
Shoot & root..........	17.8	2.0	16.3	10.5	1.2	24.8	27.4	—	3.99 ± .16
Intermediate									
Shoot only............	34.5	4.2	10.2	13.1	—	38.0	—	—	2.58 ± .18
Shoot & root..........	25.6	3.1	7.6	9.7	—	28.2	25.7	—	3.47 ± .17
Small									
Shoot only............	11.0	4.7	2.7	22.8	—	58.8	—	—	1.66 ± .04
Shoot & root..........	8.6	3.7	2.1	17.8	—	45.9	21.9	—	2.16 ± .09

TABLE II. Average production interrelations for "dominant" *Rhododendron maximum* in a hemlock-beech cove forest, with standard errors of means for sets of 10 measurements

	Shoot Only	Shoot & Root
Total production, g dry weight / Estimated stem volume increment, cm³	2.39 ± .18	3.29
Total production / Stem wood production	3.67 ± .34	5.06
Total production, g dry weight / Parabolic volume increment, cm³	1.96 ± .08	
Total production, g dry weight / Leaf blade area, m²	73	101
Stem wood production, g dry weight / Estimated stem volume increment, cm³	0.667 ± .039	
Clipping production, g dry weight / Estimated stem volume increment, cm³	0.82 ± .085	
Clipping production / Stem wood production	1.28 ± .16	
Actual wood volume increment / Estimated stem volume increment	1.14 ± .06	
Actual stem wood volume / Parabolic estimate of wood volume	.995 ± .050	
Wood specific gravity, dry g/cm³	.615	
Leaf blade area, cm² / Blade & petiole dry wt., g	76.4	

2510 larger stems per hectare was 0.736 m³/ha (73.6 cm³/m², 10.5 ft³/acre). Application of the ratios to these values yields for arborescent Rhododendron: 60 g/m² current twig and leaf production, 176 g/m²/yr total shoot production, 242 g/m²/yr total shoot and root production, and 0.42 and 1.96 m²/m² of current and older leaf surface in relation to ground surface on which this production is based. These values may be compared with those for all species together in a shrub community of a mixed heath bald on a north-facing slope of Brushy Mountain, 4900 ft elevation, Great Smoky Mountains National Park, Tennessee: 266 g/m² current twig and leaf production, 572 g/m²/yr total shoot production, about 1000 g/m²/yr shoot and root production, and 1.69 and 2.29 m²/m² of current and older leaf surface in relation to ground surface. Herb growth is sparse; and herb production less than 1% of the total, in both stands. Consumption of current shrub foliage by animals also is small in both cases, and the measured net growth should closely approach net annual production.

PRODUCTION AND BIOMASS DISTRIBUTION IN THE COMMUNITY

Such data permit a number of other approaches to community structure and characterization, among them: (a) The biomass/production or "biomass accumulation ratio," the ratio of the total dry weight of living plants, or other appropriate biomass measurement, to net annual production, may be determined. The ratio is thought to be an ecologically significant variable ranging downward from large, old-growth forests to 1.0 in some herb communities and less than 1.0 in plankton. Ratios estimated from stem sizes and numbers in the stands discussed are 10.0 for *R. maximum* only in the cove forest and 4.7 for the shrubs of the mixed heath bald. (b) A perennial problem of plant synecology is the ranking of species of different growth forms and statures, for which the "importance value" (Curtis & McIntosh 1951) and other measures are used. Production seems the best single dimension by which species may be ranked according to functional significance in the community; for some purposes biomass, or such particular expressions of production as clipping dry weight or volume increment may be appropriate. Ranked by estimated volume increment, *R. maximum* is the third most important species, after *Tsuga canadensis* and *Fagus grandifolia*, in the cove forest stand. These three species share "production dominance" in the stand. The shrubs of the mixed heath bald rank by clipping dry weight in the order—*Rhododendron catawbiense* 162, *Kalmia latifolia* 76, *Viburnum cassinoides* 9, *Vaccinium constablaei* 9, *Gaylussacia baccata* 6, and *Pyrus melanocarpa* 4 g/m². (c) Ranking of species in this manner permits approaches to relative dominance and species diversity which are independent of area relations and assumptions about "individual" plants. Of the various diversity measurements (Hairston 1959), two of the simplest may be applied to the data—Simpsons (1948) index of concentration of dominance, $\alpha = \Sigma (\frac{x}{n})^2$ (in which x is an individual species production measurement, and n is the total production for all species), and the slope r for a geometric progression, $x = ar^{(n-1)}$ (in which a is the production measurement for the first species and n the number of a species when all species are arranged in sequence from highest production to lowest, with least squares fit for r). For the cove forest and heath bald these values become—, $\alpha = 0.208$, 0.455, and $r = 0.683$, 0.477—expressing the greater species-diversity of the cove forest, the greater concentration of production dominance in a few species in the heath bald.

SUMMARY-CONCLUSION

Measurement of production in forest and shrub communities may always be difficult, but it is by no means impossible. In climates with strong seasonal fluctuations, the marking of the current year's growth by wood rings and bud-scale scars makes possible effective estimation of net primary production by woody plants. Steps in the approach to such estimation described are: (a) field measurement of community production by clipping current twigs and leaves of smaller shrubs, and stand counts of larger shrubs and trees together with increment borings and height determinations, (b) computation, for trees, of estimated volume increment, (c) analysis of major species to determine production ratios (total production to clipping production, and stem wood and total production to estimated volume increment), and (d) estimation of net production by woody plants in the community from the field measurements times the correction ratios to total production. The resulting data also permit effective approaches to species ranking, relative dominance and species diversity, and other aspects of community analysis and characterization. The techniques described here may be subject to substantial improvement; they are offered not as a perfected method but as a direction of research. Analysis of production and volume relations of trees is laborious, but such research will have to be carried out by ecologists and foresters if the method is to be applied to the productivity of forests.

REFERENCES

Burger, H. 1953. Holz, Blattmenge und Zuwaches. XIII. Fichten im gleichaltrigen Hochwald. (French summ.) Mitt. Schweiz. Anstalt forstl. Versuchswesen [Zürich] 29(1): 38-130.

Curtis, J. T. & R. P. McIntosh. 1951. An upland forest continuum in the prairie-forest border region of Wisconsin. Ecology 32: 476-496.

Hairston, N. G. 1959. Species abundance and community organization. Ecology 40: 404-416.

Ovington, J. D. 1957. Dry-matter production by *Pinus sylvestris* L. Ann. Bot., N. S. 21: 287-314.

Ovington, J. D. & H. A. I. Madgwick. 1959. The growth and composition of natural stands of birch. I. Dry-matter production. Plant & Soil 10: 271-283.

Reukema, D. L. 1959. Missing annual rings in branches of young-growth Douglas-fir. Ecology 40: 480-482.

Simpson, E. H. 1949. Measurement of diversity. Nature [London] 163: 688.

Spurr, S. H. 1952. Forest inventory. Ronald, New York. 476 pp.

17

Reprinted from pages 149-174 of *Nature and Life in Southeast Asia*, Vol. V, edited by T. Kira and K. Iwata, Kyoto: Fauna and Flora Research Society, 1967.

Comparative ecological studies on three main types of forest vegetation in Thailand
IV. Dry matter production, with special reference to the Khao Chong rain forest

Tatuo KIRA,* Husato OGAWA,* Kyoji YODA* and Kazuhiko OGINO**

Estimation of dry matter production by tropical forest vegetation has been our principal concern since we began field work in Southeast Asia in 1957 (Ogawa et al. 1961). Particular effort was made along this line during the Joint Thai-Japanese Biological Expedition to Southeast Asia 1961-62 (Iwata 1965) in three types of climax forest vegetation in Thailand, especially in a tropical rain forest at the Khao Chong Forest Reserve, Trang Province, of peninsular Thailand.

Some 500 ha of well preserved rain forest, on the west slope of the central granitic range of the Malay Peninsula at 7°35′ north latitude, comprise the Khao Chong Reserve controlled by the Royal Forest Department of Thailand. The structure and physiognomy of the forest have been described in detail by Ogawa et al. (1965a). Although somewhat inferior to equatorial rain forests of Malaysia in tree size (maximum height of dominants 35-40 m) and in floristic diversity (98 species of trees and woody climbers over 4.5 cm D.B.H. per 0.32 ha), its physiognomic features are quite typical of a tropical rain forest with its multi-layered structure, absence of seasonal aspects, development of plank-buttresses, abundance of woody lianas, and so forth. Climatic records also indicate a rain forest climate, with mean annual temperature of 27.2°C, mean annual precipitation of 2,696 mm, and only three months receiving rainfall less than 100 mm (72, 34 and 70 mm in January, February and March respectively). The leaf area index (12.3 ha/ha including ground vegetation) suggests a high level of primary production.

This paper deals first with the principle and methods employed for determining gross production. Most of the items necessary for this determination, such as biomass increment, litter-fall and amount of dying trees,

---

* Laboratory of Plant Ecology, Faculty of Science, Osaka City University, Osaka, Japan.

** The Center for Southeast Asian Studies, Kyoto University, Koyto, Japan.

were successfully measured at Khao Chong.   These measurements, together
with the estimation of annual community respiration made by Yoda(1967),
allowed fairly accurate estimations of gross and net production by the
forest.   Two other deciduous forest types, monsoon forest and savanna
forest, were investigated at Ping Kong, Chieng Mai Province, northwestern Thailand (Ogawa et al. 1965a).   Little more than the estimation
of biomass and respiration loss at the end of the rainy season was possible
in these stands (Ogawa et al. 1965b, Yoda 1967), but some tentative estimates of their primary productivity are given in this paper.   One of the
most noteworthy facts revealed by this study is the nature of dynamic
equilibrium in a climax forest in terms of the balance between input and
output of matter to and from the standing biomass.   The paper concludes
with comparison of productivity among different types of forest vegetation.

          We have already published a preliminary account of dry matter pro-
duction by the Khao Chong rain forest (Kira et al. 1964), but the results
described there must be corrected in some respects.   Later studies revealed
that the respiration in leaf and branch had been badly underestimated
(Yoda 1967), and the record of biomass increment over a longer period
of three years became available.   The results presented here should therefore replace the former account.

          We are especially indebted to Messrs. Hiroyuki Watanabe and Kyôzô
Chiba of the Laboratory of Forest Ecology, Kyoto University, who measured D. B. H. increment at Khao Chong in March of 1965.   Our thanks
are also extened to Mr. Hiromitsu Kirita of the Laboratory of Plant
Ecology, Osaka City University, for allowing us to cite some of his unpublished data on the litter production in a warm temperate forest of central
Japan, and also to Prof. C. F. Cooper of the University of Michigan for
revision of English expressions.

PRINCIPLE AND METHODS OF ESTIMATION

   1.   Principle of *the summation method* for the estimation of dry matter
        production by plant communities

          Methods based on two different principles have hitherto been proposed
for estimating the primary production by terrestrial plant communities.
One is *the harvest method* in the wide sense of the word, which involves
repeated biomass measurements at specific time intervals and attempts
to arrive at the estimate of net production by integrating the difference
of biomass between two successive measurements.   The other is *the photosynthetic method* which combines the photosynthetic rate of a single leaf

with the vertical distributions of leaf area and light intensity within the community into an estimate of dry matter acquisition by the whole community. Monsi & Saeki(1953) initiated the latter approach, which was latter elaborated and applied to various types of community including forest by several Japanese investigators (Saeki 1960, Kimura 1960, Oshima 1961, Kuroiwa & Monsi 1963, Nomoto 1964, etc.).

The Danish school led by Boysen Jensen thoroughly elaborated the harvest method with special reference to forests. Their method is based on the so-called fundamental equation of dry matter production (Boysen Jensen 1932),

> Annual increment(dry matter production) = gross production *minus* (loss by death of roots, branches, leaves, bark and fruits *plus* loss of dry matter by respiration in roots, stems, branches and leaves),

and essentially consists of various experimental and sampling techniques for independently estimating the respective terms in the equation. Gross production is obtained by summing the estimated amounts of annual biomass increments, annual losses as litter-fall and annual losses by respiration in various organs. An example of this form of productivity assessment, which we may call *the summation method*, was presented by Möller et al. (1954) in beech plantations of Denmark, and recently by Müller & Nielsen(1965) in a tropical humid forest of Ivory Coast.

The summation method is no doubt one of the most elaborate and complete means of production study. But it seems relevant to point out that the method is not always properly understood in current literature. To make the conditions for its use clearer, we derive the fundamental equation on the basis of the balance sheet of organic matter acquired and lost during the period from t_1 to t_2 (Table 1).

The amount of living plant biomass of the community at t_1 or the beginning of the period concerned is designated by y_1. The income of

Table 1 Dry matter budget in a plant community during a period t_1—t_2
    All amounts given in the table are expressed on an area basis.

Input (credit account)	Output (debit account)
Biomass at t_1 : y_1	
Total assimilation or gross production : P_g	Loss by respriration : R Loss by death : L Loss by grazing and parasitism : G
	Biomass at t_2 : y_2

organic matter to the community during the period t_1—t_2 is the gross production (P_g). No other sources of income can exist. On the other hand, a certain amount of organic matter is lost from the community biomass through various pathways. One of these is the loss due to respiration (R) by all living plants in the community. Some portions of plants may also die during the period; this loss (L) represents another pathway. And finally, some percentage of biomass may be consumed by grazing animals and other heterotrophic organisms such as parasitic plants, mycorrhizal fungi, etc. The amount of loss through the last pathway is denoted by G. Giving the symbol y_2 to the plant biomass at the end of the period or t_2, we have the equation,

$$y_1 + P_g - (R + L + G) = y_2, \tag{1}$$

or

$$\Delta y \equiv y_2 - y_1 = P_g - (R + L + G), \tag{1'}$$

where Δy stands for the difference of biomass between t_2 and t_1. Eq. (1) thus derived is the same in its construction as Boysen Jensen's formula.

The Danish workers defined the net production (P_n) as the synonym of Δy or biomass change, but the definition, $P_n = P_g - R$, is prefered to theirs from the purely biological standpoint (Odum 1959). Adopting the latter definition,

$$y_1 + P_n - (L + G) = y_2. \tag{2}$$

Furthermore, it is possible to divide both L and G into new and older components. For instance, a certain fraction of the initial biomass (y_1) dies and is shed in the period between t_1 and t_2. This is *the older component* of L or L_O, whereas *the new component* L_N is the amount of plant tissue which was newly formed during the period and which died before the end of the period ($L_O + L_N = L$). G_O and G_N may also be defined in the same way. The distinction between the older and the new component is based on whether the part of plant body concerned was formed before t_1 or after t_1. New components may receive reserve substances stored in older components, and a certain fraction of organic matter produced in new components may also be translocated into older components, but here it is assumed that any exchange of matter between the older and the new parts of a tree is balanced.

Interrelations between these amounts are graphically shown in Fig. 1. The biomass at the end of the period (y_2) consists of the new (y_{2N}) and the older components (y_{2O}).

$$\left. \begin{array}{l} y_{2O} = y_1 - (L_O + G_O) \\ y_{2N} = P_n - (L_N + G_N) \end{array} \right\} \tag{3}$$

Therefore,

$$\Delta y = y_2 - y_1 = y_{2N} - (L_O + G_O) \tag{4}$$

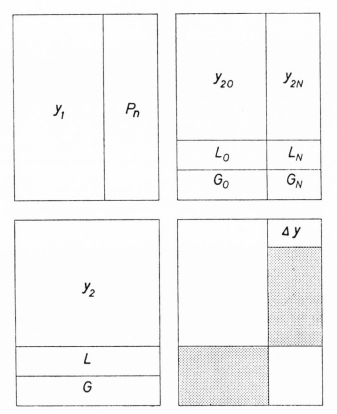

Fig. 1 Diagrammatic representation of the interrelations between the amounts related to net production

and
$$P_n = y_{2N} + L_N + G_N, \tag{5}$$
or
$$P_n = \varDelta y + L + G. \tag{6}$$

Eqs. (5) and (6) respectively represent two different principles of estimating net production. Working on deciduous hardwood forests of Denmark, Boysen Jensen (1932) and Möller et al. (1954) estimated the increment of biomass as the difference between two successive measurements of wood biomass made at a few years' interval, and by adding to it the amounts of leaf and branch litter obtained an estimate of P_n, disregarding the G-term as relatively small. This procedure, which is essentially based on Eq. (6), was later adopted by Kimura (1960) and by Müller & Nielsen (1965) for evergreen forests.

On the other hand, 'net production' has sometimes been estimated as the sum of the amounts of the current year's organs actually measured (Shidei 1960, 1964; Kimura 1963; Whittaker, Cohen & Olson 1963; Baskerville 1965). In this method, annual wood increments in trunk and older branches are estimated by stem analysis, the current year's twigs and leaves are clipped and weighed, growth of bark and older leaves, if any, is estimated, and so forth. If these measurements are made at the proper time near the end of the growing season, the current year's growth of sample trees can be fairly accurately estimated, though estimation of root growth encounters serious difficulties. It should be noted, however, that the amount of current year's growth thus obtained is not net production in the strict sense of the word, but corresponds to y_{2N} of Eq. (3). The amount y_{2N} is apparently an underestimate of the real net production, which is greater than y_{2N} by $(L_N + G_N)$ as indicated by Eq. (5), whereas the sum $(y_{2N} + L + G)$ is evidently an overestimate. Since it is generally difficult to separate litter-fall into new and older components, the procedure based on Eq. (5) is not always feasible.

Eq. (6) may therefore be prefered to Eq. (5) as the basis of net production estimation, provided that the biomass can be estimated accurately enough to assure a reliable estimate of Δy. Another advantage of the former approach is that it can be applied to such non-seasonal communities as the tropical forests discussed here, in which the stem analysis technique is impracticable owing to the incomplete formation of annual growth rings.

This study concerns the estimation of net production according to the procedure given by Eq. (6), involving repeated estimations of biomass in the same stand and the measurement of litter-fall with a litter trap. The result is combined with the estimate of community respiration obtained by Yoda (1967) to arrive at gross production. The methods employed are described in the following. More detailed discussions on the organic matter budget and balance sheet will be made elsewhere (Ogawa 1967).

2.   Methods for measuring input and output of dry matter

Increment of tree biomass

Only trees over 4.5 cm D. B. H. are treated in this paper. The amount of smaller saplings and ground vegetation in the tropical rain forest at Khao Chong was no more than 1-2% of the total biomass of the community.

Procedures for estimating the tree biomass have already been described in detail by Ogawa et al. (1965b). Based on measurements of weight, diameter and height of 119 sample trees belonging to 80 species, a method was developed to calculate the oven-dry weight of a tree based on its

D. B. H. value alone.   In a small plot (10 m × 40 m) in the Khao Chong rain forest, all trees were clear-felled and weighed. Comparing the weighed biomass of aerial shoots on the plot with the calculated value based on the D. B. H. census, it was found that the relative error of estimation was only 0.15% of the true biomass.

On the 3rd-4th of February 1962, a sample plot called Stand 5, 40 m × 40 m in area, was established in the Khao Chong rain forest for the continuous observation of D. B. H. increments. All trees larger than 4.5 cm D. B. H. were mapped, numbered and marked with a band painted in red around the trunk at a height of 130 cm above the ground (breast height). D. B. H. of every trunk was measured exactly along the band with a diameter tape. Artificial disturbances were carefully avoided in the plot, except the clipping of ground vegetation on a small area.

The census of D. B. H. was repeated by Yoda and Ogino on July 21, 1963 and by Watanabe and Chiba on March 30, 1965, using the same diameter tape. The interval between the first and the second census was 533 days and that between the second and the third census was 618 days. Records of D. B. H. increments over a total of 1,151 days (3.15 years) were thus obtained, and enabled the trends of biomass change to be calculated by the method described above. The calculations were as follows:

i)   The height of a tree (H m) was estimated from its D. B. H. (D cm) by the empirical formula,

$$\frac{1}{H} = \frac{0.543}{D} + 0.0217. \tag{7}$$

ii)   The amounts in kg dry weight of trunk (w_S), branches (w_B) and roots (w_R) per tree were respectively calculated using the following formulae.

$$\left.\begin{array}{l} w_S = 0.0396\ (D^2H)^{0.9326} \\ w_B = 0.006002\ (D^2H)^{1.027} \\ w_R = 0.0264\ (D^2H)^{0.775} \end{array}\right\} \tag{8}$$

iii)   The amount of leaves per tree (w_L in kg dry weight) was estimated from the trunk weight obtained above according to the formula,

$$\frac{1}{w_L} = \frac{13.75}{w_S} + 0.025. \tag{9}$$

iv)   The sum of w_S, w_B, w_R and w_L gave the total weight of a tree (w). The biomass of whole stand was obtained as Σw for all living trees over 4.5 cm D. B. H.

Death of standing trees

During the three years period, several trees were found standing dead on the plot. Since the exact time of death was not known, it was assumed that the trees had died at the beginning of the two time intervals; viz. the biomass of dead trees in February 1962 or in July 1963 was assigned to the loss of dry matter due to their death during the following time interval.

Litter-fall

Amount of litter-fall was measured by means of a long sheet of plastic net, $0.9\,m \times 12\,m$, stretched over the ground surface along the border of Stand 4, a sample plot of the same size adjacent to Stand 5. Daily records were kept during 27 days from January 13 to February 9, 1962. Total accumulation of litter in the trap during 15 days from February 9 to 24, 1962 was also recorded within Stand 5. Annual amount of litter-fall was estimated by simply multiplying the average rate of daily litter-fall during the 42-day period by 365.

Loss by grazing and parasitic organisms

No attempts at estimating the G-term in Eq. (6) could be made, but the resultant underestimation of net production may be small. Field observation suggested that the consumption of leaves by herbivores was not so large in amount.

Loss by respiration

That a large proportion of the gross production in mature forest communities is consumed in plant respiration has been pointed out by Kira (1965) and Yoda et al.(1965). Since this proportion normally increases with the age of the forest stand, a production estimate in forest vegetation would be incomplete without allowance for community respiration. A new method for the latter estimation was recently developed by Yoda et al.(1965), and the community respiration in three different types of Thailand forest here concerned was treated in detail in the preceding paper of this serial study(Yoda 1967).

RESULTS

Increase of biomass during three year period

On February 4, 1962, there were 214 trees and 30 woody climbers over 4.5 cm D. B. H. in Stand 5 at Khao Chong, $40\,m \times 40\,m$ in area. A number of the trees comprised two or more stems which united into a single trunk between breast height and the ground surface. Thus the total number of stems, of which D. B. H. was measured, amounted to 275 in all, including both trees and lianas.

On July 21, 1963, a tree of moderate size with a single stem was found standing dead. The remaining 274 stems were again measured with the result given in Table 2. The increment of D. B. H. ranged between 1.7 cm and -0.7 cm. Of all 274 stems, 66 (23%) neither increased nor decreased

Table 2 D. B. H. increments in all living trees and woody lianas in Stand 5 during 533 days from February 4, 1962 to July 21, 1963

D. B. H. increments mm	−7 −5	−4 −3 −2 −1 0 1 2 3 4	5 6 7 8 9 10 11	15 17
	−10	0	+10	+20
No. of trees	1 1	3 5 9 29 66 59 38 20 18	7 3 2 3 1 4 3	1 1
	2	247	23	2

their diameter, and a slight decrease(most probably the result of inaccurate measurement) was observed in 48 (18%), while the remaining 160 (59%) were found to have more or less positive increments. There was thus a general trend of D. B. H. increase in spite of the rather crude method of diameter measurement, although the increments themselves were small.

Similar measurements were repeated on March 30, 1965. During this second period of 618 days, 7 small stems died while 15 stems entered the 4.5cm D.B.H. class. As shown in Table 3, there was an apparent correlation between the increments in the first period and the corresponding values in the second period, showing that the measurements had generally been made correctly.

Based on these records, the biomass of all living trees 4.5 cm D. B. H. and over was calculated separately for the three dates, February 4, 1962,

Table 3 Correlation between D. B. H. increments in the first(533 days) and the second(618 days) period

	D. B. H. increment classes	2nd period (1963. 7. 21.—1965. 3. 30)				
		> 1.5 cm	1.4~0.5 cm	0.4~−0.4 cm	Dead	Total
1st period (1962. 2. 4.—1963. 7. 21.)	>1.5 cm	1	1	*	*	2
	1.4~ 0.5 cm	*	14	9	*	23
	0.4~−0.4 cm	2	61	177	7	247
	−0.5 cm >	*	1	1	*	2
	Total	3	77	187	7	274

Table 4 Annual rate of biomass increase (Δy) in Stand 5

Figures in parentheses indicate relative increase as percentage of the biomass at the beginning of the period concerned.

Period	Components	Stem Δy_S t/ha·yr	Branch Δy_B t/ha·yr	Root Δy_R t/ha·yr	Total wood Δy_C t/ha·yr	Leaf Δy_L t/ha·yr	Total Δy t/ha·yr	LAI ΔF ha/ha·yr
Feb. 4, 1962 Jul. 21, 1963 533 days (1.46 yrs)	Surviving trees	3.29 (1.6)	1.39 (1.8)	0.41 (1.4)	5.09 (1.6)	0.090 (1.1)	5.18 (1.6)	0.121 (1.06)
	Trees died	−1.30	−0.49	−0.19	−1.99	−0.061	−2.05	−0.85
	Total	1.99	0.90	0.22	3.10	0.029	3.13	0.035
	Trees newly reached 4.5 cm D.B.H. during the period	0	0	0	0	0	0	0
	Sum total	1.99 (0.97)	0.90 (1.13)	0.22 (0.73)	3.10 (0.98)	0.029 (0.35)	3.13 (0.97)	0.035 (0.31)
Jul. 21, 1963 Mar. 30, 1965 618 days (1.69 yrs)	Surviving trees	4.93 (2.4)	2.02 (2.5)	0.63 (2.1)	7.58 (2.4)	0.156 (1.9)	7.74 (2.4)	0.213 (1.9)
	Trees died	−0.35	−0.10	−0.09	−0.54	−0.025	−0.56	−0.037
	Total	4.58	1.92	0.54	7.04	0.131	7.18	0.176
	Trees newly reached 4.5 cm D.B.H. during the period	0.176	0.044	0.051	0.272	0.0126	0.285	0.0181
	Sum Total	4.76 (2.3)	1.96 (2.4)	0.60 (2.0)	7.32 (2.3)	0.144 (1.8)	7.46 (2.3)	0.194 (1.7)
Feb. 4, 1962 Mar. 30, 1965 1,151 days (3.15 yrs)	Surviving trees	4.08 (2.0)	1.67 (2.1)	0.53 (1.8)	6.28 (2.0)	0.125 (1.5)	6.41 (2.0)	0.176 (1.5)
	Trees died	−0.78	−0.27	−0.13	−1.18	−0.041	−1.22	−0.059
	Total	3.30	1.40	0.40	5.10	0.084	5.18	0.117
	Trees newly reached 4.5 cm D.B.H. during the period	0.095	0.024	0.028	0.146	0.0068	0.153	0.0097
	Sum Total	3.40 (1.7)	1.42 (1.8)	0.43 (1.4)	5.24 (1.7)	0.091 (1.1)	5.33 (1.6)	0.127 (1.1)

July 21, 1963 and March 30, 1965. As discussed by Ogawa et al. (1965b), the amounts of branches and leaves of a tree entangled by lianas are reduced to a considerable extent as compared with the amounts expected from Eqs.(8) and (9), but the reduction tends to be counterbalanced by the amounts of the lianas' shoots and leaves. Therefore the calculations were made for tree stems only, on the assumption that w_B and w_L in Eqs. (8) and (9) respectively include the weight of stems and leaves of the lianas climbing on the tree concerned. The rate of biomass increment obtained as the difference between measurements on two successive occasions is given in Table 4 on a hectare-year basis.

As recognized in Table 4, the net increase of biomass in a specified period was largely determined by the balance between the growth of living trees and the death of some individuals, since only a few small trees first attained 4.5 cm D.B.H. during each period. When the two periods are compared, the second period showed a higher rate of biomass increase than the first period (7.46 versus 3.13 ton/ha·yr). This difference was caused partly by the greater rate of growth in living trees during the latter period, but the smaller amount of dead trees was relatively the more influential factor. Over the whole 3.15 years from February 1962 to March 1965, the trees which survived throughout the period increased their biomass at the rate of 6.41 ton/ha·yr or 2.0% of the initial biomass, 324 ton/ha on Feb. 4, 1962. On the other hand, 8 trees died during the period and resulted in an average rate of dry matter loss of −1.22 ton/ha·yr (0.63% of the initial biomass). The net increase was therefore only 1.6% per year of the initial biomass or 5.33 ton/ha·yr. The rate of biomass increase tended to be the highest in trunk, lower in root and the lowest in foliage. It is noteworthy that even the highest rate observed (stem biomass in the second period) did not surpass 2.5% per year, whereas the lowest rate (leaf biomass in the first period) was only 0.35% per year.

Amount of litter-fall

Three fractions of litter, leaf, branch, and other minor components (including bark, fruit, flower, etc.), were separately weighed. The daily amount of leaf litter in Stand 4 ranged between 23.6 and 71.0 g (air-dry weight)/trap·day during 20-day period from January 13 to February 2, 1962. That of branch litter or minor components was much more variable, the range during the same period being 6.3-158.3 in branch and 1.0-21.1 g (air-dry weight)/trap·day in minor components.

As tabulated below, there was no significant difference between the rates of litter-fall in Stand 4 and in Stand 5. During the 42-day period,

Table 5   Rate of daily litter-fall in the Khao Chong
rain forest in oven-dry weight g/m²·day

	Period	Leaf	Branch	Others	Total
Stand 4	13/I—2/II (20 days)	3.44	2.89	0.55	6.88
Stand 5	2/II—24/II (22 days)	3.07	2.36	0.46	5.89
Mean		3.24	2.62	0.50	6.36

an average of 6.36 g/m² of oven-dry litter reached the ground each day.
Of this amount, 51%, 41% and 8% were leaf, branch and other components
respectively.   If the rate were assumed to be maintained throughout the
year, the expected annual litter-fall would amount to 6.36×365 g/m²·yr or
23.2 ton/ha·yr.

   Dry matter budget in the Khao Chong rain forest
   It is now possible to present a tentative balance sheet of dry matter
for Stand 5 at Khao Chong (Table 6), on the basis of the following assumptions:
      1)   The rate of biomass increase in living trees and that of loss
by the death of standing trees were both estimated as the average
for the three year period 1962-65.
      2)   The mean rate of litter-fall obtained by 42 days' observation
in January and February of 1962 was tentatively assumed to represent
the annual average.
      3)   Losses of dry matter due to the turnover of fine roots and to
the activity of consumer organisms could not be estimated and were
disregarded.
      4)   The estimates of Yoda (1967) were used for the annual respira-
tory consumption of dry matter.
      5)   Only the dry matter budget of the tree components of the
forest over 4.5 cm D.B.H. was considered.   Smaller plants were
excluded.
   Within the extent of these conditions, some important conclusions
may be drawn from this balance sheet.   Despite the exclusion of losses
due to the turnover of fine roots and to grazing by herbivores, the estimated annual gross production of 123.2 metric tons per hectare is
perhaps one of the highest of such estimates ever obtained in a plant community.   Nearly eighty per cent. of this gross production was found to
be consumed by the respiration of living vegetation.   Such a high value
of the respiration/gross production ratio has rarely been reported.   Of
the net production of about 29 ton/ha·yr, which is equivalent to 23% of

the gross production, nearly 4/5 were used for the turnover or renewal of leaves and twigs, while the increase of living plant biomass was very small, being less than 1/5 of the net production. The significance of these features will be discussed later.

Table 6 Summary of dry matter production by the Khao Chong
rain forest (Stand 5)

	Stem	Branch	Root	Leaf	Total
					ton/ha·yr
Biomass increase in living trees $(\Delta y')$	4.18	1.69	0.56	0.132	6.56
Death of standing trees (L')	0.78	0.27	0.13	0.041	1.22
Biomass increase in the whole stand $(\Delta y = \Delta y' - L')$	3.40	1.42	0.43	0.091	5.33
Litter-fall (L)	—	11.39*	—	11.84	23.22
Net Production $(P_n = \Delta y + L)$	3.4	12.8	0.4	11.9	28.6
Loss due to respiration (R)	13.1	19.0	5.6	57.0	94.6
Gross production $(P_g = P_n + R)$	16.5	31.8	6.0	68.9	123.2

* Including such minor components of litter as bark, fruit, etc.

DISCUSSION

1) Dry matter production by the tropical rain forest

Litter-fall

The most suitable size and form of litter trap for sampling litter-fall in a tropical rain forest were not known. Information was also lacking about the spatial distribution of litter-fall over the floor of rain forest, on which the statistical design of the sampling procedure should be based. Since we had no time to make a preliminary investigation to obtain this basic information, it was decided to use a long belt of plastic net as the trap, in the hope of reducing the error due to the uneven distribution of litter-fall. By stretching the belt immediately above the ground, the bias inevitably associated with a particular size and shape of trap could also be avoided. The smooth surface of plastic net might affect the accumulation of litter under wind-swept conditions, but there was no appreciable wind under the dense canopy of the forest.

At present we have no information on the seasonal rhythm of the rate of litter-fall in this forest. The measurement of litter-fall was made in January and February, which represent the short dry season at Khao Chong as previously stated, although light showers were not infrequent during those months. In the dry season, big emergent crowns were sporadically found in a leafless state. Two trees of *Sterculia* sp.(native name, *Pong*) and *Ficus* sp.(*Du'ai*), for instance, became leafless in Stand 4 between January 10 and 31, 1962, but the bare condition lasted for only one or two weeks. Observations in July and in January and February suggested that there was no marked difference in the frequency of bare crowns between winter and summer months. For this reason it was assumed that the rate of litter-fall remained more or less constant throughout the year, and the annual total of litter-fall was estimated by simply extrapolating the average of the observed daily rates to a one-year period.

The annual amount of leaf litter estimated in this way was 11.9 ton/ha·yr oven-dry weight. The leaf biomass in Stand 5(including undergrowth) was 8.2 ton/ha (Ogawa et al. 1965b), so that the average annual turnover of leaves is 1.45/yr. The mean longevity of leaves in this forest is thus estimated to be 0.69 years. According to Koriba(1958), the greater part of Malayan trees belong to his category of 'intermittent evergreens' characterized by periodic leafing that occurs once to several times a year. In such trees the shedding of old leaves also takes place more or less periodically following the expansion of new leaves. A relatively small number of tree species in Malaya shed their leaves before leafing and are therefore temporarily deciduous, though he found a series of intermediate habits between the intermittent evergreens and the deciduous species. In both life forms, the longevity of leaves varies greatly between three months and 2.5 years, but the interval between successive leafing is mostly less than one year. The commonnest case among Malayan trees is said to be two crops of leaves per year. These conclusions drawn by Koriba are consistent with the present estimate of the average rate of leaf turnover.

Where leaf duration is not synchronized with the annual cycle as in tropical rain forest, we cannot estimate the annual leaf production from the leaf biomass. It may equal the annual litter-fall in a mature, stabilized community, but otherwise a troublesome procedure of determining leaf longevity in each species might be needed before leaf production could be estimated accurately.

The non-leaf litter(11.5 ton/ha·yr) was nearly equivalent in amount to the leaf litter, being 49% of the total litter-fall. Bray & Gorham (1964) recently stated on the basis of a world-wide comparison of litter-fall re-

Table 7 Composition of litter-fall in some climax forest communities, in oven-dry weight

Forest type and locality	Leaf	Non-leaf components				Total	Authors and period
		Branch	Bark	Others	Total		
						ton/ha·yr	
Subalpine conifer forest (*Picea-Abies-Tsuga*) Yatugatake, Japan,	2.3 (66)	—	—	—	1.2 (34)	3.5 (100%)	Kimura 1963, 4 years
Subalpine conifer forest (*Abies* spp.) Yatugatake, Japan	2.7 (55)	—	—	—	2.2 (45)	4.9 (100%)	Kimura 1963, 4 years
Mixed hardwood forest (*Acer-Quercus*) Toronto, Canada	3.1* (72)	—	—	—	1.3 (28)	4.3 (100%)	Bray & Gorham 1964, 5 years
Evergreen laurel forest (*Cyclobalanopsis-Shiia*) Nara, Japan	3.6 (59)	1.5 (24)	0.07 (1)	0.93 (16)	2.5 (41)	6.0 (100%)	Kirita, unpublished data, 1 year
Eucalyptus forest (Mature, 200 yrs old) Victoria, Australia	4.2 (52)	—	—	—	3.9 (48)	8.1 (100%)	Ashton, after Bray & Gorham 1964
Eucalyptus forest (Virgin) Dwellingup, Australia	1.2 (52)	—	—	—	1.1 (48)	2.4 (100%)	Hatch, after Bray & Gorham 1964
Tropical rain forest Khao Chong, Thailand	11.9 (51)	9.6 (41)	1.9 (8)		11.5 (49)	23.3 (100%)	Kira et al. 1966, 2 months
Tropical rain forest** Koh Kong, Cambodia	(64)	(25)	(11)		(36)	(100%)	Hozumi & Yoda, unpublished data, 1 month
Average	(58.9)				(41.1)	(100%)	

* Including bud scales and fruits.   ** Air-dry weight basis.

cords that the average share of non-leaf components in total litter-fall
was about 25%. It seems, however, that most of the data cited by them
were concerned with secondary forests, plantations, or forests under
silvicultural management. As recognized in the table prepared by them,
the percentage of non-leaf litter components tends to be greater in older
stands of the same species. If the litter-fall records in mature climax
forests are examined, the percentage is apparently much larger, as collated
in Table 7. High values of the (non-leaf litter)/(leaf litter) ratio may
well be a characteristic of stabilized, mature forest communities.

Since litter-fall is likely to form the greater part of net production in
mature forests, litter sampling is an indispensable procedure in the study
of organic production by such communities. In view of the wide variability of annual litter-fall pointed out by various authors (e. g. Kittredge
1948), the sampling should necessarily be made continuously for at least a
few years. The present estimate in the Khao Chong rain forest may therefore involve considerable bias, probably an overestimation.

Nye(1961) reported an annual litter-fall of 10.6 ton/ha·yr in a well-
developed secondary forest of Ghana, about 40 years old and situated in
the ecotone between moist evergreen and moist semi-deciduous forest
zone. About two-thirds of this amount was leaf litter. These figures are
much smaller than the corresponding estimates in the Khao Chong forest.
Other litter-fall records in the tropics cited by the same author are more
or less similar in amount to that of the Ghana forest. Although our
present estimates might have been overestimated to a certain extent, the
Khao Chong forest seems to produce much more litter, probably owing to
the more rapid turnover of leaves and twigs as well as to the very large
foliage biomass under typical rain forest climate.

Biomass increment

Annual growth rings tend to be only incompletely formed in tropical
trees. Even where a well-defined dry season makes most tree species
deciduous as in northwestern Thailand, we cannot expect complete tree
ring formation in most species. According to our experience in the
monsoon and savanna forests of Ping Kong, growth rings were not so
distinct nor regular as to allow exact counting in woods of such purely
tropical species as *Tectona grandis, Dipterocarpus* spp., etc., though *Quercus,
Lithocarpus* and some other trees of temperate families showed somewhat
more regular ring formation. It seems that annual ring formation is under
both environmental and genetical control.

The stem analysis technique and hence the net production estimation
of Eq.(5) can not, therefore, be applied to tropical forest vegetation,
especially to rain forest, and Eq.(6) must solely be utilized for the pur-

pose. The success or failure of the latter method largely depends upon the degree of accuracy of biomass estimation. Where the biomass estimation is based entirely on D. B. H. measurements as in this study, the accuracy of measuring stem diameter increment may also influence the reliability of the result of estimation.

Though the increment of D. B. H. could be measured quite accurately by the use of a dendrometer, it would be quite impracticable to attach a dendrometer to each of a large number of trees on the sample plot. The diameter tape is a tool much inferior in its accuracy to the dendrometer, but the present result proved that an ordinary diameter tape could provide the required accuracy if the diameter were measured at exactly the same position on each trunk. Foresters working with diameter tapes normally measure D. B. H. in centimeters, since readings in mm scale are not always reliable. Increments of D. B. H. in the Khao Chong rain forest were read on the diameter tape to the nearest millimeter for the period between February 4, 1962 and July 21, 1963. The readings were later converted to round numbers in cm unit by counting fractions of more than 0.5 mm as 1 cm, and the calculations of biomass increment were made starting from D. B. H. figures expressed in both mm and cm unit. As collated in Table 8, differences between the results of the two different ways of calculation were generally within ±10% of their mean. This result seems to indicate the extent of the relative error caused by the use of diameter tape as the measuring instruments.

Table 8 Comparison between the calculated biomass increments in Stand 5 based on the D. B. H. figures measured in mm unit (A) and those measured in cm unit (B), for the period from February 4, 1962 to July 21, 1963

| | B/A | | | | | |
	Stem	Branch	Root	Leaf	Total	LAI
Increments in surviving trees	94	92	98	100	94	110
Losses by dead trees	98	98	99	99	98	100
Net increments	92	89	91	103	91	137

Dynamic equilibrium of biomass in the Khao Chong forest

It has long been believed since the establishment of the climax concept that a climax plant community should be in an apparent steady state not only in its floristic composition but also in its structure and biomass. The constancy of biomass is evidently the result of dynamic equilibrium maintained between the growth and the death of component plants, but this property of climax communities has rarely been demonstrated by concrete field data.

Whereas abundant information is available on the amount of growth or biomass increment in forest communities, the loss of biomass due to death of trees is much more difficult to estimate accurately, because the mortality rate of trees tends to vary widely in different years, especially under cool or temperate climates. Not infrequently a number of trees are killed in a year owing to such environmental catastrophes as storm and drought, while the mortality may remain low for a long spell of normal years. The biomass of a mature forest community may thus fluctuate from year to year to a considerable extent and the equilibrium could be recognized only on very long term, if it existed, even in a climax stand. Accordingly, the average amount of dying trees could not be rightly estimated unless the observation is kept continuously for a long period, probably a few decades in temperate forests.

Three years observation on the Khao Chong forest, however, revealed that the life span of trees was doubtlessly shorter there than in temperate forests, as indicated by the fairly large observed annual mortality. As pointed out by Ogawa et al.(1965b), the exuberance of woody lianas seemed to be partly responsible for the short duration of tree life. Lianas accelerate the death of their host trees by covering the surface of hosts' crowns with their leaves, reducing the foliage amount of host trees as a consequence, and eventually reducing the matter budget of the hosts. The closed canopy, which developed between 15 m and 26 m above the ground at Khao Chong, was most heavily entangled by lianas, and it seemed quite probable that only those trees which by chance escaped the attack of woody climbers could successfully grow through the closed canopy layer and become emergent individuals sporadically overtopping the forest. High mortality may also be caused by the shading of the dense canopy as well as by the vigorous activity of parasitic fungi. Only under such tropical rain forest conditions are growth and mortality likely to proceed at the same rate even over a short period, so that it is possible to demonstrate the dynamic equilibrium of biomass within a limited time of observation.

Of the 275 stems(214 individuals) larger than 4.5 cm D. B. H. which were alive on February 4, 1962 in the sample plot, 8 stems died before March 30, 1965. The average mortality was therefore 0.92% per year on a stem number basis, and corresponds to the loss of dry matter of 1.22 ton/ha·yr(Table 4) or the annual loss of 0.38% of the initial biomass(324 ton/ha). On the other hand, the biomass of all trees over 4.5 cm D. B. H. increased at the rate of 6.56 ton/ha·yr during the same period. The resultant net biomass increment amounted to 5.33 ton/ha·yr or 1.64% of the initial biomass. Since the size of the sample plot was 0.16 ha, the net biomass increment on the plot equals 0.85 ton/plot·yr. This could be counterbalanced by the death of one more tree whose D. B. H. is between 32 cm and 33 cm within the plot per annum. A tree with a diameter as great as 100 cm at breast height is expected to weigh nearly 10 tons according to Eqs.(7)-(9), so that the net biomass increment of 5.33 ton/ha·yr would be reduced to zero, if such a big tree died once every two years on each hectare. These simple calculations may suffice to show that the rain forest at Khao Chong was very near the balanced state with respect to its standing plant biomass.

Two attributes of the forest as a climax community are above all related to the realization of such an equilibrium. One is the very high percentage of respiratory consumption in the total gross production (R/P_g =0.77), and another is the large share of litter-fall in the net production (L/P_n=0.81). As 77% of annual gross production is consumed by respiration and 81% of the rest or net production is used for the renewal of leaves and twigs, only 5.3% of gross production or 6.6 ton/ha·yr is left for the growth of living trees (Table 4). The latter attribute may perhaps be the result of the full development of dense canopy, maintaining the maximum amount of leaves in equilibrium with the average intensity of incident solar radiation. Growth of individual trees cannot increase the total amount of foliage of the stand under such a situation. Instead, older leaves are successively shed as new leaves are produced. The same may also be true of fine twigs which support the leaves.

The former attribute is, on the other hand, presumably caused by the large ratio of the biomass of woody organs to that of leaves. The wood biomass in the Khao Chong rain forest is almost 40 times as much as the leaf biomass. Although the average respiration rate of woody organs (0.064 ton/ton·yr in stem, 0.238 in branch and 0.187 in root) is much smaller than that of leaves(6.951 ton/ton·yr), the vast accumulation of wood biomass exerts a great influence on the dry matter budget of the forest. The respiratory losses given in Table 6 were calculated on the basis of the biomass figures on Feb. 4, 1962. In the following three-year period, the wood biomass increased at the rate of 5.25 ton/ha per annum. The wood biomass increment is expected to increase the respiratory consumption by 0.64 ton/ha·yr. This increase in the respiration of woody organs, when maintained for eight years, is enough to match the biomass increment of 5.33 ton/ha·yr. Here is further evidence showing the nearly balanced state of biomass in the Khao Chong forest.

2) Dry matter production by other forest types of Thailand

Measurements of biomass increment and litter-fall could not be made in the other three types of decicuous forest at Ping Kong, northwestern Thailand. A crude estimation of their productivity is therefore made in the following based on the theory developed by Monsi and Saeki(Monsi & Saeki 1953, Saeki 1960). According to their theory, the daily rate of gross production(p_g) in a plant community can be calculated by the following formula.

$$p_g = \frac{1}{K}\frac{b}{a}\ln\frac{1+KaI_o}{1+KaI_o\exp(-KF)} \tag{10}$$

K is the coefficient of light extinction by leaves, I_o the average light intensity during the growing season falling over the canopy that has the leaf area index F, and a and b the parameters in the hyperbolic equation concerning the relation of daily gross photosynthetic rate per unit leaf area to light intensity. Eq.(10) means that there exists a certain optimum value of leaf area index (F_{opt}) by which the maximum amount of gross production by the whole community is realized under the given intensity of incident solar radiation. F_{opt} is given by the equation,

$$F_{opt} = \frac{1}{K}\ln\frac{KI_o(b-ar)}{r}, \tag{11}$$

in which r stands for the daily rate of leaf respiration per unit area of leaf surface. It is quite probable that the leaf area index in a stabilized natural forest is approximately equal F_{opt} corresponding to the prevailing light condition.

For the Khao Chong rain forest here concerned(Stand 5), the following values are available.

$F_{opt}=F=11.4\,m^2$(leaf area)/m^2(land area)
$K=0.44\,m^2$(land area)/m^2(leaf area)
$I_o=100\%$ (in relative unit)
$p_g=124.4$ ton/ha(land area)·yr=34.08 g/m^2·day
$r=57$ ton/11.4 ha(leaf area)·yr=1.370 g/m^2·day

Inserting these values into Eqs.(10) and (11), we have the solution,
$\begin{cases} a=1.99\ (1/\%),\ \text{and} \\ b=7.41\ (g/m^2\text{(leaf area)}\cdot day\cdot\%). \end{cases}$

If the average intensity of incident solar radiation during the growing(rainy) season in northwestern Thailand were the same as the annual average at Khao Chong, and the daily photosynthesis—light intensity curve remained the same in the leaves of both rain forest and other deciduous forests, it would be possible to calculate the daily gross production rates in the deciduous types where the values of F and K are known

(Ogawa et al. 1965b).   The results are given in Table 9.

Table 9   Estimated rates of daily gross production(p_g) in four
different forest types of Thailand, on a land area basis

Forest type	Locality	F ha/ha	K 1/F	p_g g/m²·day
Savanna forest (Stand 3)	Ping Kong	1.8	0.48	6.78
Savanna forest—monsoon forest ecotone (Stand 2)	" "	2.3	0.52	8.40
Monsoon forest (Stand 1)	" "	3.9	0.40	14.09
Rain forest (Stand 4)	Khao Chong	10.2	0.37	33.45

Since the value of K does not differ greatly among different forest
types, daily rate of gross production is more or less proportional to leaf
area index as expected from Eq. (10).   Tentatively assuming that the
duration of leafy period (growing season) is 12 months in Stand 4, 10
months in Stand 1, 9 months in Stand 2, and 8 months in Stand 3 respectively, the annual gross production is expected to be around 122, 43,
23 and 17 ton/ha·yr, or about 99, 35, 19 and 14% of that of Stand 5.   Although these estimates for Stands 1-3 are far less reliable than the direct
estimate for Stand 5 given in Table 6, they may give some indications
for the productivity of the tree components in tropical forests under
climatically limited conditions.

3)   Comparative considerations

Only a very limited number of gross production estimates for forest
communities are available for the present.   Furthermore, they were obtained
through more or less different procedures, and hence it may not always
be fair to compare them without any qualification.

Table 10 contains several estimations of gross production hitherto
obtained in various types of hardwood forest by more or less similar procedure.   In general, the annual gross production tends to be greater with
decreasing latitude as far as these forests are concerned in which the
amount of rainfall is probably not limiting to primary production throughout the growing season.   Both the increase of leaf area index and the
elongation of growing period seem responsible for this latitude-dependent
increase of productivity, as shown by the fact that the annual gross
production is more or less proportional to the product of leaf area index
and duration of growing period in months(Fig. 2).

One of the most noteworthy features recognized in Table 10 is that
the respiration/gross production ratio is consistently smaller in plantations
(40-45%) than in natural, especially climax, forests(70-75%).   Silvicultural

Table 10   Comparison of gross and net productivity in various hardwood forests of the world

	Biomass increment (Δy) ton/ha·yr	Losses by death (L) ton/ha·yr			Net production (P_n) ton/ha·yr	Losses by respiration (R) ton/ha·yr			Gross production (P_g) ton/ha·yr	R/P_g	Leaf area index	Authors
		Wood	Leaf	Total		Wood	Leaf	Total				
Fraxinus excelsior plantation, Denmark												
12 years old	4.1*	0.6	2.7	3.3	7.4			3.1	10.5	0.30	5.4	Boysen Jensen 1932
35–45 years old					13.5			7.0	21.5	0.33		Möller 1945
Fagus sylvatica plantation, Denmark												
8 years old	4.8*	0.6	2.1	2.7	7.5	2.8	3.6	6.4	13.9	0.46	4.2	Möller et al. 1954
25 ″ ″	9.6*	1.2	2.7	3.9	13.5	4.2	4.6	8.8	22.3	0.39	5.4	
46 ″ ″	9.6*	1.2	2.7	3.9	13.5	5.4	4.6	10.0	23.5	0.43	5.4	
85 ″ ″	7.4*	1.2	2.7	3.9	11.3	5.5	4.6	10.1	21.4	0.47	5.4	
Warm temperate broad-leaf evergreen forest, Kyûsyû, Japan												
Castanopsis cuspidata, coppice wood, 11 years old	14.1*	0.9	3.7	4.6	18.7	2.9	23·7	26.6	45.3	0.59	8.0	Tadaki 1965
Climax forest dominated by *Distylium racemosum*	9.2*	—	11.4	11.4	21.6	28.3	24.1	52.4	73.1	0.72	8.8	Kimura 1960
Tropical humid forest, Ivory Coast	9.0*	2.3	2.1	4.4	13.4	22.2	16.9	39.1	52.5	0.74	3.2	Müller & Nielsen 1965
Tropical rain forest, Khao Chong, Thailand	5.3	12.6	11.9	24.4	29.8	37.6	57.0	94.6	124.4	0.76	11.4	Kira et al. 1966

* Not including the increment of leaf biomass.

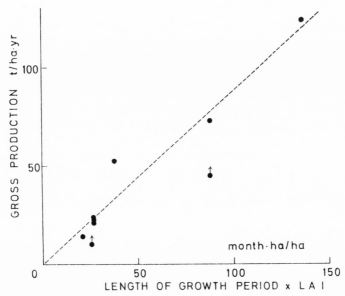

Fig. 2 Estimates of gross production in different types of hardwood forest as related to leaf area index multiplied by number of months of annual growth period

Arrows indicate apparent underestimation in the original productivity data.

practices, thinning in particular, might reduce the ratio by removing suppressed, slender individuals, in which the photosynthetic efficiency is expected to be lower than in dominant trees while the respiratory activity is not, as suggested by Boysen Jensen (1932).   Tadaki et al. (1965), however, obtained the R/P_g values as high as 70-80% in plantations of *Cryptomeria* in Kyñsyñ, southern Japan.   It is to be left for future study which of the following factors should be assigned to the cause of the difference of R/P_g ratio between European plantations and Asian or tropical forests; different procedures used, different sylvicultural practices, different length of growing season, or difference of tree habits (deciduous versus evergreen).

The wide difference of productivity between the two tropical rain forests of Thailand and Ivory Coast may at first seem astonishing, but it is doubtful whether the forest of Ivory Coast studied by Müller & Nielson could properly be called a rain forest.   According to the climate diagram of Walter (1955), the climate of Abidjan, 25 km to the east of the forest, has two months of dry season.   In six months December-March and August-September, the average monthly rainfall is less than 100 mm there.   At Khao Chong, on the other hand, the dry season in Walter's sense is only one month in the average year, and three months from January to March receive average rainfall less than 100 mm. As judged by the monthly summation of Ångström's coefficient of humidity(Ogawa et al. 1961), the hydrothermal climate of Abidjan corresponds to that of the deciduous monsoon forest zone of Southeast Asia, although the precipitation is more evenly distributed throughout the year at Abidjan than in Southeast Asia. These climatic records suggest that the water supply may at least seasonally limit organic production in the Ivory Coast forest.

The small leaf area index in the Ivory Coast forest, 3.2 ha/ha excluding undergrowth vegetation, as well as the low productivity figure may be caused by this climatic limitation. The leaf area index in the typical monsoon forest(Stand 1) of northwestern Thailand was in fact very near the value obtained by Müller & Nielsen in Ivory Coast, amounting to 3.9 ha/ha(Table 9).

SUMMARY

1) The principle, procedures and results of a trial to estimate the primary organic production by a tropical rain forest in the Khao Chong Forest Reserve, Trang Province, of southern Thailand were presented.

2) Empirical formulae proposed by Ogawa et al.(1965b), which describe the relation between D. B. H. and individual tree biomass, were combined with the result of D. B. H. increment observations during three years period in all trees over 4.5 cm D. B. H. on a 40 m × 40 m plot(Stand 5), for the purpose of estimating the annual rate of biomass increment. The increase of biomass of living trees amounted to 6.56 ton/ha·yr in average, while the loss of biomass due to the death of trees was found to be 1.22 ton/ha·yr. The net biomass increase was therefore 5.33 ton/ha·yr in oven-dry weight or only 1.6% of the initial tree biomass.

3) The amount of litter-fall was observed by means of a long, broad strip of plastic net stretched over the ground for 1.5 months at the beginning of 1962. By extrapolating the observed mean daily rate, the annual amount of litter-fall was estimated to be 23.3 ton/ha·yr, of which 51% were leaves and the rest comprised branches, bark and other minor components.

4) Thus the net production by the tree components of the Khao Chong rain forest was estimated to be 28.6 ton/ha·yr, the consumption by heterotrophic organisms and the loss of fine roots being excluded. Of this amount, nearly 4/5 are used for the turnover of leaves and twigs, and only 1/5 is available for the growth of trees.

5) Adding the amount of annual respiratory consumption calculated

by Yoda(1967), the gross production by the Khao Chong rain forest, with its large leaf area index(11.2 ha/ha for tree components), was found to amount to 123.2 ton/ha·yr. About 77% of this vast amount of organic matter is consumed by the community respiration.

6) Respiratory consumption of a large proportion of gross photosynthetic production, and a very small rate of net biomass increase, were suggested as characteristic properties of a climax forest community. It was also shown that the biomass of the Khao Chong rain forest was very near the balanced state or dynamic equilibrium on the basis of the dry matter budget of the community.

REFERENCES

BASKERVILLE, G. L. 1965. Dry matter production in immature balsam fir stands. *Forest Sci. Monogr.* 9, 42 pp.

BOYSEN JENSEN, P. 1932. *Die Stoffproduktion der Pflanzen.* Gustav Fischer, Jena.

BRAY, J. R. & E. GORHAM 1964. Litter production in forests of the world. *Adv. Ecol. Res.* 2: 101–157.

IWATA K. 1965. The Joint Thai-Japanese Biological Expedition to Southeast Asia 1961–62. *Nature & Life in SE Asia* 4: 1–12.

KIMURA, M. 1960. Primary production of the warm temperate laurel forest in the southern part of Ôsumi Peninsula, Kyûshû, Japan. *Misc. Rep. Res. Inst. Nat. Resources* 52–53: 36–47.

———— 1963. Dynamics of vegetation in relation to soil development in northern Yatsugatake Mountains. *Jap. J. Bot.* 18: 255–289.

KIRA T., H. OGAWA, K. YODA & K. OGINO 1964. Primary production by a tropical rain forest of southern Thailand. *Bot. Mag. Tokyo* 77: 428–429.

KIRA, T. 1965. A reasonable method for estimating the total amount of respiration of trees and forest stands. Ist Intern. Symposium on Ecosystems, Copenhagen.

KITTREDGE, J. 1948. *Forest influences.* McGraw-Hill, New York.

KORIBA, K. 1958. On the periodicity of tree-growth in the tropics. *Gard. Bull. Singapore* 17: 11–81.

KUROIWA, S. & M. MONSI 1963. Theoretical analysis of light factor and photosynthesis in plant communities. (1) & (2). *J. Agr. Met.* 18: 143–151, 19: 15–21. In Japanese with English summary.

MÜLLER, Car Mar 1945. Untersuchungen über Laubmenge, Stoffverlust und Stoffproduktion des Waldes. *Forstl. Forsögsv. Danmark* 17: 1–287.

————, D. MÜLLER & J. NIELSEN 1954. The dry matter production of European beech. *Ibid.* 21: 253–335.

MONSI, M. & T. SAEKI 1953. Über die Lichtfaktor in den Pflanzengesellschaften und seine Bedeutung für die Stoffproduktion. *Jap. J. Bot.* 14: 22–52.

MÜLLER, D. & J. NIELSEN 1965. Production brute, pertes par respiration et production nette dans la forêt ombrophile tropicale. *Forstl. Forsögsv. Danmark* 29: 60–160.

NOMOTO, N. 1964. Primary productivity of beech forest in Japan. *Jap. J. Bot.* 18: 385–421.

NYE, P. H. 1961. Organic matter and nutrient cycles under moist tropical forest. *Pl. Soil* 13: 333–346.

ODUM, E. P. 1959. *Fundamentals of ecology.* 2nd ed. Saunders, Philadelphia & London.

OGAWA, H. 1967. Balance sheet of organic matter in a plant community as the basis for determining primary production—a theoretical consideration. Manuscript.

——, K. YODA & T. KIRA 1961. A preliminary survey on the vegetation of Thailand. *Nature & Life in SE Asia* 1: 21-157.

—— et al. 1965a. Comparative ecological studies on three main types of forest vegetation in Thailand. I. Structure and floristic composition. *Ibid.* 4: 13-48.

——, K. YODA, K. OGINO & T. KIRA 1965b. Ibid. II. Plant biomass. *Ibid.* 4: 49-80.

OSHIMA, Y. 1961. Ecological studies of *Sasa* communities. IV. Dry matter production and distribution of products among various organs in *Sasa kurilensis* community. *Bot. Mag. Tokyo* 74: 473-479.

SAEKI, T. 1960. Interrelationships between leaf amount, light distribution and total photosynthesis in a plant community. *Ibid.* 73: 55-63.

SHIDEI, T. (ed.) 1960. *Studies on the productivity of the forest. I. Essential needle-leaved forests of Hokkaido.* Kokusaku Pulp Ind. Co., Tokyo, In Japanese.

—— (ed.) 1964. *Ditto. II. Larch(Larix leptolepis GORD.) forests of Shinshu District.* Nippon Ringyô-Gizyutu Kyôkai, Tokyo. In Japanese.

TADAKI, Y. 1965. Studies on production structure of forests. VII. The primary production of a young stand of *Castanopsis cuspidata. Jap. J. Ecol.* 15: 142-147.

——, N. OGATA & Y. NAGATOMO 1965. The dry matter productivity in several stands of *Cryptomeria japonica* in Kyushu. *Bull. Gov. For. Exp. Sta.* 173: 45-66. In Japanese with English summary.

WALTER, H. 1955. Klimagramme als Mittel zur Beurteilung der Klimaverhältnisse für ökologische, vegetationskundliche und landwirtschaftliche Zwecke. *Ber. d. deutsch. bot. Ges.* 68: 331-344.

WHITTAKER, R. H., N. COHEN & J. S. OLSON 1963. Net production relations of three tree species at Oak Ridge, Tennessee. *Ecology* 44: 806-810.

YODA, K. 1967. Comparative ecological studies on three main types of forest vegetation in Thailand. III. Community respiration. *Nature & Life in SE Asia* 5: 83-148.

——, K. SHINOZAKI, H. OGAWA, K. HOZUMI & T. KIRA 1965. Estimation of the total amount of respiration in woody organs of trees and forest communities. *J. Biol. Osaka City Univ.* 16: 15-26.

Authors' Note added in June 1976

According to the recent investigation of a West Malaysian rain forest by Kira, Ogawa, and Yoda (1971-1974), the rate of community respiration given in this paper seems to have been considerably overestimated, owing to inadequate treatments of plant samples. The litterfall rate was also obviously overestimated as mentioned in the text. The most probable estimates of net and gross production rates in Khao Chong Forest would be 20-25 t/ha·yr and 70-80 t/ha·yr. The outline of our Pasoh Forest study has been reported to the Malaysian IBP Synthesis Meeting at Kuala Lumpur in August 1974, and will be published in the near future.

18

Reprinted from *Ecology* 42(3):581–584 (1961)

ENERGY VALUES OF ECOLOGICAL MATERIALS

Frank B. Golley

AEC, Savannah River Project, University of Georgia

Since Lindeman (1942) formulated his concept of trophic dynamics, American ecologists have become increasingly interested in the energy relationships of ecosystems. Even though the application of thermodynamic theory to ecology has progressed rapidly (see Odum 1956, Odum and Pinkerton 1955, Patten 1959, Sobodkin 1960), understanding of the energy dynamics of individual populations has been hampered by incomplete knowledge of the energy content of many plants and animals. In energy flow studies it is often necessary to convert biomass to energy by using caloric equivalents obtained from the literature. However, many workers, reluctant to depend upon equivalents, have made their own energy determinations in the oxygen bomb calorimeter. This paper summarizes my analyses of over 400 wild plants and animals collected from the field and over 200 analyses of other workers. These values were previously listed in a mimeographed publication (Golley 1959).

I am indebted to a number of persons who have aided these studies. In 1956-57 equipment for caloric analyses in the laboratories of the Department of Foods and Nutrition, Michigan State University, was used through the courtesy of Dr. Evelyn Jones. Later studies were made in the ecological laboratory of Dr. E. P. Odum, Univ. of Ga. and the work was supported by the U. S. Atomic Energy Commission, contract At(07-2)-10. I am grateful to Dr. Odum for his critical comments during these investigations. I am also grateful to Drs. Lawrence Bliss, Univ. of Illinois; Francesco Trama, Rutgers Univ.; Edward Kuenzler, Woods Hole Oceanographic Institute; Clyde Connell, Valdosta State College, Ga.; and J. D. Ovington, The Nature Conservancy, London for providing unpublished data from their studies.

Methods

The caloric value of ecological materials was determined by burning samples in a Parr oxygen bomb calorimeter. The standard procedure for determining energy values is fully discussed in the Parr Manual (1948).

Since the purpose of the measurements is to provide estimates of energy content of biomass that can be applied to populations of plants and animals in the field, the samples must reflect the variations inherent in field populations. Therefore, vegetation samples are collected from random plots on study areas, dried at 100°C for 24 hours and ground in a Wiley Mill fitted with a 40 mesh-to-the-inch screen. When only the small mill is available, portions from pooled samples from all plots are selected for grinding. The ground tissue is then thoroughly mixed in jars before samples are removed for calorimetry.

Since the amount of body fat (which has a high energy value) varies with age and season in many animals, each age group must be sampled at each season to obtain a caloric estimate that can be applied to the populations at specific seasons. For many groups of animals, e.g., insects collected by sweep-netting, the entire dried sample can be run through the mill. For fleshy animals, such as birds or mammals, specimens are minced in a Waring Blender, the slurry of water and tissue dried in a vacuum oven, and the dried material ground in the Wiley Mill. Extremely fat animals are

difficult to handle even with this procedure. With these it is best to extract the fat after vacuum drying then redry and combust the nonfat residue.

The calorimeter is restandardized (the water equivalent determined) every third month, especially when the laboratory temperatures fluctuate widely. The magnesium fuse wire and the total acid corrections are determined for each individual analysis, rather than assumed to be constant, since the amounts of fuse wire used and the acid produced vary widely. For instance, insect samples may require more than 9.0 ml of base to neutralize the acid formed during combustion, while plant material usually requires only 5.0 ml or less. The difference of 4 ml is equivalent to 4 g cal. The amount of fuse wire burned has varied from 9.3-3.6 cm in these studies. Corrections are made by measuring the unburned wire with a ruler (1 cm of burned wire = 2.8 g cal) and by titrating the bomb washings with a sodium carbonate solution (3.658 g $Na_2CO_3/1H_2O$, 1 ml solution = 1 g cal).

The per cent ash in each sample has been determined since 1958. This merely requires drying and weighing the tarred calorimeter cup containing the ash after the tests are completed. The caloric value per gram ash-free weight more accurately reflects differences between materials than the value per gram dry weight because of contamination. Samples of roots and ground litter, in particular, often contain soil which cannot be dislodged by routine washing and if it is included the caloric value will be lowered.

The Parr Manual (1948) states that the American Society for Testing Materials requires 0.3% accuracy for tests of the same material made in the same laboratory and 0.5% for tests made in different laboratories. I have not achieved this degree of accuracy with all types of ecological materials and have accepted as satisfactory a variation of not more than 3.0% between 3 tests on a given sample. All of my analyses have been run in triplicate.

The caloric value of plant materials was tabulated under 3 categories: plant parts, month collected, and ecological community. Comparisons between members of a category were tested by an analysis of variance procedure. Mr. James Fortson, Institute of Experimental Statistics, Univ. of Ga., transformed the data using matrices and the statistics were calculated on an IBM 650 computer. Significance of the difference between values in each separate category was determined with the F-test. Animal caloric values were subjected to an analysis of variance using a desk calculator. Since many values were obtained from the literature or were made before the ash-correction became routine, all are reported as gram calories per gram dry weight.

Results

Table I presents the mean energy values for parts of plants. The differences are significant at the 99% level (F = 27.72, d.f. = 4 and 462). Seeds, which often contain high proportions of fat, have the highest values. Long (1934), one of the first to apply calorimetric methods in ecological research, also found that energy values varied for parts of the plant. In the sunflower he found

TABLE I. Average energy values for parts of plants, based on determinations from 57 species
(Value g cal/g dry wt)

Part	Number Samples	Average Value	Coefficient Variation
Leaves...............	260	4229	.116
Stems and Branches...	51	4267	.081
Roots...............	52	4720	.092
Litter...............	82	4298	.104
Seeds...............	22	5065	.219

that the values ranged from 4308 g cal/g dry wt for the seed head to 3435 g cal/g dry wt for one of the oldest leaves. The highest values he obtained were for seeds of conifers, which ranged from 5625-7117 g cal/g dry wt.

The seasonal analysis includes vegetative samples from 3 old-field communities: a blue-grass field in Mich., a broomsedge field and young pine stand in Ga. (Table II). Values for the dominant plants in these 3 communities were grouped so that the comparison is between months, irrespective of the species of plants. The analysis of variance showed that significant differences at the 99% level (F = 2.36, d.f. = 11 and 272) existed between months. The highest caloric values occurred in the fall and winter, presumably from storage of energy in the roots, culms, and seeds. Morrison (1949) reports that in many plants the per cent of crude protein in the green foliage decreases, that of crude fiber and nitrogen-free extract (mostly carbohydrate) increases, while that of ether extract (partly fat) remains constant through the growing season. Accordingly, the caloric value of the foliage should be higher in the spring than in the fall. However, when considering all parts of the vegetation growing in the community the caloric value per gram total vegetation (including roots and seeds) appears to be greater in the fall and winter.

Caloric data are available for the dominant plants in 9 ecological communities, ranging from tropical rain-forest to alpine tundra. The pine community data furnished by J. D. Ovington are from England. The alpine tundra data obtained by L. Bliss are from New Hampshire. The Spartina (analysed by C. Connell), the Andropogon, and the old-field herb communities were studied in Ga., the Poa community in Mich., and the rain-forest and mangrove forest in Puerto Rico. The values in Table

TABLE II. Average energy values for dominant species in three old-field communities collected at different seasons
(Value g cal/g dry wt)

Season	Number Samples	Average Value	Coefficient Variation	Seasonal Average
January.......	18	4039	.152	
February......	17	4225	.022	
March........	3	4034	.028	4099
April..........	27	3900	.129	
May..........	24	4127	.070	
June..........	24	3917	.088	3981
July..........	21	4072	.065	
August........	41	3919	.079	
September.....	38	4197	.061	4063
October.......	20	4192	.066	
November.....	33	4151	.097	
December.....	18	3907	.215	4083

III are mixed-species averages of all the data available for the dominant species; they are not average weighted by the importance of individual species in the phytosociology or biomass composition of the communities. The statistical analysis showed that the average caloric value per gram total vegetation (roots, leaves, and stems) in these communities differed significantly at the 99% level (F = 11.3, d.f. = 8 and 343). The tundra estimates were considerably higher than those for the other communities indicating that communities with a long period of nonproduction accumulate a greater energy store than those which grow throughout the year. However, the pine community, with its high resin and turpentine content, also had a high energy value.

TABLE III. Average energy values of dominant vegetation in ecological communities
(Value g cal/g dry wt)

Community	Number Samples	Average Value	Coefficient Variation
Tropical rain-forest.......	15	3897	.060
Mangrove forest.........	11	3764	.082
Spartina marsh.........	14	4072	.042
Andropogon field.........	143	3905	.104
Herb old-field............	35	4177	.096
Poa old-field.............	115	4075	.064
Pinus sylvestris stand.....	14	4787	.078
Alpine meadow..........	3	4711*	.005
Alpine Juncus dwarf heath	2	4790*	.003

* Bliss (pers. comm.) recently reported that the average value for tundra, based on 32 determinations, is 4709 g cal/g dry wt.

The 3 analyses described are not as precise as desired because it was necessary to compare caloric values per gram dry weight rather than values per gram ash-free weight. The large amount of ash-free weight data available for the Andropogon virginicus community (Table IV) shows how ash may influence the differences between categories. When the caloric values by plant part and month uncorrected for ash are compared a significant difference exists between parts at the 95% level (F = 2.74, d.f. = 4 and 32) and months at the 99% level (F = 4.07, d.f. = 8 and 32). However, the same comparison but using calories per gram ash-free weight (Table IV) shows significant differences at the 95% level between seasons only. The average value for parts for all seasons, shown below, illustrates the differences when ash is considered.

Part	cal/g dry wt	cal/g ash-free wt
Green Broomsedge	4231	4377
Standing-dead vegetation	4116	4290
Litter	3902	4139
Roots	3607	4169
Green herbs	3634	4288

ENERGY VALUES OF ANIMALS

Only limited caloric data are available for animals. Seven taxa, including invertebrates and vertebrates, are in Table V. The crabs were especially low in energy content, probably because the calcareous exoskeleton was not separated from the soft parts. E. P. Odum has provided unpublished data on the ash content of crabs from Sapelo Island, Ga., which indicate that the ash is about 51% of the total dry weight. This means that the caloric value per gram ash-free weight for crabs is about 4400 cal. The analysis of variance showed that

Table IV. Energy values in an *Andropogon virginicus* Old-field Community in Georgia
(Value g cal/g dry organic matter)

Part	April	May	June	July	Sept.	Oct.	Nov.	Dec.	Jan.	Ave.
Green grass	4254	4372	4325	4187	4256	4529	4508	4505	4422	4373
Standing dead	4435	4338	4201	4208	4281	4205	4429	4190	4325	4290
Litter	3928	4369	4104	4225	4124	4126	4264	4029	4078	4139
Roots	4387	4056	4344	4270	4104	4074	4137	4236	3891	4167
Green herbs	4477	4157	4088	4212	4399	4429	4265	4375	4193	4288
Average	4296	4258	4212	4220	4233	4273	4321	4267	4182	4251

Table V. Energy values for animal taxa
(Value g cal/g dry wt)

Taxa	Number Samples	Average Value	Coefficient Variation	Authority
Crustacea				
Daphnia	18	4419	.115	Richman (1958)
Stenonema	29	5596	.048	Trama (1957)
Uca and other crabs	8	2248	.188	Connell (unpubl.)
Mollusca				
Modiolus	3	4600	—	Kuenzler (unpubl.)
Insecta				
Schistocerca	8	5363	.048	Connell (unpubl.)
Annelida				
Earthworms	3	4617	.030	French *et al* (1957)
Mammalia				
Mice	8	5163	.157	Golley (1960)

the energy value of the animals, excluding the crabs, did not differ significantly (F = 1.84, d.f. = 5 and 63). In general, the values for the animals are about 1000 g cal/g dry wt higher than the plant values.

Discussion

Examination of over 600 records of plants has shown that significant differences in caloric value exist between plant parts, between vegetation collected in different months, and between vegetation growing in different ecological communities. When the variation in chemical composition of various cultivated and noncultivated crop plants reported in Morrison (1949) is considered (e.g., fat ranges from 1.0% in *Lespedeza* stems to 38.8% in wild mustard seed) and the fact that Long (1934) found that caloric values varied with light intensity, length of day, amount of nutrients, and type of soil, the observed differences are not unexpected.

Richman and Slobodkin (1960) have emphasized the constancy of the caloric value of animal tissue. Except under starvation or storage conditions animal tissue averages about 5000 g cal/g dry wt. The data in this report are less extensive but support their conclusions. Richman and Slobodkin (1960) point out that when an animal is storing food material before hibernation or a nonfeeding portion of the life cycle, the energy value of the body may increase to 6000 or 7000 g cal/g dry wt. This condition in animals is analogous to the seed stage in the life history of the plant and to the fall condition of many perennial plants which store food in the root, tuber, or rhizomes. This analogy helps to explain why significant differences exist between the plant categories.

The caloric value of a plant or animal is a function of its genetic constitution, nutritive condition, and life his-

tory. Because these factors may vary with species, seasons, and environmental conditions the ecologist making intensive measurements of energy flow through natural systems cannot depend on caloric constants or equivalents. This study shows that the ecologist must determine the energy content under the specific conditions of his particular study. However, those engaged in extensive surveys are probably justified in converting biomass to energy by using the average caloric values in the tables.

Summary

This report summarizes and evaluates the variation between caloric values of plants and animals. Analysis of over 600 records shows significant differences between plant parts, between vegetation collected in different seasons, and between vegetation from different ecological communities. Differences between animal taxa were not significant. It was concluded that ecologists should directly determine the energy content of ecological materials when studying energy flow through natural systems. It is also hoped, however, that the average values presented may be useful for energy estimation in some types of ecological research.

References

French, C. E., S. A. Liscinsky, and D. R. Miller. 1957. Nutrient composition of earthworms. J. Wildlife Management, 21: 348.

Golley, F. B. 1959. Table of caloric equivalents. Mimeo. Univ. of Georgia, 7p.

———. 1960. Energy dynamics of a food chain of an old-field community. Ecological Monog. 30: 187-206.

Lindeman, R. L. 1942. The trophic-dynamic aspect of ecology. Ecology 23: 399-418.

Long, F. L. 1934. Application of calorimetric methods to ecological research. Plant Physiol. 9: 323-337.

Morrison, F. B. 1949. Feeds and Feeding. Morrison Publ. Co. Ithaca. 1207p.

Odum, H. T. 1956. Efficiencies, size of organisms, and community structure. Ecology 37: 592-597.

——— and R. C. Pinkerton. 1955. Time's speed regulator: the optimum efficiency for maximum power output in physical and biological systems. Am. Sci. 43: 331-343.

Parr Instrument Company. 1948. Oxygen bomb calorimetry and oxygen bomb combustion methods. Manual No 120. Moline, Ill. 80p.

Patten, B. C. 1959. An introduction to the cybernetics of the ecosystem; the trophic-dynamic aspect. Ecology 40: 221-231.

153

Richman, S. 1958. The transformation of energy by *Daphnia pulex*. Ecological Monog. **28**: 273-291.

—— and B. Slobodkin. 1960. A micro-bomb calorimeter for ecology. Bull. Ecol. Soc. of Am., **41**(3): 88-89.

Slobodkin, L. B. 1960. Ecological energy relationships at the population level. Am. Naturalist **94** (876): 213-236.

Trama, F. B. 1957. The transformation of energy by an aquatic herbivore, *Stenonema pulchellum*. Ph.D. Thesis, Univ. of Mich.

19

Reprinted from pages 602, 604, and 605–608 of *Science* **130**:602–608 (1959)

POTENTIAL PRODUCTIVITY OF THE SEA

John H. Ryther

Under ideal conditions for photosynthesis and growth, what is the maximum potential rate of production of organic matter in the sea? Is this potential ever realized, or even approached? How does the sea compare with the land in this respect? These questions may be approached empirically with some measure of success but, aside from the time and effort required by this method, one can never be certain how close to the optimum a given environment may be and, hence, to what extent the biotic potential is realized.

However, we do know with some degree of certainty the maximum photosynthetic efficiency of plants under carefully controlled laboratory conditions; and there is a considerable literature concerning the effects of various environmental conditions on photosynthesis, respiration, and growth, particularly with respect to the unicellular algae. From such information it should be possible to estimate photosynthetic efficiencies and, for given amounts of solar radiation, organic production under natural conditions. This indirect and theoretical approach cannot be expected to provide exact values, but it does furnish a supplement to the empirically derived data which may help substantiate our concepts both of the environmental physiology of the plankton algae and the level of organic production in the sea.

An attempt has been made to use this joint approach for the marine environment in the following discussion. The only variable considered is light, and the assumption is made that virtually all of the light which enters the water (and remains) is absorbed by plants. Such situations are closely approximated in plankton blooms, dense stands of benthic algae, eelgrass, and other plants. For the rest, it is assumed that temperature, nutrients, and other factors are optimal, or at least as favorable as occur under ideal culture conditions. Given these conditions, I have attempted to calculate the organic yields which might be expected within the range of solar radiation incident to most of the earth. These data are then compared with maximal and mean observed values in the marine environment and elsewhere, and an attempt is made to explain discrepancies.

The calculations which appear below are based, for the most part, upon experimentally derived relationships between unicellular algae and the environment, and are therefore applicable only to this group. This must be kept in mind when, later in the discussion, comparisons are drawn between the theoretical yields and observed values of production by larger aquatic and terrestrial plants.

The values for the efficiency of photosynthesis under natural conditions are based on the utilization of the visible portion of the solar spectrum only (400 to 700 mμ), or roughly half of the total incident radiation. In converting these efficiencies to organic yields, it is assumed that the heat of combustion of the dry plant material is 5.5 kcal per gram, which closely approximates values for unicellular algae reported by Krogh and Berg (*1*), Ketchum and Redfield (*2*), Kok (*3*), Aach (*4*), Wassink *et al.* (*5*), and others.

Reflection and Backscattering

Of the sunlight which strikes the surface of the ocean, a certain fraction is reflected from its surface and never enters the water. The remainder penetrates to depths which depend upon the concentration of absorbing and scattering particles or dissolved colored substances. While scattering may be as important as absorption in the vertical attenuation of the light, it makes little difference as far as the biological utilization of the radiation is concerned, since the scattered light is eventually absorbed, with the exception of a small fraction which is backscattered up out of the water. The combined reflected and backscattered light is lost to the aquatic system; the rest remains in the water, where, under the ideal conditions postulated, it is absorbed entirely by plants.

The fraction of the incident radiation which is reflected and backscattered has been studied by Powell and Clarke (*6*), Utterback and Jorgenson (*7*), and Hulburt (*8*). The two factors have been treated separately, but they may be considered together here. Their combined effect is rather small, ranging from about 3 to 6 percent, depending somewhat upon who made the measurements and the conditions under which the measurements were observed. The highest values were observed when the sky was overcast. Sea states, ranging from flat calm to whitecap conditions, made surprisingly little difference. Reflection and backscattering were also found by Hulburt to be independent of the sun's angle, despite the fact that reflection increases greatly with the angle (from the zenith) of the incident light, particularly at angles above 60° The explanation for this apparent contradiction lies in the fact that as the sun approaches the horizon, indirect sky light becomes increasingly important, and it eventually exceeds the intensities of the sun itself.

[Editor's Note: Material has been omitted at this point.]

The author is on the staff of the Woods Hole Oceanographic Institution, Woods Hole, Mass. This article is based on a paper presented by the author at the AAAS meeting in Washington, D.C., December 1958.

J. H. Ryther

By extrapolating the lower, exponential portion of the photosynthesis curve to the surface, one may create a hypothetical curve of photosynthesis if the latter maintained the same efficiency at all depths. The ratio of the actual photosynthesis curve to this hypothetical exponential curve will then show the reduction in efficiency caused by light intensities above saturation in the upper waters. This has been done in Fig. 4 for a series of photosynthesis curves on days of varying incident radiation. Since photosynthesis at the various depths is a function of light intensity and not of depth per se, the units on the ordinate of Fig. 4 are natural logarithms of I_0/I and thus represent the depths to which given fractions of the incident radiation penetrate. The curve for the day with lowest radiation (20 gcal/cm² day) is exponential all the way to the surface, indicating that on such a day there is no reduction in photosynthetic efficiency from the effects of light intensity. On days of progressively higher light intensity, the photosynthesis curve departs more and more from the exponential curve illustrating the increasing reduction in efficiency.

[*Editor's Note:* Material has been omitted at this point.]

Net Production

Returning to Fig. 5, gross production may be reduced by the respiratory loss (Fig. 6), giving the curve of net production, which begins at 100 g cal/cm² day and reaches a value of 25 g/m² day under radiation of 600 g cal/m² day (the lower broken line in Fig. 5).

Table 1. Gross and net organic production of various natural and cultivated systems in grams dry weight produced per square meter per day.

System	Gross	Net
A. Theoretical potential		
Average radiation (200 to 400 g cal/cm² day)	23–32	8–19
Maximum radiation (750 g cal/cm² day)	38	27
B. *Mass outdoor* Chlorella *culture (26)*		
Mean		12.4
Maximum		28.0
C. Land (*maximum for entire growing seasons*) (18)		
Sugar cane		18.4
Rice		9.1
Wheat		4.6
Spartina marsh		9.0
Pine forest (best growing years)		6.0
Tall prairie		3.0
Short prairie		0.5
Desert		0.2
D. Marine (*maxima for single days*)		
Coral reef (27)	24	
Turtle grass flat (28)	20.5	(9.6)
Polluted estuary (29)	11.0	(11.3)
Grand Banks (Apr.) (30)	10.8	(8.0)
Walvis Bay (23)	7.6	(6.5)
Continental Shelf (May) (19)	6.1	
Sargasso Sea (Apr.) (31)	4.0	(3.7)
		(2.8)
E. Marine (*annual average*)		
Long Island Sound (32)	2.1	0.9
Continental Shelf (19)	0.74	(0.40)
Sargasso Sea (31)	0.88	0.40

Although the annual range of daily incident radiation is extremely wide, even for a given latitude, this short-term variability is probably not very significant in affecting the general level of organic production of a given area. If one examines the tables compiled by Kimball (*17*) showing mean monthly radiation for different latitudes, it appears that over 80 percent of the data (including all latitudes and seasons) fall within a range of 200 to 400 g cal/cm² day. Thus, over most of the earth for most of the year a potential production of organic matter of some 10 to 20 g/m² day may be expected, while for shorter periods of fine summer weather, a net production of 25 g/m² day or slightly more may occur.

Comparison of Theoretical and Observed Production Rates

We may now compare the production rates which were calculated in the preceding sections with some values which have been observed empirically. Since the former are based on hypothetical situations in which all light entering the water is absorbed by plants, the observational data, to be comparable, must be restricted to natural environments in which these conditions are at least closely approximated (for example, in dense plankton blooms, thick stands of benthic algae and rooted plants). In addition to these maximal values, the theoretical potential may be contrasted with average oceanic productivity rates.

We may also extend this comparison to the terrestrial environment, including some of the better agricultural yields, bearing in mind, however, that the physiology and hence, perhaps, the biotic potential of land plants may differ significantly from those of algae.

Finally, we may include the yields of *Chlorella* grown in outdoor mass culture, drawing here upon the excellent, continuing studies of H. Tamiya and his collaborators. These are of particular interest, since the conditions of these experiments were as optimal as possible and since the physiology of *Chlorella* is identical or closely similar to that of the organisms upon which our calculations are based. Thus the *Chlorella* yields will serve as a check for the theoretical production rates.

It is important, in making these comparisons, to keep in mind the distinction between gross and net production as defined above. Some of the data refer to true photosynthesis measurements (gross

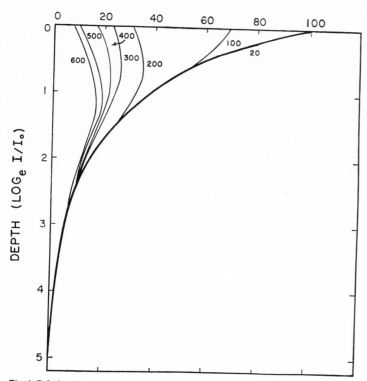

RELATIVE PHOTOSYNTHESIS

Fig. 4. Relative photosynthesis as a function of water depth for days of different incident radiation. Numbers beside curves show gram calories per square centimeter per day.

157

production) while others, such as the *Chlorella* experiments and the agricultural yields, are based on the actual harvest of organic matter (net production). In those cases in which only gross production values are available and where radiation data are given, net production has been obtained from Figure 5 and is shown in parentheses.

The theoretical production potential for average and maximal radiation, and the observational data for both marine and terrestrial environments, are given in Table 1. In each case the original source is given, except for the land values, where reference is made to the recent compilation by Odum (*18*). The various methods by which the values were obtained will not be discussed here except in the case of the unpublished data, in which gross production was calculated from chlorophyll and light, according to the method of Ryther and Yentsch (*19*) and net production was measured by the C^{14} method, uncorrected for respiration as this method is interpreted by Ryther (*20*). Where gross production (photosynthesis) was originally reported as oxygen evolution, this has been converted to carbon assimilation, using an assimilatory quotient

$$\left(\Delta \frac{+ O_2}{- CO_2} \right)$$

of 1.25 (see Ryther, *20*). Carbon uptake, in turn, has been converted to total organic production by assuming that the latter is 50 percent carbon by weight.

The maximal values for the marine environment represent the seven highest such values known to me. In addition to these, data are given for three regions (one inshore, one coastal, and one offshore) which have been studied over long enough periods of time to justify the calculation of annual means.

Discussion

The mean yield of *Chlorella* obtained by the Japanese workers is almost identical to the mean theoretical production for days of average radiation (12.4 versus 13.5 g/m² day). These yields of *Chlorella* were produced only during the warmer part of the year, presumably owing to the poor growth of *Chlorella* at low temperatures. The highest yields of *Chlorella* (up to 28 g/m² day) were, according to Tamiya, "obtained on fair days in the warmer months." This maximum is approximately the same as the theoretical net production for days of

maximum radiation. Thus, the *Chlorella* yields agree very well with the theoretical productive potential of the sea.

The land values for net production quoted from Odum's tables range from 18.4 g/m² day for the highest yields of sugar cane to 0.2 g/m² day for deserts.

The best agricultural yields are generally of the same order of magnitude as the theoretical net production of the sea, as are the values for the salt marsh and the pine forest (during its years of best growth). Uncultivated grasslands range from 3.0 for tall prairie to 0.2 for desert

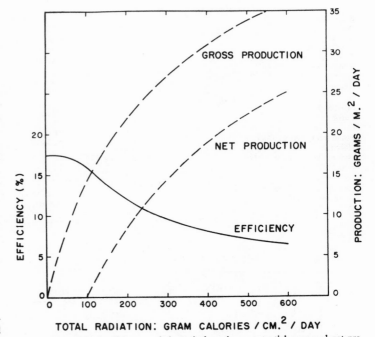

Fig. 5. Photosynthetic efficiency and theoretical maximum potential gross and net production as a function of incident radiation.

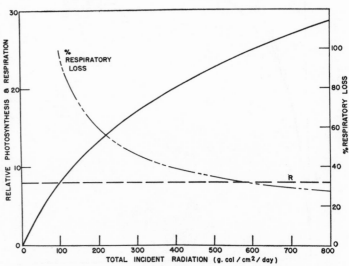

Fig. 6. Relative photosynthesis, respiration, and percentage of respiratory loss as a function of incident radiation.

conditions. Because of the extreme contrasts among terrestrial environments, mean values for the land as a whole are difficult to determine and would have little meaning. It is interesting, however, that Schroeder's estimate (21) of the annual production of all the land is equivalent to a mean daily production of 0.55 g/m², roughly the same as the value given in Table 1 for short prairie grass.

With regard to the marine data, it is perhaps surprising that net production rates differ by less than a factor of 2 in such diverse environments as a coral reef, a turtle grass flat, a polluted creek, and the Grand Banks. This alone would indicate that production in each case is limited by the same basic factor, the photosynthetic potential of the plants, and indeed these and the other high values in *D* in Table 1 all closely approach the theoretical potential.

Seasonal studies have been made of three marine areas, Long Island Sound, the continental shelf off New York, and the Sargasso Sea off Bermuda. In each case temporary rates of production were observed during the spring flowering which approached the theoretical maximum, but the annual means were more than an order of magnitude lower (*E* in Table 1). True, these regions do not, throughout the year, satisfy the postulated conditions necessary to obtain this maximum, namely, that all light entering the water be absorbed by plants. For example, in his Long Island Sound studies, Riley (22) found that no more than one-third of the incident radiation was utilized by plants, the remainder presumably being absorbed by nonliving particulate and dissolved materials. Using Riley's techniques, I estimated that only 25 to 40 percent of the light penetrating the continental shelf waters was absorbed by the phytoplankton. This alone, however, is insufficient to account for the discrepancy between observed and potential production rates. In the clear waters of the Sargasso Sea only 10 to 20 percent of the light is absorbed by the phytoplankton during most of the year. But there is little if any other particulate matter present; the remainder of the light is absorbed by the water itself. This is not a cause but an effect of low production. The underlying reason for low production rates here and in most parts of the ocean is the limitation of essential nutrients in the upper, euphotic layers and the inadequacy of

vertical mixing processes in bringing deep, nutrient-rich water to the surface.

With the exception of the three planktonic communities which have been discussed, the seasonal cycles of marine production are largely unknown and can only be surmised. Probably high levels may be maintained throughout the year in benthic populations such as the coral reef, the turtle grass flats (see *D* in Table 1) and in thick beds of seaweeds, provided that seasonal temperature extremes do not impair growth. While the concentrations of nutrients in the surrounding waters may be very low, the fact that they are continually being replenished as the water moves over the plants probably prevents their ever being limiting. Plankton organisms, on the other hand, suspended as they are in their milieu, can probably never maintain high production rates in a given parcel of water, for their growth rapidly exhausts the nutrients from their surrounding environment and any mixing process which enriches the water must, at the same time, dilute the organisms. However, high plankton production may be sustained in a given geographic area (a polluted estuary, a region of permanent upwelling of deep water, and so forth), which is continually replenished with enriched water. In these situations, the productive capacity of the sea may be sustained for long periods, perhaps permanently.

For most of the ocean, as stated above, no such mechanism for nutrient replenishment is available. The combined meteorological and hydrographic conditions which produce the typical spring flowering of the phytoplankton over much of the oceans have been adequately described elsewhere and need not be discussed here. Suffice it to say that, in the oceans as a whole, as seasonal studies have demonstrated, high production approaching the theoretical maximum under optimal conditions is restricted to periods of a few days or, at most, weeks, per year.

Steemann Nielsen (23) has recently estimated the net production of the entire hydrosphere as 1.2 to 1.5 × 10¹⁰ tons of carbon per year, roughly one-tenth the earlier estimates made by Riley (24) and others, and about comparable to Schroeder's figure (21) for the land. Our production estimates are somewhat higher than those of Steemann Nielsen, the annual mean net production of organic matter for the Sargasso Sea (0.40

g/m² day) being about 6 times as great as his value for the same area, and twice his average for the oceans as a whole. This discrepancy appears to be largely due to the fact that Steemann Nielsen's values are based on single observations which probably seldom included seasonal maxima. His observations in the Sargasso Sea, for example, were made in June and did not differ greatly from our June values, which were the seasonal minima. If the Sargasso Sea is one of the less fertile parts of the ocean, as is generally believed, then our data would indicate that the seas are more than twice as productive as the land (25).

References and Notes

1. A. Krogh and K. Berg, *Intern. Rev. ges. Hydrobiol. Hydrog.* 25, 205 (1931).
2. B. H. Ketchum and A. C. Redfield, *J. Cellular Comp. Physiol.* 33, 281 (1949).
3. B. Kok, *Acta Botan. Neerl.* 1, 445 (1952).
4. H. G. Aach, *Arch. Mikrobiol.* 17, 213 (1952).
5. E. C. Wassink, B. Kok, J. L. P. van Oorschot, "The efficiency of light-energy conversion in *Chlorella* cultures as compared with higher plants," in "Algal Culture from Laboratory to Pilot Plant," *Carnegie Inst. Wash. Publ. No. 600* (1953), pp. 55–62.
6. W. M. Powell and G. L. Clarke, *J. Opt. Soc. Am.* 26, 111 (1936).
7. C. L. Utterback and W. Jorgensen, *ibid.* 26, 257 (1936).
8. E. O. Hulburt, *ibid.* 35, 698 (1945).
9. R. Emerson and C. M. Lewis, *Am. J. Botany* 30, 165 (1943).
10. T. Tanada, *ibid.* 39, 276 (1951).
11. P. Moon, *J. Franklin Inst.* 230, 583 (1940).
12. K. Kalle, *Ann. Hydrog. mar. Meteor.* 66, 1 (1938).
13. J. H. Ryther *et al.*, *Biol. Bull.* 115, 257 (1958).
14. E. Steemann Nielsen and E. A. Jensen, *Galathea Repts.* 1, 49 (1957).
15. J. H. Ryther, *Deep-Sea Research* 2, 134 (1954).
16. E. I. Rabinowitch, *Photosynthesis and Related Processes* (Interscience, New York, 1956), vol. 2, part 2, pp. 1925–1939.
17. H. H. Kimball, *Monthly Weather Rev.* 56, 393 (1928).
18. E. P. Odum, *Fundamentals of Ecology* (Saunders, Philadelphia, ed. 2, 1959).
19. J. H. Ryther and C. S. Yentsch, *Limnol. Oceanog.* 2, 281 (1957).
20. ——, *ibid.* 1, 72 (1956).
21. H. Schroeder, *Naturwissenschaften* 7, 8 (1919).
22. G. A. Riley, *Bull. Bingham Oceanog. Coll.* 15, 15 (1956).
23. E. Steemann Nielsen, *J. conseil, Conseil permanent intern. exploration mer* 19, 309 (1954).
24. G. A. Riley, *Bull. Bingham Oceanog. Coll.* 7, 1 (1941).
25. This paper is contribution No. 1016 of the Woods Hole Oceanographic Institution. The work was supported in part by research grant G-3234 from the National Science Foundation and under contract AT (30-1)-1918 with the Atomic Energy Commission.
26. H. Tamiya, *Ann. Rev. Plant Physiol.* 8, 309 (1957).
27. H. T. Odum and E. P. Odum, *Ecol. Monographs* 25, 291 (1955).
28. H. T. Odum, *Limnol. Oceanog.* 2, 85 (1957).
29. J. H. Ryther *et al.*, *Biol. Bull.* 115, 257 (1958).
30. J. H. Ryther and C. S. Yentsch, unpublished data.
31. J. H. Ryther and D. W. Menzel, unpublished data.
32. G. A. Riley, *Bull. Bingham Oceanog. Coll.* 15, 324 (1956).

20

Reprinted from pages 147 and 149-166 of *Primary Productivity in the Biosphere*, edited by H. Lieth and R. H. Whittaker, New York: Springer-Verlag, 1975, 339pp.

Methods of Assessing the Primary Production of Regions

David M. Sharpe

Most published work on primary production has been done at the local level. Estimates of production for regions have been based upon extrapolations from small samples of stand productivities. Recently, regional production rates have been studied in a more integrated manner, for example, in the Biome and Regional Analysis Program of the Eastern Deciduous Forest Biome Program, which is one of the contributions of the United States to the International Biological Program (IBP). The objective of this chapter is to review the methods used in estimating primary production rates for specific regions, and to suggest directions for improvement.

[*Editor's Note:* Material has been omitted at this point.]

Methods of Estimating Regional Production

Three sources of data have been used in IBP studies to provide information with different degrees of spatial resolution. The first method draws upon continuous forest-inventory plot data and allometric relations to develop the equivalent of a network of intensive plot studies with several plots per county. A second method involves the use of published data on areas in counties devoted to various land uses and data on agricultural and forest yields as provided by the *Census of Agriculture* and published forest inventories. Both methods assume that primary production can be extrapolated from these data through the use of appropriate conversion factors, but obtaining these factors is an unfinished task that is discussed subsequently. A third method relates regressions of the most appro-

priate index of production on environmental variables, for example, from networks of weather stations, to map production over large regions (R2 and R3)
and the world.

Each of the methods considered in the next section utilizes a large existing
data base, because it is virtually impossible to collect new data for so many
points in the field. Unfortunately, data sets that have been developed for other
purposes require major adjustments to derive regional production estimates.

Continuous forest inventories

The forest resources of the United States are censused periodically by the
Forest Service in a program of continuous forest inventory (CFI). Other agen-
cies have CFI programs in their regions, such as the Tennessee Valley
Authority (TVA).

The CFI program involves sampling of forests by establishing forest inventory
plots that are resurveyed at intervals (e.g., 5 or 10 years). Plots usually are
located on a grid system; the Forest Service locates plots on a 3-mile- (4.8-km-)
square grid, and the TVA uses an 8.5-mile- (13.7-km-) square grid. A variable-
radius plot (Forest Service), or a fixed-radius plot, usually of 0.08 ha (TVA)
may be established. Tallied trees on each plot are identified so that the basic
record of the CFI program relates measurements to specific trees. Each tree is
classified by species, and by diameter-size class as sapling, poletimber, and sawtimber. Measurements to assess both the quantity and quality of forest products
in trees of commercial species and merchantable size (12.7 cm DBH and
larger) are made; saplings of commercial species are censused to assess the
potential for forest products; and noncommercial species are measured. The
DBH (diameter breast height) of each measured tree is recorded, along with
other attributes of commercial trees of merchantable size.

Records of CFI surveys are available in two forms: (1) as published summaries that provide information on forest types, merchantable standing crop, and
in some cases growth to the merchantable growing stock, as discussed in the
next section, and (2) as unpublished data for each tree on each plot on the
CFI program, available on punched cards or magnetic tape. The plot records
provided by the TVA are the data base for the study discussed here. The method
is generally applicable to other CFI data as well.

Records of 224 plots in Tennessee are being used to test and revise this
method. Some of these plots were installed in 1960 and were resurveyed in
1965 and 1970. Others were installed in 1966 and were resurveyed in 1970.
The 1965 (or 1966) to 1970 period was used to compute average annual net
primary production for each plot, and 1970 was chosen to compute biomass
of the plot (DeSelm *et al.*, 1971).

The average annual net primary production and biomass of each plot are the
summations of the production and biomass of each stand component. The biomass and primary production of poletimber, sawtimber, and saplings were computed; then adjustments were made to account for biomass and production of
undergrowth and roots and for insect consumption. The components of the stand

Table 7–1   Outline for computing net primary production and standing crop of Tennessee Valley Authority forest inventory plots

Components of the stand	1965–1970 net primary productivity (g/m²/year)	1970 standing crop (kg/m²)
Undergrowth (shoots and roots)	30—Average of local research	0.135—Average of local research
Saplings	Foliage only: foliage biomass × turnover rate	Bole and branch, foliage (1970 survey)
Poles and sawtimber	Bole and branch: 1970 biomass minus 1965 biomass Foliage: average biomass × turnover rate	Bole and branch, foliage (1970 survey)
Roots (saplings, poles, sawtimber)	25% of shoot production	25% of shoot biomass
Insect consumption (saplings, poles, sawtimber)	3% of shoot production	None

under consideration and the general procedure for computing net primary production and biomass standing crop of each component are shown in Table 7–1.

The biomass of each measured tree of any species and size was computed from the recorded DBH of the tree by using allometric relations between DBH and bole and branch biomass, and DBH and foliage biomass. The equations were developed by Sollins and Harris (personal communication) from stem analyses of conifers and deciduous species in Georgia, North Carolina, and Tennessee computed by Sollins and Anderson (1971). The equations were adjusted to remove the bias inherent in logarithmic transformation of data in regression analysis discussed by Beauchamp and Olson (1972). The equations (with the original in brackets, preceded by the adjustment factor) are

$$BB = (1.15)[0.119D^{2.393}]$$
$$FB = (1.37)[0.03D^{1.695}]$$

where BB is the biomass of bole and branches in kilograms for dry weight; FB is the biomass of foliage in kilograms for dry weight; and D is the diameter breast height (expressed as DBH throughout this volume) in centimeters.

The biomass of poletimber and sawtimber for 1965 (or 1966) and 1970 and for saplings in 1970 was computed as the sum of bole and branch plus foliage biomass. Net primary production of each poletimber and sawtimber tree was computed as (1) the difference between bole and branch biomass in 1970 and 1965 (or 1966) divided by the 5- or 4-year interval; and (2) the average foliage biomass for 1965 (or 1966) and 1970 multiplied by the foliage turnover rate (once every year for deciduous species and an assumed 3 years for evergreens). Because the TVA did not measure saplings until the 1970 survey, only their foliage production could be computed as the 1970 foliage biomass multiplied by the appropriate foliage turnover rate. Neglect of bole and branch production of saplings was compensated for by the production computed for trees that grew to poletimber size between 1965 (or 1966) and 1970. These trees had a DBH of zero in 1965 (or 1966) that was arbitrarily assigned to them by TVA, and a correspondingly high bole and branch production in this study.

Unfortunately, no forest inventory program takes into account all of the ecologically significant components of the forest stand. Saplings were not measured by the TVA in the initial survey, and the biomass and growth of seedlings, herbs, shrubs, and root systems are not measured. The adjustments shown in Table 7–1 were made to account for this additional production. Above- and belowground production of herbs and shrubs was considered as a constant 30 g/m²/year. Root production of trees was assumed to be 25% of aboveground production, and animal consumption to be 3% of aboveground production.

One uncertainty of this method results from the extensive use of one set of allometric relations. Stem analyses of trees from West Tennessee are not included in the data set of Sollins and Anderson, and the difficulty of stem analysis of large trees biases the sample toward saplings and small poletimber. As more stem analyses are made we shall gain confidence in the allometric relations.

Moreover, the adjustments for undergrowth, roots, and insect consumption are recognized as arbitrary. Table 7–2 shows the estimated net primary production and biomass for the TVA CFI plots in Knox County, Tennessee. Net primary production increases generally as basal area increases and stocking improves from 36 g/m²/year for plot 360 to 1230 g/m²/year for plot 365. Basal area in this case takes account of pole- and sawtimber, but not saplings. Stocking is defined as follows (TVA, 1967):

Overstocked. 100% crown closure or more than 700 seedlings and saplings per acre (1750 per hectare)

Good stocking. 70–99% crown closure or 550 seedlings and saplings per acre (1360 per hectare)

Fair. 40–69% crown closure or 300–549 seedlings and saplings per acre (1040–1359 per hectare)

Poor. 10–39% crown closure or 100–300 well-distributed seedlings and saplings per acre (247–1039 per hectare)

Other. Less than 10% crown closure or less than 100 well-distributed seedlings and saplings per acre (247 per hectare).

Plot 360 was a dense pole stand of *Pinus echinata* and *P. virginiana*, which was cut between 1960 and 1965; only an estimated 100 saplings per hectare remained in 1970. Plot 365, by contrast, is in a yellow pine–hardwood stand with a large number of rapidly growing pole- and sawtimber trees and no evidence of recent cutting.

The extremely low production for plot 360 results from assigning to each plot a constant undergrowth production of 30 g/m²/year, which is more representative of closed stands in Tennessee than of abandoned and recently distributed land. Similarly, recent studies suggest that root production is significantly higher

Table 7–2   Net primary productivity (g/m²/year) and biomass (kg/m²), dry matter, for selected plots in Knox County, Tennessee

Plot	Forest type	Basal area (m²/ha)[a]	Stocking class	Net primary production (trees)			Biomass		
				DBH 2.5-12.4 cm	DBH ≧ 12.5 cm	Total	DBH 2.5-12.4 cm	DBH ≧ 12.5 cm	Total
360	391[b]	0.0	Poor	6	0	36	0.08	0	0.21
361	380[c]	1.44	Fair	213	260	503	5.02	0.92	6.08
363	380	1.73	Fair	57	278	365	2.01	1.29	3.44
365	380	27.74	Good	43	1153	1226	1.02	22.84	23.99

[a] Trees DBH ≧ 12.5 cm.
[b] Cedar–pine–hardwood.
[c] Yellow pine–hardwood.

164

than was hitherto suspected (Harris and Todd, 1972). Better measures of these frequently ignored or hard-to-measure stand components will certainly raise the estimates of forest production.

Table 7–3 shows the primary production and biomass as computed from the TVA plots located in each physiographic province (after Fenneman, 1938). If only the plots that have overstocking and good stocking are considered, estimated productivity varies from 899 g/m²/year in the Cumberland Plateau to 1419 g/m²/year in West Tennessee. If plots comprised of all stocking classes are included, computed productivity and biomass are decreased, as shown in Table 7–3. This reduction is less extreme than is shown in Table 7–2 because 134 of the 224 TVA plots in Tennessee were classified as overstocked or well stocked and 79 as having fair stocking; most had high production reflected in computed production of pole- and sawtimber.

Censuses and conversion factors

An alternative approach to the study of productivity is shown by the productivity profiles of North Carolina, Tennessee, New York, Massachusetts, and Wisconsin. These studies were conducted by four teams that coordinated their work, but that, in some respects, used different techniques and sources of data (Art *et al.*, 1971; DeSelm *et al.*, 1971; Stearns *et al.*, 1971; Whigham and Lieth, 1971). The productivity profiles also tapped reservoirs of data collected by federal and state agencies, again with purposes other than regional productivity. The general strategy of each profile was to determine the area in each county that was devoted to each of a number of land-use categories, to establish an average primary production value for each land-use category, and by multiplying area by average production and summing across all land uses, to estimate county-level productivity.

Table 7–3   Tennessee productivity profile: Forest biomass and net primary productivity by physiographic region for two stocking categories in Tennessee

Region	Well stocked		All stocking classes	
	1965–1970 net primary productivity (g/m²/year)	1970 biomass (kg/m²)	1965–1970 net primary productivity (g/m²/year)	1970 biomass (kg/m²)
Appalachian Mountains	1203	19.5	1081	17.6
Great Valley	1108	15.6	940	12.5
Cumberland Plateau	899	14.0	831	12.5
Highland Rim	1001	15.4	894	13.4
Nashville Basin	1086	21.4	970	17.0
West Tennessee	1419	20.6	1074	15.3
Average:	1091	16.7	936	13.8

Land-use categories, and the area per county in each category, were determined from state or federal sources. The 1964 *Census of Agriculture* (U.S. Bureau of the Census, 1967) was a major source of information for the New York–Massachusetts profiles as was the Crop Reporting Service in North Carolina and Tennessee. Forest yields for North Carolina and Wisconsin are based on published records of net annual growth to growing stock (the increment to pole- and sawtimber, plus ingrowth to these size classes, minus losses incurred by mortality). These data have been collected by the Forest Service in the CFI program of each state. Yields are published by species in some states (e.g., Wisconsin), or by forest type (e.g., North Carolina).

In the Wisconsin and Tennessee productivity profiles no published statistical data were available on production of wetlands and water bodies, urban areas and rights-of-way, and the catchall category of open land (for abandoned farmland, nonstocked forest land, farmstead, and county roads). Consequently, indirect evidence of a limited number of ecologic studies had to be relied upon (Stearns *et al.*, 1971; DeSelm *et al.*, 1971).

The major issue of the productivity profiles has been how to convert yields of agricultural crops and net annual growth to growing stock of forests to a more complete net primary production budget for these land-use categories. This involves more than converting the units in which yield is reported, for example, bushels, to dry weight of the yield; accounting for the unharvested or uneconomic components of production that are not included in yield figures has been treated to date only as an approximation. For agricultural crops, the unharvested biomass of the plant, such as roots, stalk, husk, and leaves for corn, must be accounted for, along with any of these components that are lost during the growing season. For forests, growth in roots, unharvested portions of merchantable boles, branches, and foliage, and unmerchantable trees is needed, along with mortality of individuals and parts (e.g., branch pruning and root sloughing); and herbaceous and woody undergrowth all must be added to net annual growth to growing stock.

The following quotation details how the conversion factor for wheat was determined for the North Carolina study (Whigham and Lieth, 1971). The same logic was used for other crops and other states, but the values probably can be improved in all cases:

> Extant data were given as yield in bushels per acre (bu/ac). Each bushel of wheat weighs approximately 60 pounds (lb); the data were initially multiplied by 60 to convert yield in bu/ac to yield in lb/ac. Finally, yield in lb/ac was converted to yield in t/ha. The formula:
>
> $$\frac{\text{yield in lb/ac}}{892.2} = \text{yield in t/ha} \ (100 \ g/m^2)$$
>
> Yield rates (t/ha) were then converted to total plant productivity rates by using a conversion factor (total plant production/plant yield). Adjusting that ratio (2.42) for water content (12%), it became 2.13 and the rate of yield $\times$ 2.13 = corrected production rate. Total county wheat produc-

tion was then calculated by multiplying the corrected production rate by the hectares of wheat in the county (i.e., tons of wheat per hectare multiplied by hectares = tons of wheat).

A similar procedure was used to convert annual growth to growing stock of forests, expressed in cords or cubic feet per acre, to dry weight production per hectare. Detailed information about the conversion from yield to total productivity and the assessment of production for individual counties is given in Chapter 6 of this volume.

Conversion factors used to date to extrapolate yield data for selected crops, forest, urban, open land, and water land-use categories are shown in Table 7–4. In all cases except for hay, commercial yields are no more than 50% of total production, that is, conversion factors are 2.0 or more. The production of urban areas, open land, and water was computed as a proportion of the hay production of a county, or a constant production was assigned, except as noted in footnote *c* of Table 7–4. The choice of conversion factor is therefore critical to the accuracy of estimated primary production. The teams working on the productivity profiles collaborated on this issue, so the similarity of these trial conversion factors is not surprising. Conversion factors remain a major issue for estimating total production, as well as the fraction of that total that can be used by man for particular purposes.

Table 7–4 Factors used in productivity profiles to convert yields for selected agricultural and forest crops to dry matter net primary productivity[a]

Cover type	New York–Massachusetts	Wisconsin	Tennessee	North Carolina
Corn (grain)	2.14	2.12	3.68	2.03
Small grains	2.32	2.45	3.46	2.64
Hay	1.12	1.48	1.48	1.12
Forest				
Pine	—	3.5	—	2.0
Aspen, ash	—	3.0	—	2.0
Urban	—	160 g/m²/year	0.33 × hay production	25 g/m²/year
Open	—	Avg. other categories	0.5 × hay production	0.6 × hay production
Water	—	12 g/m²/year[b] 62 g/m²/year[d]	213–841 g/m²/year[c]	500 g/m²/year

[a] (primary productivity = conversion factor × yield dry weight); conversion factors and constants for urban, open, and water–land-use categories.

[b] North of tension zone (Curtis, 1959).

[c] Based on production measured for selected TVA reservoirs by M. P. Taylor (*personal communication*), Environmental Biology Branch, TVA, Norris, Tennessee.

[d] South of tension zone.

Table 7-5   Range in primary productivity for selected vegetation
(dry matter g/m²/year)

State	Maize		Small grain		Hay		Forest		Average	
	Low[a]	High[b]	Low	High	Low	High	Low	High	Low	High
New York–Massachusetts	260	590	120	580	280	580	—	—	230	800
Wisconsin	280	1260	360	670	590	1340	220	490	230	800
Tennessee	660	910	220	560	300	510	720[c]	1050[c]	420	900
North Carolina	490	1080	260	810	250	450	100[d]	430[d]	150	500
							765[e]	1500[e]		

[a] Production of county with lowest production.

[b] Production of county with highest production.

[c] Based on TVA plot data.

[d] Based on Forest Service statistics, Whigham and Lieth (1971).

[e] Based on plantation survey data; see Table 6–4.

In spite of these uncertainties, two practical types of information came from the productivity profiles. The first type is variation in primary production of selected crops from one state to another. Table 7–5 shows, for each state, the productivity of the least productive and most productive county for selected land-use categories and for the average of all land uses. For example, the lowest average corn production for the least productive county for corn in the New York–Massachusetts profile (Franklin County, Massachusetts) was 260 g/m²/ year, and the highest (for Chautauqua County, New York) was 590 g/m²/year. The lack of any sharp distinction between agricultural and forest production, except in Tennessee, should be noted. The higher forest production in Tennessee, computed from CFI plot data as discussed previously, is probably more a result of conservative initial conversion factors for forests in North Carolina and Wisconsin, than of higher actual productivity in the forests of Tennessee.

Another comparison of interest is the ranking of production by land-use category, as shown for Tennessee in Table 7–6. Forests rank highest, perhaps because of the close (but still incomplete) accounting made of forest production. Row crops can produce nearly as much as forests, but only by investing more management input. The computed production of 110 g/m²/year for urban land may be very conservative. The aboveground production of one residential landscape studied in Madison, Wisconsin, exceeds the production of adjacent woodland per unit area of vegetated surface (perhaps the result of the inputs of fertilizer and supplemental irrigation and of decreased competition), and is equal when paved and roofed surfaces are included (Lawson *et al.*, 1972). This may be found to be true generally (except, of course, for urban cores) when urban vegetation is subjected to the same close accounting as natural vegetation.

D. M. Sharpe

Table 7–6   Median and range in primary productivity by
land-use categories for counties in Tennessee
(dry matter $g/m^2/year$)

Land-use category	Productivity		
	Low	Median	High
Forest	720	900	1050
Agricultural row crops	660	750	920
Hay crops	300	400	510
Small grains	220	390	560
Pasture	200	250	300
Urban	080	110	190
Open land	120	160	280
Lakes and rivers	220	560	900

Relating Productivity to Environment

Weather records constitute another widespread source of data that can be brought to bear on the problems of regional productivity. A number of models for agricultural and natural vegetation are available from the work of agronomists, silviculturists, climatologists, and ecologists (see Chang, 1968a; Lowry, 1969; Munn, 1970), so that the task becomes one of selecting models that satisfy the needs of regional ecosystem productivity analysis. Some of these are (1) the weather data needed must be available for a large number of stations, and of uniform quality; (2) the model derives production for a variety of plant communities in a region, and is not restricted to single crops or species; and (3) the model reflects major changes in production at a particular scale, for example, differences in annual production between such R3 regions as states.

Most models relate to a limited number of species (e.g., Currie and Peterson, 1966; Zahner and Stage, 1966; Albrecht, 1971); rely on data with high resolution (e.g., deWit, 1958; Monteith, 1965); ignore seasonality of energy or moisture resources, which becomes important in interregional comparisons (e.g., Drozdov, 1971; Lieth, 1971b); or assume either energy or moisture to be in adequate supply in all climatic conditions (e.g., Chang, 1968b, 1970).

Models that relate growth to a component of a water balance, usually actual evapotranspiration but sometimes deficit as well, avoid these shortcomings. A water balance is a budget of water in response to an estimated demand for moisture imposed by an energy load versus precipitation and available supply of soil moisture. The accounting period may vary from a day to a month. Many water-balance schemes have been devised, but the simplest of these requires only air temperature and precipitation data, the geodetic coordinates of the weather station, and a measured or assumed moisture-storage capacity for the root zone (Thornthwaite and Mather, 1957).

Actual evapotranspiration has been related to yields of agricultural crops (Arkley and Ulrich, 1962; Arkley, 1963), diameter growth of trees (Zahner

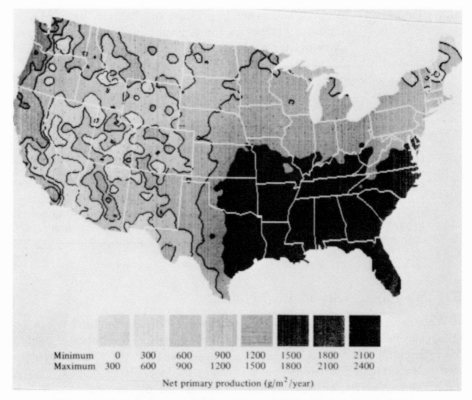

Minimum   0   300   600   900   1200   1500   1800   2100
Maximum   300   600   900   1200   1500   1800   2100   2400

Net primary production (g/m^2/year)

FIGURE 7–2. Average annual net primary production in conterminous United States after C. W. Thornthwaite Memorial model developed by Lieth and Box, and average annual water balances computed by C. W. Thornthwaite Associates.

and Stage, 1966; Zahner and Donnelly, 1967; Manogaran, 1972), and to the net primary production of ecosystems (Rosenzweig, 1968; Lieth and Box, 1972). Maps of the primary production of the United States were developed using the Lieth–Box and Rosenzweig models, as shown in Figures 7–2 and 7–3. The Rosenzweig and Lieth–Box models have the same general logic; each considers primary production as a function of actual evapotranspiration. However, there are differences between the data sets of measured primary production and of actual evapotranspiration upon which each is based, and the mathematical function chosen to relate production to evapotranspiration. This has been discussed in detail (Lieth and Box, 1972; see also Chapter 6, this volume); some general comments on the methods used will clarify disparities between the two maps.

Each model uses measured values of primary production as a data set; Lieth and Box use a data set of about 50 values of aboveground and belowground production from North America, South America, Eurasia, and Africa. Rosenzweig's data set of 25 values of aboveground production only derives largely from the Great Smoky Mountains in Tennessee (15 points from Whittaker, 1966) with a selection of other data from desert, grassland, tundra, and tropical

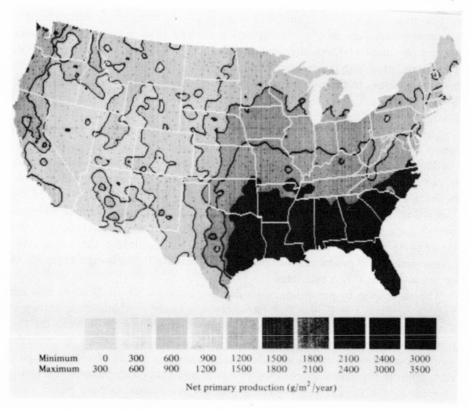

Minimum   0   300   600   900   1200   1500   1800   2100   2400   3000
Maximum   300   600   900   1200   1500   1800   2100   2400   3000   3500

Net primary production (g/m²/year)

FIGURE 7–3. Average annual net primary production in conterminous United States after M. L. Rosenzweig and average annual water balances computed by C. W. Thornthwaite Associates.

forest. The maximum net primary production value in each data set is about 2900 g/m²/year. Evapotranspiration values in Rosenzweig's model are from published water budgets computed by the Thornthwaite method (C. W. Thornthwaite Associates, 1964). Evapotranspiration values in the Lieth–Box data set are derived from the map *Annual Effective Evapotranspiration* (scale 1 : 30,000,000) (Geiger, 1965) by estimating evapotranspiration for each site in the production data set.

Each model results from a least-squares fit of a curve to the respective data sets. Rosenzweig's model is a linear regression of the logarithms of the variables. The Lieth–Box model is a saturation curve with 3000 g/m²/year as asymptote. The equations are

$$NPP(\text{aboveground}) = 0.0219E^{1.66} \text{ (Rosenzweig)}$$
$$NPP(\text{total}) = 3000[1 - e^{-0.0009695(E-20)}] \text{ (Lieth–Box)}$$

where *NPP* is net primary production (g/m²/year) and *E* is actual evapotranspiration (millimeters per year).

The maps in Figures 7–2 and 7–3 were produced using these equations and a common data base of evapotranspiration derived from average annual water budgets for about 1100 weather stations computed by the Thornthwaite method. The maps show similar trends in production across the United States with maximum values in the Southeast, and a minimum in the Intermountain West. The estimates are most similar in the Southern Appalachian Mountains, and diverge toward places having higher and lower production.

The evapotranspiration data base used for these maps and the logic of each model account for their differences. Brief review of the Geiger map indicates that the Thornthwaite estimates of evapotranspiration in the eastern United States are higher than the Geiger map shows. An adjustment to account for the overestimate of evapotranspiration (as viewed from the perspective adopted for the Lieth–Box model) would reduce the Lieth–Box production estimate. Rosenzweig's data set accounts for aboveground production only; adjustment of the Rosenzweig or Lieth–Box model, so that each accounts for the same stand components in the production estimate, would increase Rosenzweig's estimates or decrease the Lieth–Box estimates.

The sharp rise in net primary production between North Georgia and the Florida panhandle in Figure 7–3 (Rosenzweig's model), which is not shown in Figure 7–2 (Lieth–Box model) identifies the major difference between the two models. The data set of production values used by Rosenzweig is generally more conservative than the Lieth–Box model for given values of evapotranspiration, and the maximum values of production and evapotranspiration are nearly coincident. Yet Rosenzweig considers primary production to be a power function of evapotranspiration, which imposes no limit on production as evapotranspiration increases, whereas the Lieth–Box model is a saturation curve that imposes an upper limit to production of 3000 g/m²/year. Tests of alternative curves on Rosenzweig's data show that a simple linear regression of production on evapotranspiration has the same correlation coefficient ($r = 0.95$) as the linear regression of the logarithms of the variables. A logistic curve with an asymptote of 3000 g/m²/year has a slightly poorer least-square fit. Alternative curve forms could be fitted to the Lieth–Box data sets, as well, perhaps with some decrease in the goodness-of-fit.

Lieth and Box (1972) justify their use of the saturation curve on the ground that it conforms to Mitscherlich's yield law. Other lines of evidence indicate a need for a ceiling on primary production. Stanhill (1960), Black (1966), and Chang (1968b) point to the increasing toll taken by respiration on the gross photosynthesis of agricultural crops, pasture, and forests as temperatures increase when moisture is in adequate supply. Drozdov (1971) models primary production as a saturation-curve function of net radiation in subhumid and humid environments. The ecologic reasoning expressed in the Lieth–Box model, appears to be superior to that of the Rosenzweig model. The values of production associated with evapotranspiration and the specific form of the curve are likely to change as more measurements of production and studies of the physiology of net photosynthesis provide further insight into plant–environment relationships in production.

Discussion

Two goals for studies of regional production are confidence in the methods and results of each study, and as a corollary, convergence of the results of different studies. Considerable progress toward this goal is being made in the local ecosystem analysis programs of IBP; the methods and results of the study of regional production discussed here deserve less confidence. Comparison of several results illustrates this point.

The Lieth–Box production map shows production for Tennessee in the range of 1500–1800 g/m²/year; the summary for Tennessee in Table 7–6 shows production for forests as computed from CFI data as 720–1050 g/m²/year, and an agricultural production of 660–920 g/m²/year for row crops and 300–510 g/m²/year for hay. By contrast, the production for Wisconsin in the Lieth–Box map is 1200–1500 g/m²/year, whereas Table 7–5 shows a forest production of 220–490 g/m²/year, a corn production of 280–1260 g/m²/year, and a hay production of 590–1340 g/m²/year. The trend for Tennessee is: Lieth–Box > forest ≧ row crops > pasture; for Wisconsin it is: Lieth–Box ≧ hay ≧ corn ≧ forest. Does this ordering indicate real differences in the absolute and relative production of these land–use categories within and between Tennessee and Wisconsin, or does it indicate inaccuracies in the methods? The Wisconsin forest production values are quite low in relation to production estimates for a range of eastern forests, some of them in cool mountain climates, in the Great Smoky Mountains (Whittaker, 1966). No firm answer can be given, however, and the uncertainties that underlie these orderings indicate some themes for further study.

[*Editor's Note:* Material has been omitted at this point.]

Acknowledgments

This research was supported by the Deciduous Forest Biome Project, International Biological Program, funded by the National Science Foundation under Interagency Agreement AG–199, 40–193–69 with the Atomic Energy Commission, Oak Ridge National Laboratory, Oak Ridge, Tennessee.

References

Albrecht, J. C. 1971. A climatic model of agricultural productivity in the Missouri River Basin. *Publ. Climatol.* 24(2):1–107. Centerton (Elmer), N.J.: Thornthwaite Lab. Climatol.

Arkley, R. J. 1963. Relationships between plant growth and transpiration. *Hilgardia* 34:559–584.

————, and R. Ulrich. 1962. The use of calculated actual and potential evapotranspiration for estimating potential plant growth. *Hilgardia* 32:443–468.

Art, H. W., P. L. Marks, and J. T. Scott. 1971. Progress report—Productivity profile of New York. 30 pp. (mimeogr.) US-IBP EDFB Memo Rep. 71-12.

Beauchamp, J. J., and J. S. Olson. 1972. Estimates for the mean and variance of a lognormal distribution where the mean is a function of an independent variable. 22 pp. (mimeogr.) US-IBP EDFB Memo Rep. 71-101.

Black, J. N. 1966. The utilization of solar energy by forests. *Forestry Suppl.* 98–109.

Chang, J.-H. 1968a. *Climate and Agriculture: An Ecological Survey*, 304 pp. Chicago, Illinois: Aldine.

———. 1968b. The agricultural potential of the humid tropics. *Geograph. Rev.* 58: 333–361.

———. 1970. Potential photosynthesis and crop productivity. *Ann. Assoc. Amer. Geogr.* 60:92–101.

Currie, P. O., and G. Peterson. 1966. Using growing-season precipitation to predict crested wheatgrass yields. *J. Range Mgt.* 19:284–288.

Curtis, J. T. 1959. *The Vegetation of Wisconsin*, 657 pp. Madison, Wisconsin: Univ. of Wisconsin Press.

Dansereau, P. 1962. The barefoot scientist. *Colo. Quart.* 12:101–115.

DeSelm, H. R., D. Sharpe, P. Baxter, R. Sayres, M. Miller, D. Natella, and R. Umber. 1971. Final report; Tennessee productivity profile; 182 pp. (mimeogr.) US-IBP EDFB Memo Rep. 71-13.

deWit, C. T. 1959. Potential photosynthesis of crop surfaces. *Neth. J. Agr. Sci.* 7:141–149.

Drozdov, A. V. 1971. The productivity of zonal terrestrial plant communities and the moisture and heat parameters of an area. *Sov. Geogr. Rev. Transl.* 12:54–59.

Evans, F. C. 1956. Ecosystem as the basic unit in ecology. *Science* 123:1127–1128.

Fenneman, N. A. 1938. *Physiography of the Eastern United States*, 714 pp. New York: McGraw-Hill Book Co.

Friederichs, K. 1958. A definition of ecology and some thoughts about basic concepts. *Ecology* 39:154–159.

Geiger, R. 1965. *The Atmosphere of the Earth* (12 wall maps and text). Darmstadt, Germany: Justus Perthes.

Goff, F. G., F. P. Baxter, and H. H. Shugart, Jr. 1971. Spatial hierarchy for ecological modeling, 12 pp. (mimeogr.) US-IBP EDFB Rep. 71-41.

Golley, F. B., and J. B. Gentry. 1965. A comparison of variety and standing crop of vegetation on a one-year and a twelve-year abandoned field. *Oikos* 15:185–199.

Harris, W. F., and D. E. Todd. 1972. Forest root biomass production and turnover, 17 pp. (mimeogr.) US-IBP Memo Rep. 72-156.

Kira, T., and T. Shidei. 1967. Primary production and turnover of organic matter in different forest ecosystems of the western Pacific. *Jap. J. Ecol.* 17:70–87.

Lawson, G. J., G. Cottam, and O. Loucks. 1972. Structure and primary productivity of two watersheds in the Lake Wingra Basin, 51 pp. (mimeogr.) US-IBP EDFB Memo Rep. 72-98.

Lieth, H. 1970. Phenology in productivity studies. In *Analysis of Temperate Forest Ecosystems*, D. E. Reichle, ed., *Ecological Studies* 1:29–46. New York: Springer-Verlag.

———. 1971. The phenological viewpoint in productivity studies. In *Productivity* of Forest Ecosystems: Proc. Brussels Symp. 1969, P. Duvigneaud, ed., *Ecology and Conservation*, Vol. 5, 71–84. Paris: UNESCO.

———. 1973. Primary production: Terrestrial ecosystems. *Human Ecol.* 1:303–332.

———, and E. Box. 1972. Evapotranspiration and primary productivity: C. W. Thornthwaite Memorial Model. In *Papers on Selected Topics in Climatology*, J. R. Mather, ed. (*Thornthwaite Mem.* 2:37–46.) Elmer, New Jersey: C. W. Thornthwaite Associates.

Loucks, O. L. 1970. Evolution of diversity, efficiency and community stability. *Amer. Zool.* 10:17–25.

Lowry, W. P. 1969. *Weather and Life: An Introduction to Biometeorology*, 305 pp. New York: Academic Press.

Monogaran, C. 1972. Climatic limitations on the potential for tree growth in southern forests. Unpublished dissertation. Carbondale, Illinois: Southern Illinois Univ.

Monteith, J. L. 1965. Light distribution and photosynthesis in field crops. *Ann. Bot. N.S.* 29:17–37.

Munn, R. E. 1970. *Biometeorological Methods*, 336 pp. New York: Academic Press.

Novikoff, A. B. 1945. The concept of integrative levels and biology. *Science* 101: 209–215.

Odum, E. P. 1960. Organic production and turnover in old field ecosystems. *Ecology* 41:34–49.

Rosenzweig, M. L. 1968. Net primary productivity of terrestrial communities: Prediction from climatological data. *Amer. Natur.* 102:67–74.

Rowe, J. S. 1961. The level-of-integration concept and ecology. *Ecology* 42:420–427.

Sharp, D. D., H. Lieth, and D. Whigham. Chapter 6, this volume.

Sollins, P., and R. M. Anderson. 1971. Dry-weight and other data for trees and woody shrubs of the Southeastern United States. ORNL-IBP-71-6, 80 pp. Oak Ridge, Tennessee: Oak Ridge National Laboratory.

Stanhill, G. 1960. The relationship between climate and the transpiration and growth of pastures. In *Proc. 8th Int. Grassland Congress*, Tel Aviv: 293–296.

Stearns, F., N. Kobriger, G. Cottam, and E. Howell. 1971. Productivity profile of Wisconsin, 82 pp. (mimeogr.) US-IBP EDFB Memo Rep. 71-14.

Tennessee Valley Authority. 1967. TVA forest inventory field manual for county-wide units and watersheds in the Tennessee Valley, 20 pp. (mimeogr.) Norris, Tennessee: Forest Survey Sect., Forest Products Branch, Div. of Forestry, Fisheries and Wildlife Development.

Thornthwaite, C. W., and J. R. Mather. 1957. Instructions and tables for computing potential evapotranspiration and the water balance. *Publ. Climatol.* 10(3):181–311. Centerton, N.J.: Thornthwaite Lab. Climatol.

Thornthwaite, C. W. Associates. 1964. Average climatic water balance data of the continents, Part 7: United States. *Publ. Climatol.* 17(3):415–615. Centerton, N.J.: Thornthwaite Lab. Climatol.

U. S. Bureau of the Census, Census of Agriculture. 1967. Statistics for the State and Counties, Tennessee. Washington, D. C.: U. S. Government Printing Office.

Whigham, D., and H. Lieth. 1971. North Carolina Productivity Profile 1971, 143 pp. (mimeogr.) US-IBP EDFB Memo Rep. 71-9.

Whittaker, R. H. 1966. Forest dimensions and production in the Great Smoky Mountains. *Ecology* 47:103–121.

Woodwell, G. M. 1967. Radiation and the patterns of nature. *Science* 156:461–470.

————. 1970. The energy cycle of the biosphere. *Sci. Amer.* 223:64–74.

Zahner, R., and J. R. Donnelly. 1967. Refining correlations of water deficits and radial growth in young red pine. *Ecology* 48:525–530.

————, and A. R. Stage. 1966. A procedure for calculating daily moisture stress and its utility in regressions of tree growth on weather. *Ecology* 47:64–74.

21

Reprinted from pages 3–32 of *Forest Biomass Studies,* Sec. 25: Growth and Yield, 15th Internat. Union Forest Res. Organ. Congr., 1971, 205pp.

A SUMMARY TABLE OF BIOMASS AND NET ANNUAL PRIMARY PRODUCTION IN FOREST ECOSYSTEMS OF THE WORLD

H. W. Art
Williams College

P. L. Marks
Cornell University

ABSTRACT

A working table lists by major species, by location and by stand age, the oven dry weights of biomass and net annual primary production as well as the leaf area index for over 280 forest stands. Specific stands are arranged according to origin (plantations, natural stands, stands of uncertain origin and forest type averages).

Methods used both for conversion of sample tree data to a land area basis and for estimating annual primary production are indicated.

INTRODUCTION

The purpose of this paper is to compile published biomass and productivity data into a relatively simple compendium, unencumbered by excessive detail. As such, the table best serves as a lead into the literature and a rapid means of comparing data accumulated from different forest stands. With the current interest in biomass and productivity studies intensified by the International Biological Program, the present paper is a timely updating of the numerous data on forest biomass and productivity.

This table lists by major species, by location, and by stand age, the oven dry weights of biomass and net annual primary production as well as the Leaf Area Index for various forest stands. Unless otherwise noted, data given are for above-and below-ground portions of living trees, shrubs and herbs exclusive of dead material and soil organic matter.

Specific stands are arranged according to origin (plantations, natural stands, stands of uncertain origin and forest type averages). Natural stands are further grouped as coniferous, temperate broad-leaved evergreen, temperate deciduous and tropical forests. Stands of uncertain origin include those whose origin was not clear in the reference cited, as well as the synthetic pure stands of Young & Carpenter (1967).

Although it is beyond the scope of this paper to give a detailed, critical analysis of the many different methods used in estimating forest biomass and production, it was felt some basis for distinguishing between approaches would be helpful. There are several general categories of methods used to convert field and sample tree data onto a land area basis:

177

Mean tree technique (m). Biomass and production of an entire stand are estimated from the harvesting and analysis of a tree or trees of mean dimension. Sample tree data are converted to a land area basis by multiplying the biomass and production of sample trees by the average number of trees per unit area. This technique has been used most widely in plantations and even-aged stands, but thought inappropriate for uneven-aged stands (Ovington, et al., 1967; Baskerville, 1965 b; Whittaker & Woodwell, 1969 b).

Stratified tree technique (s). This method, a variant of the mean tree technique, is used in uneven-aged stands. The stand is stratified by diameter size classes, each of which is sampled by a tree of mean dimensions for that particular stratum (Peterken & Newbould, 1966).

Regression estimation technique (r). This method has also been called "dimension analysis of woody plants" (Whittaker, 1961, 1962, 1965; Whittaker & Woodwell, 1968, 1969 a) and "allometry" (Kira & Shidei, 1967; Satoo, 1970). This technique involves three phases: 1) Trees are tallied by species and diameters at breast height on a quadrat basis; 2) for sample trees, difficultly measured allometric growth parameters are related by regression analysis to more easily measured parameters (ultimately to diameters at breast height); 3) the conversion to a land area basis is then accomplished by applying the regressions of phase 2 to the quadrat measurements of phase 1. In tropical stands with high species diversities, multi-species regression equations are generally calculated.

Unit area method (e). All individuals within small randomly located plots are completely harvested (Ovington, et al., 1967). Alternatively, all individuals are harvested in a large plot representative of a larger vegetational complex (Satoo, 1970).

Canopy-area method (c). Conversions to a land area basis are made by dividing the sample tree biomass and production by the effective canopy projection of the sample tree (Bray & Dudkiewicz, 1963).

Basal area proportion method (d). Biomass and production of the stand are estimated from sample trees by the ration of stand basal area to sample tree basal area (Ando, 1965).

Estimation of total net annual primary production involves a wide variety of techniques, especially for branch and root tissues (Newbould, 1967, 1968; Whit aker & Woodwell, 1969 b). It is helpful to distinguish between the three basic approaches to the estimation of stem wood production: 1) Current annual increment (ζ) in which production is based upon the most recent annual radial increment of the stem; 2) current periodic annual increment (p) in which production is based upon the mean of usually the last five or ten annual radial increments; 3) mean annual increment ($\int$) in which the woody biomass is divided by the age of the sample trees. The mean annual increment generally underestimates the current woody production.

Leaf area indices represent the leaf surface area per unit of land surface area. For broad-leaved forest this index is calculated from one side of the leaf blade only, while the total leaf surface is used in needle-leaved stands. In mixed stands the leaf area index is a combination of the two.

Analyses of the stands in the table were conducted using a wide variety of plot sizes and sampling intensities. The mean tree technique frequently utilizes only a single sample tree for the estimation of biomass and productivity of an entire stand. Furthermore, oven dry weights for the various stands were calculated using subsample drying temperatures ranging between 70°C and 105°C, although these differences may not be significant. Consequently, differences in approaches to biomass and productivity estimation make difficult accurate comparisons of the many stand data in the following table.

178

ENTRY NO.	SPECIES	LOC.	AGE	BIOMASS: TREES, SHRUBS, & HERBS g/m^2	NET PRIMARY PRODUCTION: TREES, SHRUBS, & HERBS $g/m^2/yr$	LEAF AREA INDEX m^2/m^2	REFERENCE*
PLANTATIONS							
1	Abies grandis	UK	21	35541†γ	1840§	n	Ovington, 1956 [m]
2	Abies grandis	UK	24	16485†γ	925§	n	Ovington, 1956 [m]
3	Abies sacharinensis	JAP	26	n	1340γ§	n	Satoo, 1966 [r]
4	Alnus incana	UK	22	12470†γ	630§	n	Ovington, 1956 [m]
5	Alnus incana	UK	22	9754†γ	506§	n	Ovington, 1956 [m]
6	Betula alba	UK	22	6284†γ	286§	n	Ovington, 1956 [m]
7	Betula alba	UK	22	5881†γ	267§	n	Ovington, 1956 [m]
8	Betula ermanii	JAP	22	6252γ	n	5.6	Satoo, 1970 [e]
9	Castanea sativa	UK	47	11661†γ	375§	n	Ovington, 1956 [m]
10	Chamaecyparis lawsoniana	UK	21	20687†γ	1057§	n	Ovington, 1956 [m]
11	Cinnamomum camphora	JAP	46	19600–20000γ	1500–1660ζ	n	Satoo, 1966 [r]
12	Cinnamomum camphora	JAP	52	n	1526γ§	n	Satoo, 1966 [r]
13	Cryptomeria japonica	JAP	5	9730†	2910ζ	17.2	Tadaki & Kawasaki, 1966 [d]
14	Fagus sylvatica	UK	39	13333†γ	505§	n	Ovington, 1956 [m]
15	Fagus sylvatica	SWITZ	80	n	967†γ§	n	Burger, 1940 [φ]
16	Larix decidua	UK	46	18942†γ	598§	n	Ovington, 1956 [m]
17	Larix decidua	SWITZ	50	n	492†γ§	n	Burger, 1945 [φ]
18	Larix decidua	SWITZ	105	n	426†γ§	n	Burger, 1945 [φ]
19	Larix decidua	SWITZ	220	n	76†γ§	n	Burger, 1945 [φ]
20	Larix eurolepis	UK	23	14776†γ	964§	n	Ovington, 1956 [m]
21	Larix leptolepis	JAP	21	n	1944§	n	Satoo, 1966 [r]
22	Larix leptolepis	UK	22	8601†γ	473§	n	Ovington, 1956 [m]
23	Larix leptolepis	UK	22	7845†γ	439§	n	Ovington, 1956 [m]
24	Nothofagus obliqua	UK	22	8077†γ	431§	n	Ovington, 1956 [m]
25	Picea abies	UK	20	21832†γ	1125§	n	Ovington, 1956 [m]
26	Picea abies	UK	47	13980†γ	745§	n	Ovington, 1956 [m]
27	Picea abies	UK	47	26273†γ	939§	n	Ovington, 1956 [m]
28	Picea abies	JAP	39	16870γ	n	n	Ovington, 1962 (Satoo, T., unpubl. data) [φ]
29	Picea abies	JAP	46	8820γ	n	n	Ovington, 1962 (Satoo, T., unpubl. data) [φ]
30	Picea abies	JAP	46	16920γ	n	n	Ovington, 1962 (Satoo, T., unpubl. data) [φ]
31	Picea abies	SWED	52	13220†γ	n	n	Ovington, 1962 (Tamm & Carbonnier, 1961) [φ]
32	Picea abies	SWITZ	98	n	820§	n	Burger, 1953 [φ]
33	Picea abies	JAP	37	n	1328γ§	n	Satoo, 1966 [r]
34	Picea abies	JAP	45	n	873γ§	n	Satoo, 1966 [r]
35	Picea abies	JAP	46	n	1277γ§	n	Satoo, 1966 [r]
36	Picea abies	JAP	46	n	1300γ§	n	Satoo, 1966 [r]
37	Picea abies	JAP	47	n	1401γ§	n	Satoo, 1966 [r]
38	Picea omorika	UK	21	29893†γ	1427§	n	Ovington, 1956 [m]
39	Pinus densiflora	JAP	13	n	1222γ§	n	Satoo, 1966 [r]
40	Pinus densiflora	JAP	13	n	1101γ§	n	Satoo, 1966 [r]
41	Pinus densiflora	JAP	13	n	1254γ§	n	Satoo, 1966 [r]
42	Pinus densiflora	JAP	13	n	1503γ§	n	Satoo, 1966 [r]
43	Pinus densiflora	JAP	16	5920γ	n	n	Ovington, 1962 (Satoo, 1955) [φ]
44	Pinus densiflora	JAP	16	10470γ	n	n	Ovington, 1962 (Satoo, 1955) [φ]
45	Pinus nigra	UK	18	17387†γ	1033§	n	Ovington, 1956 [m]
46	Pinus nigra	UK	22	12238†γ	587§	n	Ovington, 1956 [m]
47	Pinus nigra	UK	22	14187†γ	676§	n	Ovington, 1956 [m]

ENTRY NO.	SPECIES	LOC.	AGE	BIOMASS: TREES, SHRUBS, & HERBS g/m²	NET PRIMARY PRODUCTION: TREES, SHRUBS, & HERBS g/m²/yr	LEAF AREA INDEX m²/m²	REFERENCE*
PLANTATIONS							
48	Pinus nigra	UK	46	24220†γ	926§	n	Ovington, 1956 [m]
49	Pinus radiata	NZ	18	30531†	3920 - 4480ζ	n	Will, 1966 [φ]
50	Pinus radiata	AUSTL	3	630γ	375ρ	n	Forrest & Ovington, 1970 [r]
51	Pinus radiata	AUSTL	5	1040γ	720ρ	n	Forrest & Ovington, 1970 [r]
52	Pinus radiata	AUSTL	7	5250γ	2560ρ	n	Forrest & Ovington, 1970 [r]
53	Pinus radiata	AUSTL	9	7460γ	1270ρ	n	Forrest & Ovington- 1970 [r]
54	Pinus radiata	AUSTL	12	11920γ	1570ρ	n	Forrest & Ovington 1970 [r]
55	Pinus strobus	JAP	41	10470†γ	n	n	Ovington, 1962 (Senda & Satoo, 1956) [φ]
56	Pinus strobus	JAP	41	20420†γ	n	n	Ovington, 1962 (Senda & Satoo, 1956) [φ]
57	Pinus sylvestris	UK	47	15655†γ	787§	n	Ovington, 1956 [m]
58	Pinus sylvestris	UK	3	4†	2200§	-	Ovington, 1957 [m]
59	Pinus sylvestris	UK	7	746†	2200§	2.2	Ovington, 1957 [m]
60	Pinus sylvestris	UK	11	2598†	2200§	6.0	Ovington, 1957 [m]
61	Pinus sylvestris	UK	14	3331†	2200§	6.9	Ovington, 1957 [m]
62	Pinus sylvestris	UK	17	4811†	2200§	9.3	Ovington, 1957 [m]
63	Pinus sylvestris	UK	20	6535†	2200§	10.8	Ovington, 1957 [m]
64	Pinus sylvestris	UK	23	9170†	2200§	5.3	Ovington, 1957 [m]
65	Pinus sylvestris	UK	31	12720†	2200§	8.5	Ovington, 1957 [m]
66	Pinus sylvestris	UK	35	16336†	2200§	10.1	Ovington, 1957 [m]
67	Pinus sylvestris	UK	55	15072†	2200§	7.5	Ovington, 1957 [m]
68	Pinus sylvestris	UK	33	18591	n	n	Ovington & Madgwick, 1959a [s]
69	Pinus thunbergii	JAP	10	5651†γ	n	n	Ando, 1965 [d]
70	Pinus thunbergii	JAP	10	6777†γ	n	n	Ando, 1965 [d]
71	Pinus thunbergii	JAP	10	3783†γ	n	n	Ando, 1965 [d]
72	Pinus thunbergii	JAP	10	2044†γ	n	n	Ando, 1965 [d]
73	Pinus thunbergii	JAP	10	4409†γ	n	n	Ando, 1965 [d]
74	Pinus thunbergii	JAP	10	2262†γ	n	n	Ando, 1965 [d]
75	Pseudotsuga menziesii	USA	36	20554	999ζ	n	Cole, Gessel, & Dice, 1967 [e]
76	Pseudotsuga mensiesii	UK	21	11475†γ	720§	n	Ovington, 1956 [m]
77	Pseudotsuga mensiesii	UK	22	16821†γ	864§	n	Ovington, 1956 [m]
78	Pseudotsuga mensiesii	UK	22	18144†γ	924§	n	Ovington, 1956 [m]
79	Pseudotsuga mensiesii	UK	47	25242†γ	977§	n	Ovington, 1956 [m]
80	Quercus	UK	44	9238†γ	324§	n	Ovington, 1956 [m]
81	Quercus petraea	UK	21	4243†γ	202§	n	Ovington, 1956 [m]
82	Quercus robur	UK	47	12830†γ	394§	n	Ovington, 1956 [m]
83	Quercus rubra	UK	21	4306†γ	205§	n	Ovington, 1956 [m]
84	Tsuga heterophylla	UK	23	24960†γ	1294§	n	Ovington, 1956 [m]
85	Thuja plicata	UK	22	6832†γ	311§	n	Ovington, 1956 [m]
NATURAL STANDS (coniferous)							
86	Abies balsamea	CAN	40-50	13756†γ	944ρ	n	Baskerville, 1965,1966 [r]

180

ENTRY NO.	SPECIES	LOC.	AGE	BIOMASS: TREES, SHRUBS, & HERBS g/m²	NET PRIMARY PRODUCTION: TREES, SHRUBS, & HERBS g/m²/yr	LEAF AREA INDEX m²/m²	REFERENCE*
	NATURAL STANDS (coniferous)						
87	Abies balsamea	CAN	40-50	13301†γ	935ρ	n	Baskerville, 1965, 1966 [r]
88	Abies balsamea	CAN	40-50	14904†γ	962ρ	n	Baskerville, 1965, 1966 [r]
89	Abies balsamea	CAN	40-50	16702†γ	1058ρ	n	Baskerville, 1965, 1966 [r]
90	Abies balsamea	CAN	40-50	18272†γ	1164ρ	n	Baskerville, 1965, 1966 [r]
91	Abies balsamea	CAN	40-50	20009†γ	1258ρ	n	Baskerville, 1965, 1966 [r]
92	Abies fraseri	USA	n	21001γ	566ρ	n	Whittaker, 1966 [r]
93	Abies fraseri	USA	n	20010γ	653ρ	n	Whittaker, 1966 [r]
94	Abies spectabilis	NEPAL	n	50100	n	n	Yoda, 1968 [r]
95	Abies spectabilis	NEPAL	n	40500	n	n	Yoda, 1968 [r]
96	Abies spectabilis	NEPAL	n	42600	n	n	Yoda, 1968 [r]
97	Abies spectabilis	NEPAL	n	33800	n	n	Yoda, 1968 [r]
98	Abies-Tsuga	NEPAL	n	52000	n	n	Yoda, 1968 [r]
99	Picea	N.TAIGA USSR	n	10000	450§	n	Rodin & Bazilevich, 1967 [φ]
100	Picea	TAIGA USSR	n	26000	700§	n	Rodin & Bazilevich, 1967 [φ]
101	Picea	S.TAIGA USSR	n	33000	850§	n	Rodin & Bazilevich, 1967 [φ]
102	Picea abies	SWED	58	10860†γ	n	n	Ovington, 1962 (Tamm & Carbonnier, 1961) [m]
103	Picea mariana	CAN	65	9350γ	156ʃ	n	Weetman & Harlan, 1964 [r]
104	Picea - Abies	USA	n	34096Y	1024ρ	14.8	Whittaker, 1966 [r]
105	Picea - Abies	USA	n	31010Y	944ρ	n	Whittaker, 1966 [r]
106	Picea - Abies	USA	n	30000Y	1402ρ	n	Whittaker, 1966 [r]
107	Picea - Rhododendron	USA	n	32100Y	812ρ	n	Whittaker, 1966 [r]
108	Pinus	N.TAIGA USSR	n	8070	n	n	Rodin & Bazilevich, 1967 [φ]
109	Pinus	S.TAIGA USSR	n	28000	610§	n	Rodin & Bazilevich, 1967 [φ]
110	Pinus densiflora	JAP	15	6396	1578§	n	Satoo, 1967 [r]
111	Pinus densiflora	JAP	15	n	1486§	n	Satoo, 1966 [r]
112	Pinus echinata	USA	n	18140Y	875ρ	7.8	Whittaker, 1966 [r]
113	Pinus echinata	USA	n	13120Y	991ρ	n	Whittaker, 1966 [r]
114	Pinus nigra var. calabrica	UK	18	2576†γ	n	n	Wright & Will, 1958 [m]
115	Pinus nigra var. calabrica	UK	28	6832†γ	n	n	Wright & Will, 1958 [m]
116	Pinus nigra var. calabrica	UK	48	11200†γ	n	n	Wright & Will, 1958 [m]
117	Pinus sylvestris	UK	18	5488†γ	n	n	Wright & Will, 1958 [m]
118	Pinus sylvestris	UK	28	9408†γ	n	n	Wright & Will, 1958 [m]
119	Pinus sylvestris	UK	64	11872†γ	n	n	Wright & Will, 1958 [m]
120	Pinus sylvestris	UK	11	5163Y	n	n	Ovington, 1957 [m]
121	Pinus sylvestris	UK	14	4939Y	n	n	Ovington, 1957 [m]
122	Pinus sylvestris	UK	11	5163Y	n	n	Ovington, 1957 [m]
123	Pinus sylvestris	Uk	14	4939Y	n	n	Ovington, 1957 [m]

181

ENTRY NO.	SPECIES	LOC.	AGE	BIOMASS: TREES, SHRUBS, & HERBS g/m²	NET PRIMARY PRODUCTION: TREES, SHRUBS, & HERBS g/m²/yr	LEAF AREA INDEX m²/m²	REFERENCE*
NATURAL STANDS (coniferous)							
124	Pinus - Quercus	USA	n	10192	1189ρ	3.8	Whittaker & Woodwell, 1969a [r]
125	Pinus - Quercus	USA	n	18152γ	867ρ	7.8	Whittaker & Woodwell, 1969a [r]
126	Pinus - Quercus	USA	n	11370γ	442ρ	n	Whittaker & Woodwell, 1969a [r]
127	Pinus heath	USA	n	9199γ	561ρ	5.4	Whittaker & Woodwell, 1969a [r]
128	Pseudotsuga menziesii	USA	30	7500	n	n	Riekerk, 1967 (Heilman, 1961) [φ]
129	Pseudotsuga menziesii	USA	32	6000	n	n	Riekerk, 1967 (Heilman, 1961) [φ]
130	Pseudotsuga menziesii	USA	38	10800	n	n	Riekerk, 1967 (Heilman, 1961) [φ]
131	Pseudotsuga menziesii	USA	38	17800	n	n	Riekerk, 1967 (Heilman, 1961) [φ]
132	Pseudotsuga menziesii	USA	52	22800	n	n	Riekerk, 1967 (Heilman, 1961) [φ]
133	Pseudotsuga menziesii	USA	12	8000ψ	n	n	Riekerk, 1967 (Paddock, 1962) [φ]
134	Pseudotsuga menziesii	USA	28	15800ψ	n	n	Riekerk, 1967 (Paddock, 1962) [φ]
135	Pseudotsuga menziesii	USA	30	22400ψ	n	n	Riekerk, 1967 (Paddock, 1962) [φ]
136	Pseudotsuga menziesii	USA	39	23500ψ	n	n	Riekerk, 1967 (Paddock, 1962) [φ]
137	Pseudotsuga menziesii	USA	75	26400ψ	n	n	Riekerk, 1967 (Paddock, 1962) [φ]
138	Pygmy Conifer-Quercus	USA	n	1878γ	183ρ	n	Whittaker & Woodwell, 1969a [r]
139	Tsuga canadensis	USA	n	61006γ	1183ρ	n	Whittaker, 1966 [r]
140	Tsuga - Fagus	USA	n	19300γ	1333ρ	n	Whittaker, 1966 [r]
141	Tsuga - Rhododendron	USA	n	51100γ	1022ρ	13.1	Whittaker, 1966 [r]
142	Tsuga dumosa	NEPAL	n	64000	n	n	Yoda, 1968 [r]
143	Tsuga - Quercus	NEPAL	n	68200	n	n	Yoda, 1968 [r]
144	Cove Forests	USA	n	50065γ	1051ρ	6.2	Whittaker & Woodwell, 1969a [r]
NATURAL STANDS (temperate broad-leaved evergreen)							
145	Acacia harpophylla	AUSTRL	n	15950†	n	n	Moore, et al., 1967 [e]
146	Acacia mollissima	JAP	4	8060†	3530ζ	9.7	Tadaki, 1965b [r]
147	Camellia japonica	JAP	70	18500†γ	1500ζ	6.2	Kan, et al., 1965 [r]
148	Camellia japonica	JAP	70	15200†γ	1400ζ	5.6	Kan, et al., 1965 [r]
149	Camellia japonica	JAP	70	17700†γ	1600ζ	6.8	Kan, et al., 1965 [r]
150	Castanopsis cuspidata	JAP	11	6620	1870ζγ	8.0	Tadaki, 1965a [r,e]
151	Castanopsis cuspidata	JAP	40	23900†γ	3100ζ	9.7	Kan, et al., 1965 [r]
152	Castanopsis cuspidata	JAP	40	19800†γ	2600ζ	8.5	Kan, et al., 1965 [r]

182

ENTRY NO.	SPECIES	LOC.	AGE	BIOMASS: TREES, SHRUBS, & HERBS g/m^2	NET PRIMARY PRODUCTION: TREES, SHRUBS, & HERBS g/m^2/yr	LEAF AREA INDEX m^2/m^2	REFERENCE*
NATURAL STANDS (temperate broad-leaved evergreen)							
153	Castanopsis cuspidata	JAP	12	4700†γ	1800ζ	5.2	Kan, et al., 1965 [r]
154	Castanopsis cuspidata	JAP	12	6600†γ	2600ζ	7.2	Kan, et al., 1965 [r]
155	Ilex aquifolium	UK	82	21100γ	1540§	6.6	Peterken & Newbould, 1966 [s]
156	Ilex aquifolium	UK	100	13100γ	970§	5.6	Peterken & Newbould, 1966 [s]
157	Ilex aquifolium	UK	92	3800γ	220§	5.7	Peterken & Newbould, 1966 [s]
158	Ilex aquifolium	UK	94	7200γ	380§	4.9	Peterken & Newbould, 1966 [s]
159	Ilex aquifolium	UK	80	6000γ	360§	5.3	Peterken & Newbould, 1966 [s]
160	Ilex opaca	USA	150	17830	1075ρ	5.9	Art, unpubl. data [r]
161	Quercus - Camellia	JAP	70	22800†γ	2000ζ	7.5	Kan, et al., 1965 [r]
162	Quercus - Camellia	JAP	70	21200†γ	1800ζ	7.0	Kan, et al., 1965 [r]
163	Quercus - Camellia	JAP	70	19800†γ	1800ζ	6.2	Kan, et al., 1965 [r]
164	Quercus - Cinnamomum	NEPAL	n	57500	n	6.9	Yoda, 1968 [r]
165	Quercus - Machilus	NEPAL	n	54900	n	5.6	Yoda, 1968 [r]
166	Quercus - Rapanaea	JAP	80	21800†γ	2300ζ	6.9	Yoda, 1965 [r]
167	Quercus - Rapanaea	JAP	80	26500†γ	2800ζ	9.4	Yoda, 1965 [r]
168	Quercus - Rapanaea	JAP	80	26400†γ	2500ζ	7.7	Kan, et al., 1965 [r]
NATURAL STANDS (temperate deciduous)							
169	Acer spicatum	CAN	1	74†γ	74ʃ	n	Post, 1970 [r]
170	Acer spicatum	CAN	3	239†γ	80ʃ	n	Post, 1970 [r]
171	Acer spicatum	CAN	8	1293†γ	162ʃ	n	Post, 1970 [r]
172	Acer spicatum	CAN	11	1587†γ	144ʃ	n	Post, 1970 [r]
173	Acer spicatum	CAN	13	2319†γ	178ʃ	n	Post, 1970 [r]
174	Acer spicatum	CAN	16	2553†γ	160ʃ	n	Post, 1970 [r]
175	Acer spicatum	CAN	18	2241†γ	124ʃ	n	Post, 1970 [r]
176	Acer spicatum	CAN	21	3350†γ	159ʃ	n	Post, 1970 [r]
177	Acer spicatum	CAN	23	3128†γ	136ʃ	n	Post, 1970 [r]
178	Acer spicatum	CAN	26	4045†γ	156ʃ	n	Post, 1970 [r]
179	Alnus rubra	USA	50	25000†γ	n	n	Zavitovski, 1970[φ]
180	Alnus rubra	USA	10-15	n	2600§	n	Zavitovski, 1970[φ]
181	Betula	USSR	n	22000	1200§	n	Rodin & Bazilevich, 1967 [φ]
182	Betula maximo-wicziniana	JAP	44	n	607γ	n	Satoo, 1966 [r]
183	Betula maximo-wicziniana	JAP	47	n	751γ	n	Satoo, 1966 [r]
184	Betula maximo-wicziniana	JAP	47	n	632γ	n	Satoo, 1966 [r]
185	Betula maximo-wicziniana	JAP	47	9060γ	n	n	Ovington, 1962 (Satoo, unpubl. data)[φ]
186	Betula maximo-wicziniana	JAP	47	14300γ	n	n	Ovington, 1962 (Satoo, unpubl. data)[φ]

183

ENTRY NO.	SPECIES	LOC.	AGE	BIOMASS: TREES, SHRUBS, & HERBS g/m^2	NET PRIMARY PRODUCTION: TREES, SHRUBS, & HERBS $g/m^2/yr$	LEAF AREA INDEX m^2/m^2	REFERENCE*
NATURAL STANDS (temperate deciduous)							
187	Betula maximo- wicziniana	JAP	47	11700Y	n	n	Ovington, 1962 (Satoo unpubl. data)[φ]
188	Betula verrucosa	USSR	20	8030†	n	n	Ovington, 1962 (Smirnova & Gorodentseva, 1958)[φ]
189	Betula verrucosa	USSR	40	24760†	n	n	Ovington, 1962 (Smirnova & Gorodentseva, 1958)[φ]
190	Betula verrucosa	USSR	67	21390†	n	n	Ovington, 1962 (Smirnova & Gorodentseva, 1958)[φ]
191	Betula verrucosa	SWED	25	4740†Y	n	n	Ovington, 1962 (Tamm)[φ]
192	Betula verrucosa	UK	6	170	n	n	Ovington & Madgwick, 1959b [m]
193	Betula verrucosa	UK	24	7990	n	n	Ovington & Madgwick, 1959b [m]
194	Betula verrucosa	UK	27	7830Y	n	n	Ovington & Madgwick, 1959b [m]
195	Betula verrucosa	UK	32	6380Y	n	n	Ovington & Madgwick, 1959b [m]
196	Betula verrucosa	UK	38	7050Y	n	n	Ovington & Madgwick, 1959b [m]
197	Betula verrucosa	UK	42	9450	n	n	Ovington & Madgwick, 1959b [m]
198	Betula verrucosa	UK	46	12080Y	n	n	Ovington & Madgwick, 1959b [m]
199	Betula verrucosa	UK	53	17270	n	n	Ovington & Madgwick, 1959b [m]
200	Betula verrucosa	UK	55	21380	n	n	Ovington & Madgwick, 1959b [m]
201	Fagus	EUROPE	n	37000	1300§	n	Rodin & Bazilevich, 1967[φ]
202	Fagus grandifolia	USA	n	13001Y	668ρ	2.9	Whittaker, 1966 [r]
203	Fagus grandifolia	USA	n	17010Y	906ρ	n	Whittaker, 1966 [r]
204	Liriodendron tulipifera	USA	n	22020Y	2408ρ	n	Whittaker, 1966 [r]
205	Nothofagus truncata	NZ	110	30870†Y	n	n	Ovington, 1962 (Satoo, unpubl. data)[φ]
206	Populus david- iana	JAP	40	11860Y	n	n	Ovington, 1962 (Satoo, et al., 1956)[φ]
207	Populus david- iana	JAP	40	n	1237Y	n	Satoo, 1966 [r]
208	Populus tremu- loides	USA	41	20401Y	n	n	Bray & Dudkiewicz, 1963 [c]
209	Populus tremu- loides & gran- identata	USA	38	5953Y	n	n	Bray & Dudkiewicz, 1963 [c]
210	Prunus pensyl- vanica	USA	4	2398	1309†ζ	n	Marks, unpubl. data [r]
211	Prunus pensyl- vanica - Pop - ulus tremuloides	USA	6	3905	1658†ζ	6.1	Marks, unpubl. data [r]
212	Prunus pensyl- vanica	USA	14	6812	1264†ζ	5.4	Marks, unpubl. data [r]

ENTRY NO.	SPECIES	LOC.	AGE	BIOMASS: TREES, SHRUBS, & HERBS g/m^2	NET PRIMARY PRODUCTION: TREES, SHRUBS, & HERBS $g/m^2/yr$	LEAF AREA INDEX m^2/m^2	REFERENCE*
NATURAL STANDS (temperate deciduous)							
213	Prunus pensyl-vanica	USA	1	162	102[†]ζ	1.1	Marks, unpubl. data [r]
214	Quercus (closed)	USA	n	n	730ρ	n	Bray, 1962 [c]
215	Quercus (open)	USA	n	n	440ρ	n	Bray, 1962 [c]
216	Quercus (savan.)	USA	n	n	340ρ	n	Bray, 1962 [c]
217	Quercus	USSR	n	40000	900§	n	Rodin & Bazilevich, 1967 [φ]
218	Quercus borealis	USA	n	13700γ	828ρ	n	Whittaker, 1966 [r]
219	Quercus robur	SWED	125-190	24000	1560ρ	n	Andersson, 1970 [r]
220	Quercus prinus	USA	n	42250γ	1465ρ	6.3	Whittaker, 1966 [r]
221	Quercus – Carya	USA	n	15700γ	542[†]ζ	4.0	Monk, et al., 1970 [r]
222	Quercus – Carya	USA	n	37005γ	1203ρ	n	Monk, et al., 1970 [r]
223	Quercus – Fagus	BELG	70-75	15600	1440	6.8	Duvigneaud & Denaeyer – DeSmet, 1970 [r]
224	Quercus – Fraxinus	BELG	115-160	38000	1430	n	Duvigneaud & Denaeyer – DeSmet, 1970 [r]
225	Quercus (mixed)	USA	n	8770γ	568ρ	n	Whittaker, 1966 [r]
NATURAL STANDS (tropical)							
226	Mixed Dipterocarp	THAI	n	6690γ	n	n	Ovington, 1962 (Owaga, et al., 1961) [φ]
227	Evergreen seasonal	CAMB	n	41500	n	7.4	Hozumi, et al., 1969 [r]
228	Evergreen seasonal	CAMB	n	34800	n	7.3	Hozumi, et al., 1969 [r]
229	Melaleucia leucadendron	CAMB	n	17166	n	7.1	Hozumi, et al., 1969 [r]
230	Evergreen Gallery	THAI	n	38370γ	n	n	Ovington, 1962 (Owago, et al., 1961) [φ]
231	Tropical Forest	GHANA	50	38974	n	n	Greenland & Kowal, 1960 [γ]
232	Tropical Forest	CONGO	18	19154	n	n	Greenland & Kowal, 1960 [φ]
233	Tropical Rain Forest	THAI	n	36500	2850ζ	12.3	Kira, et al., 1964, [e,r]
234	Tropical Rain Forest	THAI	n	40400[†]γ	n	10.7	Ogawa, et al., 1965 [e]
235	Tropical Rain Forest	THAI	n	32400[†]	2860ρ	11.4	Kira, et al., 1967 [e]
STANDS OF UNCERTAIN ORIGIN							
236	Abies balsamea	USA	48	7749[†]	n	n	Young & Carpenter, 1967 [c]
237	Picea abies	USSR	24	10120	n	n	Ovington, 1962 (Sonn, 1960) [φ]
238	Picea abies	USSR	38	16080[†]	n	n	Ovington, 1962 (Sonn, 1960) [φ]
239	Picea abies	USSR	60	27140[†]	n	n	Ovington, 1962 (Sonn, 1960) [φ]

185

ENTRY NO.	SPECIES	LOC.	AGE	BIOMASS: TREES, SHRUBS, & HERBS g/m^2	NET PRIMARY PRODUCTION: TREES, SHRUBS, & HERBS $g/m^2/yr.$	LEAF AREA INDEX m^2/m^2	REFERENCE*
STANDS OF UNCERTAIN ORIGIN							
240	Picea abies	USSR	93	32490[†]	n	n	Ovington, 1962 (Sonn, 1960)[φ]
241	Picea rubra	USA	67	9346[†]	n	n	Young & Carpenter, 1967 [c]
242	Pinus strobus	USA	34	4849[†]	n	n	Young & Carpenter, 1967 [c]
243	Thuja occidentalis	USA	90	13451[†]	n	n	Young & Carpenter, 1967 [c]
244	Tsuga canadensis	USA	27	3124[†]	n	n	Young & Carpenter, 1967 [c]
245	Acer rubrum	USA	44	1523[†]	n	n	Young & Carpenter 1967 [c]
246	Betula papyrifera	USA	32	11808[†]	n	n	Young & Carpenter, 1967 [c]
247	Fagus	NETH?	46	n	1350[ψ][§]	n	Lieth,1962 (Moller, et al., 1954)[φ]
248	Populus	USA	13	275[†]	n	n	Young & Carpenter, 1967 [c]
249	Populus	USA	28	9410[†]	n	n	Young & Carpenter, 1967 [c]
250	Quercus	USSR	22	9020[†]	n	n	Ovington, 1962 (Sonn. 1960)[φ]
251	Quercus	USSR	42	16980[†]	n	n	Ovington, 1962 (Sonn, 1960)[φ]
252	Quercus	USSR	56	23210[†]	n	n	Ovington, 1962 (Sonn, 1960)[φ]
253	Quercus	USSR	200	44970[†]	n	n	Ovington, 1962 (Sonn, 1960)[φ]
FOREST TYPE AVERAGES							
254	Boreal Forest	n	n	20000	800	n	Whittaker, 1970
255	Beech	n	n	18450	n	n	Ovington, 1965
256	Temperate Betula	n	n	14840	n	n	Ovington, 1965
257	Temperate Picea	n	n	16480	n	n	Ovington, 1965
258	Temperate Pinus	n	n	16200	n	n	Ovington, 1965
259	Pseudotsuga menziesii	n	n	13910	n	n	Ovington, 1965
260	Temperate coniferous (probable maximum	n	n	n	2800 ±25%	n	Westlake, 1963
261	Temperate Quercus	n	n	17460	n	n	Ovington, 1965
262	Temperate Deciduous (probale maximum)	n	n	n	1200 ±25%	n	Westlake, 1963
263	Temperate Forest	n	n	30000	1300	n	Whittaker, 1970
264	Sub-tropical deciduous	n	n	41000	2450	n	Rodin & Bazilevich, 1967
265	Tropical Forest	n	n	45000	2000	n	Whittaker, 1970
266	Tropical Rain Forest	n	n	750000	3250	n	Rodin & Bazilevich, 1967 [φ]
267	Tropical Rain Forest	n	n	27090	n	n	Ovington, 1965
268	Tropical Rain Forest (probable maximum)	n	n	n	5000 ±20%	n	Westlake, 1963

KEY TO TABLE

† = Trees only
γ = Above ground mass only
ψ = Unknown as to whether below ground
    mass included

ρ = Current periodic annual increment
∫ = Mean annaul increment
ζ = Current annual increment
    (net annual production)

§ = Method of estimating production not
    specified

m = Mean tree technique
s = Stratified tree technique
r = Regression estimation technique
e = Unit area method
c = Canopy-area method
d = Basal area proportion method
φ = Not clear how individual tree data
    translated to land area basis

n = Data not given
* = When review paper consulted, original
    work is given is parentheses

LITERATURE CITED

Andersson, F., 1970, Ecological studies in a Scanian woodland
    and meadow area, southern Sweden. II. Plant biomass,
    primary production and turnover of organic matter.
    Botaniska Notiser 123: 8-51.

Ando, T., 1965, Estimation of dry-matter and growth analysis of
    the young stand of Japanese black pine (Pinus thunbergii).
    Adv. Front. Plnt. Sci. 10: 1-10.

Baskerville, G.L., 1965a, Dry matter production in immature fir
    stands. For. Sci. Monog. 9: 362-478.

Baskerville, G.L., 1965b, Estimation of dry weight of tree com-
    ponents and total standing crop in conifer stands. Ecol.
    46: 867-869.

Bray, J.R., 1962, The primary productivity of vegetation in
    central Minnesota, U.S.A., and its relationship to chloro-
    phyll content and albedo. pp. 102-108. In: Lieth, H.,
    ed., Die Stoffproduktion der Pflanzendecke. Fischer-Ver-
    lag. Stuttgart.

Bray, J.R., & L.A. Dudkiewicz, 1963, The composition, biomass
    and productivity of two Populus forests. Bull. Torr. Bot.
    Club 90: 298-308.

Burger, H., 1940, Holz, Blattmenge, und Zuwachs. Iv. Ein 80
    jahriger Buchenbestand. Mitt. Schweiz. Anst. forstl.
    Versuchsw. 21: 307-348.

Burger, H., 1945, Holz, Blattmenge, und Zuwachs. VII. Die Larche.
    Mitt. Schweiz. Anst. forstl. Versuchsw. 24 (1): 7-103.

Burger, H., 1953, Holz, Blattmenge, und Zuwachs. XIII. Fichten
    im gleichalterigen Hochwald. Mitt. Schweiz. Anst. forstl.
    Versuchsw. 29 (1): 38-130.

Cole, D.W., S.P. Gessel, & S.F. Dice, 1967, Distribution and
    cycling of nitrogen, phosphorous, potassium and calcium in
    a second-growth Douglas fir ecosystem. pp. 197-213. In:
    Symposium on primary productivity and mineral cycling in
    natural ecosystems. AAAS.

Duvigneaud, P. & S. Denaeyer-DeSmet, 1970, Biological cycling
    of minerals in temperate deciduous forests. pp. 199-225.
    In: Analysis of temperate forest ecosystems. ed. D.E. Reichle.
    Springer-Verlag.

187

Forrest, W.G., & J.D. Ovington, 1970, Organic matter changes in an age series of Pinus radiata plantations. <u>J. Appl. Ecol.</u> <u>7</u>: 177-186.

Greenland, D.J., & J.M.L. Kowal, 1960, Nutrient content of the moist tropical forest of Ghana. <u>Plant and Soil</u> <u>12</u>: 154-174.

Hozumi, K., Y. Koyji, S. Kokawa, & T. Kira, 1969, Production ecology of tropical rain forests in southwest Cambodia. I. Plant biomass. <u>Nature and Life in S.E. Asia</u> <u>6</u>: 1-51.

Kan, M., H. Saito, & T. Shidei, 1965, Studies of the productivity of evergreen broad leaved forests. <u>Bull. Koyoto Univ. For.</u> <u>37</u>: 55-75. (In Japanese)

Keay, J., & A.G. Turton, 1970, Distribution of bio-mass and major nutrients in a maritime pin plantation. <u>Austrl. For.</u> <u>34</u>: 39-48.

Kimura, M., 1960, Primary production of the warm temperate laurel forest in the southern part of Osumi Peninsula, Kyushu, Japan. <u>Misc. Rep. Res. Inst. Nat. Resour.</u> <u>52-53</u>: 36-47.

Kira, T., H. Ogawa, K. Yoda, & K. Ogino, 1964, Primary production by a tropical rain forest of southern Thailand. <u>Bot. Mag.</u> <u>77</u>: 428-429.

Kira, T., H. Ogawa, K. Yoda, & K. Ogino, 1967, Comparitive ecological studies on three main types of forest vegetation in Thailand. IV. Dry matter production, with special reference to the Khao Chong rain forest. <u>Nature and Life in S.E. Asia</u> <u>5</u>: 149-174.

Kira, T., & T. Shidei, 1967, Primary production and turnover of organic matter in different forest ecosystems of the Western Pacific. <u>Jap. J. Ecol.</u> <u>17</u>: 70-87.

Lieth, H., 1962, Abschnitt 4, Stoffproduktionsdaten. <u>Die Stoffproduktion der Pflanzendecke.</u> Fischer-Verlag, Stuttgart. pp. 117-133.

Monk, C.D., G.I. Child, & S.A. Nicholson, 1970, Biomass, litter and leaf surface area estimates of an oak-hickory forest. <u>Oikos</u> <u>21</u>: 138-141.

Moore, A.W., J.S. Russell, & J.E. Coaldrake, 1967, Dry matter and nutrient content of a subtropical forest of <u>Acacia harpophylla</u>. F. Muell. (Brigalow). <u>Austrl. J. Bot.</u> <u>15</u>: 11-24.

Newbould, P.J., 1967, <u>Methods for Estimating the Primary Production of Forests</u>. International Biological Programme, London (Blackwell Sci. Publ.) 62 pp.

Newbould, P.J., 1968, Methods of estimating root production, pp. 187-190. In: Eckhardt, F.E.,ed., <u>Fuctioning of Terrestrial ecosystems at the Primary Production Level</u>. (UNESCO Copenhagen Symposium).

Ogawa, H., K. Yoda, K. Ogino, & T. Kira, 1965, Comparative ecological studies on three main types of forest vegetation in Thailand. II. Plant biomass. <u>Nature and Life in S.E. Asia</u> <u>4</u>: 49-80.

Ovington, J.D., 1956, The form weights and productivity of tree species grown in close stands. <u>New Phytol.</u> <u>55</u>: 289-304.

Ovington, J.D., 1957, Dry matter production by <u>Pinus sylvestris</u>. <u>Ann. Bot.</u> <u>4</u>: 5-58.

Ovington, J.D., 1962, Quantitative ecology and the woodland ecosystem concept. Adv. Ecol. Res. 1: 103-192.

Ovington, J.D., 1965, Organic production, turnover and mineral cycling in woodlands. Biol. Rev. 40: 295-336.

Ovington, J.D., W.G. Forrest, & J.S. Armstrong, 1967, Tree biomass estimation. pp. 4-31. In: Young, H.E., ed., Symposium on Primary Productivity and Mineral Cycling in Natural Ecosystems. (Ecol. Soc. Amer.)

Ovington, J.D., & H.A.I. Madgwick, 1959a, Distribution of organic matter and plant nutrients in a plantation of Scots pine. For. Sci. 5: 344-355.

Ovington, J.D., & H.A.I. Madgwick, 1959b, The growth and composition of natural stands of birch. I. Dry-matter production. Plant and Soil 10: 271-283.

Peterken, G. F. and P. S. Newbould, 1966, Dry matter production by Ilex aquifolium L. in the New Forest. J. Ecol. 54: 143-150.

Post, L. J., 1970, Dry-matter production of mountain maple and balsam fir in northwestern New Brunswick. Ecol. 51: 548-550.

Riekerk, H., 1967, The movement of phosphorus, potassium and calcium in a Douglas-fir forest ecosystem. Ph.D. Dissert. Univ. Wash. 142 pp.

Rodin, L. E. and N. I. Bazilevich, 1967, Production and Mineral Cycling in Terrestrial Vegetation. Oliver and Boyd. Edinburgh. 288 pp.

Satoo, T., 1966, Production and distribution of dry matter in forest ecosystems. Misc. Info. Tok. U. For. 16: 1-15.

Satoo, T., 1967, Primary production relations in woodlands of Pinus densiflora. pp.52-80. In: Symposium on primary productivity and mineral cycling in natural ecosystems. AAAS.

Satoo, T., 1968, Material for the studies of growth in stands VII. Primary production and distribution of produced dry matter in a plantation of Cinnamomum camphora. Bull. Tokyo U. For. 64: 241-275.

Satoo, T., 1970, A synthesis of studies by the harvest method: Primary production relations in the temperate deciduous forests of Japan. pp 55-72. In: Reichle, D. E. ed. Analysis of Temperate Forest Ecosystems, Springer-Verlag, N.Y.

Tadaki, T., 1965a, Studies on production structure of forests (VII). The primary production of a young stand of Castanopsis cuspidata. Jap. J. Ecol. 15: 142-147.

Tadaki, Y., 1965b, Studies on production structure of forest. VIII. Productivity of an Acacia mollissima stand in higher stand density. J.Jap. For. Soc. 47: 384-391. (In Japanese)

Tadaki, Y. and Y. Kawasaki, 1966, Studies on the production structure of forest. IX. Primary productivity of a young Cryptomeria plantation with excessively high stand density. J. Jap.For. Soc. 48: 55-61.

Tadaki, Y., N. Ogata & Y. Nagatomo, 1963, Studies on production structure of forest. V. Some analyses on productivities of artificial stand of Acacia mollissima. J.Jap.For. Soc. 45: 293-301. (In Japanese).

Tadaki, Y., N. Ogata & T. Takagi, 1962, Studies on production structure of forest. III. Estimation of standing crop and some analyses on productivity of young stands of Castanopsis cuspidata. J.Jap.For.Soc. 44: 350-359 (In Japanese).

Tadaki, Y., T. Shidei, T. Sakasegwa & K. Ogino, 1961, Studies on productive structure of forest. II. Estimation of standing crop and some analyses on productivity of young birch stand (Betula platyphylla). J.Jap. For. Soc. 43: 19-26 (In Japanese).

Weetman, G.F. & R. Harland, 1964, Foliage and wood production in unthinned black spruce in northern Quebec. For.Sci. 10: 80-88.

Westlake, D.F., 1963, Comparisons of plant productivity. Biol. Rev. 38: 385-425.

Whittaker, R.H., 1961, Estimation of net primary production of forest and shrub communities. Ecol. 42: 177-180.

Whittaker, R.H., 1962, Net production relations of shrubs in the Great Smokey Mountains. Ecol. 43: 357-377.

Whittaker, R.H., 1965, Branch dimensions and estimation of branch production. Ecol. 46:365-370.

Whittaker, R.H., 1966, Forest dimensions and production in the Great Smokey Mountains. Ecol. 47: 103-121.

Whittaker, R.H. & G.M. Woodwell, 1968, Dimensions and production relations of trees and shrubs in the Brookhaven Forest, New York. J.Ecol. 56: 1-25.

Whittaker, R.H. & G.M. Woodwell, 1969a, Structure, production and diversity of the oak-pine forest at Brookhaven, New York. J. Ecol. 57: 155-174.

Whittaker, R.H. & G.M. Woodwell, 1969b, Measurement of net primary production of forests. 43pp. In: Conference on Forest Productivity of the World. International Biological Program. Bruxelles, Belgium, October 27-31, 1969.

Whittaker, R.H., 1970, Communities and Ecosystems. Macmillan, London. 162pp.

Will, G.M., 1966, Root growth and dry matter production in a high producing stand of Pinus radiata. N.Z. For. Res. Note 44: 15pp.

Wright, T.W. & G.M. Will, 1958, The nutrient content of Scots and Corsican pines growing on sand dunes. Forestry 31: 13-25.

Yòda, K., 1968, A preliminary survey of forest vegetation of eastern Nepal. II. Plant biomass in the sample plots chosen from different vegetation zones. J.Coll. Arts & Sci. Chiba (Nat.Sci.Ser.) 5: 277-302.

Young, H.E. & P.M. Carpenter, 1967, Weight, nutrient element and productivity studies of seedlings and saplings of eight tree species in natural ecosystems. Me.Agr.Exp.Sta. Tech. Bull. 28: 39pp.

Zavitovski, J. 1970, Biomass and primary productivity of red alder (Alnus rubra Bong.) ecosystem. I. Tree layer. Bull. Ecol. Soc.Amer. 51(2):30-31.

ENTRY NO.	SPECIES	LOC.	AGE	BIOMASS: TREES, SHRUBS, & HERBS g/m^2	NET PRIMARY PRODUCTION: TREES, SHRUBS, & HERBS $g/m^2/yr$	LEAF AREA INDEX m^2/m^2	REFERENCE *

ADDENDA

PLANTATIONS

269	Acacia mollissima	JAP	3	4401γ†	2549ζ	n	Tadaki, et al., 1963 [r]
270	Acacia mollissima	JAP	3	3671γ†	2129ζ	n	Tadaki, et al., 1963 [r]
271	Acacia mollissima	JAP	5	9386γ†	3168ζ	n	Tadaki, et al., 1963 [r]
272	Acacia mollissima	JAP	5	12100γ†	4031ζ	n	Tadaki, et al., 1963 [r]
273	Acacia mollissima	JAP	5	7172γ†	2383ζ	n	Tadaki, et al., 1963 [r]
274	Acacia mollissima	JAP	7	7693γ†	2408ζ	n	Tadaki, et al., 1963 [r]
275	Acacia mollissima	JAP	7	10023γ†	3113ζ	n	Tadaki, et al., 1963 [r]

NATURAL STANDS (coniferous)

276	Abies - Rhododendron	NEPAL	n	19800	n	n	Yoda, 1968 [r]

NATURAL STANDS (temperate broad-leaved evergreen)

277	Temperate Evergreen	THAI	n	23530	n	n	Ovington, 1962 (Ogawa, et al., 1961)[φ]
278	Distylium racemosum	JAP	n	22340γ	1877ρ	8.75	Kimura, 1960 [r]

NATURAL STANDS (temperate deciduous)

279	Betula platyphylla	JAP	10	1900-2200γ	500ζ	3.5	Tadaki, et al, 1961 [r]

STANDS OF UNCERTAIN ORIGIN

280	Castinopsis cuspidata	JAP	10	4260γ†	1650ζ	12.5	Tadaki, et al., 1962 [r]

PLANTATIONS

281	Pinus pinaster	AUSTL	14	10080 †γ	720ƒ	14.6	Keay & Turton, 1970 [s,r]
282	Pinus pinaster	AUSTL	14	12260 †γ	880ƒ	17.8	Keay & Turton, 1970 [s,r]
283	Cryptomeria japonica	JAP	34	17640	960ζ	9.4	Tadaki, et al., 1965 [d,r]
284	Cryptomeria japonica	JAP	34	18250	1400ζ	12.1	Tadaki, et al., 1965 [d,r]
285	Cryptomeria japonica	JAP	34	19170	1600ζ	13.0	Tadaki, et al., 1965 [d,r]
286	Cryptomeria japonica	JAP	24	17050	1510ζ	14.7	Tadaki, et al., 1965 [d.r]

191

ENTRY NO.	SPECIES	LOC.	AGE	BIOMASS: TREES, SHRUBS, & HERBS g/m^2	NET PRIMARY PRODUCTION: TREES, SHRUBS, & HERBS $g/m^2/yr$	LEAF AREA INDEX m^2/m^2	REFERENCE*

ADDENDA

PLANTATIONS

287	Cryptomeria japonica	JAP	34	15400	1230ζ	11.5	Tadaki, et al., 1965 [d,r]
288	Cryptomeria japonica	JAP	49	17450	890ζ	10.0	Tadaki, et al., 1965 [d,r]
289	Cryptomeria japonica	JAP	11	6360	970ζ	10.7	Tadaki, et al., 1965 [d,r]
290	Cryptomeria japonica	JAP	22	13570	1400ζ	11.8	Tadaki, et al., 1965 [d,r]
291	Cryptomeria japonica	JAP	31	15980	1670ζ	13.5	Tadaki, et al., 1965 [d,r]

ADDITIONAL LITERATURE CITED

Tadaki, Y., N. Ogata & Y. Nagatomo, 1965, The dry matter productivity in several stands of Cryptomeria Japonica in Kyushu. Bull. Gout. For. Exp. Sta. 173: 45-66. (In Japanese).

22

Reprinted with permission from pages 2, 3, 4-6, and 10-12 of *Productivity of the World's Main Ecosystems*, edited by D. E. Reichle, J. F. Franklin, and D. W. Goodall, Washington, D.C.: National Academy of Sciences, 1975, 166pp.

PRODUCTIVITY OF TUNDRA ECOSYSTEMS

F. E. Wielgolaski

[*Editor's Note:* In the original, material precedes this excerpt.]

PRIMARY PRODUCTION

Most work on tundra ecosystems concerns biomass and production of plants, especially the aboveground biomass of higher plants (Bliss, 1962, 1966, 1970; and Andreev, 1966; Khodachek, 1969; Alexandrova, 1970; Wielgolaski, 1972; Bliss and Wielgolaski, 1973).

Biomass

The biomass of higher plants may be very small in zones classified as polar desert by Alexandrova (1970), e.g., 6 g/m² in moss-lichen polygons at Franz Josef Land (Table 1). Low biomass values are also found in other areas with only patchy vegetation, while higher values, 15-50 g/m², may be found in *Salix herbacea* snow beds in Norway at 1,300 m above sea level and 60°N latitude [Wielgolaski, 1972, in press (a)].

Even in the polar deserts the amount of cryptogams may be rather high. Alexandrova (1970) reports 123 g/m², while Andreev (1966) reports only 9 g/m² in polar semideserts, a northern variant of arctic tundra with more than 50 percent bare soil. In the Norwegian snow bed mentioned above the cryptogam biomass was ~40 g/m². The highest amounts of living aboveground biomass of cryptogams reaches 1,345 g/m² (Alexandrova, 1970) in shrub tundra, 800 g/m² for a polygonal bog (Khodachek, 1969; Shamurin *et al.*, 1972), and about 400 g/m² for a flat palsa bog (Pospelova, 1972).

When both above- and belowground biomass of phanerogams are summed with cryptogams (Alexandrova, 1970) a total biomass of ~150 g/m² occurs in the polar desert and ~200 g/m² in alpine snow beds in Norway. This suggests that even in closed vegetation in alpine areas, at a low latitude the total amount of biomass may be nearly as low as in polar deserts at a high latitude.

[*Editor's Note:* Material has been omitted at this point.]

193

TABLE 1 Total Live Aboveground Vascular Plant Biomass (g/m^2 dry weight) at the Time of Maximum Aboveground Biomass

Country	Area	Site Types	g/m^2	References
U.S.S.R.	Franz Josef Land	Polar desert	6	Alexandrova, 1970
Norway	Hardangervidda	Alpine snow bed	15–50	Wielgolaski, 1972, in press (a)
Canada	Devon Island	Plateau	20	Bliss, 1972
Norway	Hardangervidda	Lichen heath	30–60	Wielgolaski, 1972, in press (a)
U.S.S.R.	Northeastern Europe	Polar semidesert	40	Andreev, 1966
Canada	Devon Island	Sedge meadow	60–90	Muc, 1973
U.S.S.R.	New Siberian Island	Arctic tundra	71	Alexandrova, 1970
Canada	Devon Island	Raised beach ridge	90–130	Svoboda, 1973
Norway	Hardangervidda	Dry and wet meadow	90–140	Wielgolaski, 1972, in press (b)
U.S.S.R.	Taimyr	Moss-*Dryas*-sedge tundra	100	Shamurin et al., 1972
U.S.A.	Point Barrow	Wet sedge meadow	100	Tieszen, 1972
Sweden	Abisko	Bog	110	Rosswall, 1972
U.S.S.R.	Kola	Spotted eutrophic alpine tundra	110	Chepurko, 1972
Finland	Kevo	Subalpine heath	182	Kallio and Kärenlampi, 1971
U.S.S.R.	Western Taimyr	Dwarf shrub-sedge moss-lichen tundra	188	Pospelova, 1972
U.S.S.R.	Taimyr	Polygonal bog	190	Shamurin et al., 1972
U.S.S.R.	Salekhard	Moss-shrub-hummock	204	Gorchakovsky and Andreyashkina, 1972
Ireland	Glenamoy	Open bog	214	Moore, personal communication, 1970
U.S.S.R.	Western Taimyr	Flat palsa bog	513	Pospelova, 1972
U.S.S.R.	Kola	Alpine meadow	528	Chepurko, 1972
Australia	Macquarie Island	Herbfield	615	Jenkin and Ashton, 1970
Norway	Hardangervidda	Willow thicket	800	Kelvik and Kärenlampi, in press
U.S.S.R.	Eastern Europe	Shrub tundra	817	Alexandrova, 1970
United Kingdom	Moor House	Blanket bog	846	Forrest, 1971
Australia	Macquarie Island	Grassland	1,138	Jenkin and Ashton, 1970
Austria	Patseherkofel	Loiseleurietum	1,150	Larcher et al., 1973

Production

Primary production can be determined by repeated harvestings, preferably of both above- and belowground biomass. Estimates of the production can also be developed from phytosynthesis-respiration values when translocation within the plant is taken into account. Chlorophyll measurements and carbohydrate analyses are a third method used in primary production studies. Harvesting has been the major method used in tundra production studies.

Primary production is often described as the difference between biomass at the time of peak living aboveground vascular plant biomass and the biomass of the same parts before the growth season begins. This gives only a very rough estimate of the plant production, however. While the green parts of vascular plants have the greatest increment in biomass in the early summer, the root biomass most often decreases in spring because of translocation of food reserves for new green growth. To a certain extent the same pattern exists in tundra areas for nongreen, living, aboveground parts during periods of high respiration in spring, before photosynthesis of green parts is high enough to compensate for respiration. Usually, tundra root mass increases most in the autumn when the green parts decrease. Lichens and, to some extent, bryophytes also continue growth until relatively late autumn in tundra areas.

Decreases in biomass between two summer harvestings may be found for all plant compartments. This can result from harvesting errors, but mortality of plant parts, decomposition, and animal consumption may also be responsible. In the IBP tundra studies, for example, decreases in living biomass are sometimes found for belowground material between the time of maximum aboveground living material and the start of the growing season at some wet sedge meadow sites. This indicates simply that most belowground growth takes place in relatively late autumn after aboveground biomass peaks. "Production" of green components may be taken as the increase in green biomass, plus any increase in mass of standing dead aboveground and of litter. This assumes that the increased weight of these dead parts comes mostly from the green material; if this mortality had not occurred during the growth period, the green parts would have increased accordingly. When dead parts of the plants have lower weight at later harvestings, decomposition can be considered responsible.

Annual primary productivity is normally low in tundra which is frequently a consequence of a very short growing season—less than two months in extreme cases. In Taimyr the growing season lasts about 80 days, at Hardangervidda in Norway about 100 days, and at Moor House in the United Kingdom about 180 days. The highest yearly dry matter production is found at tundra sites in the lowest latitudes, i.e., the Austrian IBP sites in the northern hemisphere and at Macquarie Island in the southern hemisphere, as well as on moorlands in Ireland and the United Kingdom. At those sites the green vascular plant production ranged from 100 to 400 g/m². When belowground production was added, the total biomass of vascular plants at Moor House increased annually by the order of 600 to 700 g/m², with about half of the productivity belowground (Forrest, 1971). Bryophyte production at Moor House was also considerable, i.e., up to 300 g/m² in *Sphagnum* (Clymo, 1970). Relatively high primary productivity was also found at some sites in the U.S.S.R., such as in relatively dry alpine meadows on the Kola Peninsula (Chepurko, 1972), as well as in Norway [Wielgolaski, in press (b)]. The total yearly production (vascular plants) was about 500 g/m², 225 g/m² aboveground and 275 g/m² belowground. Even higher primary production have been calculated for wet eutrophic alpine meadows in Norway (vascular plants about 650 g/m²). Relatively high values were also found at a marshy brush-sedge-moss site in Western Taimyr (Pospelova, 1972); total yearly production was about 400 g/m², but only 60 g/m² was aboveground.

Many tundra sites have an aboveground vascular plant production of 40 to 100 g/m² and a total vascular plant accretion of 100 to 200 g/m². For example, such values were attained at many Russian tundra sites (Andreev, 1966;

Chepurko, 1972), on sedge meadows at Devon Island in Canada (Muc, 1973) and at some Finnish sites in the understory of sub-alpine woodlands (Kallio and Kårenlampi, 1971).

Low aboveground vascular plant production (less than 30 g/m²) was found in snow beds in Norway (Wielgolaski, unpubl.), the northern arctic tundra of the U.S.S.R. (Andreev, 1966), some spotted tundras in the U.S.S.R. (Chepurko, 1972; Pospelova, 1972), and in beach ridges and a plateau at Devon Island (Svoboda, 1973). Including lichens and bryophytes, a lichen heath in Norway showed a production of more than 150 g/m² aboveground, however (Kjelvik and Kårenlampi, in press). Cryptogams also contributed significantly to primary production at other sites, 30 to above 200 g/m² by mosses in meadows in Norway and at Devon Island, Canada, for example [Pakarinen and Vitt, 1973; Wielgolaski, in press (b)].

Primary production may vary considerably from year to year for several reasons including lemming cycles and climatic variations. Dennis (1968) found variations in aboveground dry matter production from 60 to 97 g/m² in 1964 and from 3 to 48 g/m² in 1965, when lemming populations were high. Several years are therefore necessary for productivity estimates. The values cited earlier are mostly from only a few years of IBP-tundra studies; they are, however, mostly within the 50 to 200 g/m² productivity range found in other tundra investigations (Bliss, 1970). The extremely low yearly production at a *Salix artica*-dominated barren site (3 g/m²) on Cornwallis (Warren Wilson, 1957) lies below the values reported in this paper.

The daily aboveground primary productivity may be rather high in tundra areas, Bliss (1970) having recorded up to 3 g/m²/day. Incoming radiation may be high and the energy balance is often positive during the whole 24-hour period in polar areas during parts of the growing season, e.g., until August 6th in Taimyr, U.S.S.R. (Zalenskij *et al.*, 1972). Tieszen (1972) found positive photosynthesis over 24-hour periods at Point Barrow, Alaska, on most days up to August 2nd. Bliss (1972) reported that *Dryas* is photosynthetically active within a few days of snow melt. Photosynthetic values for *Dryas* were quite comparable to temperate zone grasses and tree seedlings. On clear days most *Dryas* production took place at night because of high temperatures during the day.

Still, effective utilization of solar energy by plants may be rather low in tundra areas, e.g., 0.7 percent on a spotted *Dryas*-moss tundra and 1.8 percent on a marshy tundra in Western Taimyr (Vassiljevskaya and Grishina, 1972). At the latter site daily total primary production was 5 to 6 g/m² (Pospelova, 1972), but only about 1 g/m² was aboveground productivity. Daily production ranged to as low as 2 g/m² at other sites in the same area, i.e., in a spotted *Dryas*-moss tundra (0.25 g/m²/day in aboveground production). Based on maximum values for aboveground living biomass and biomass of similar components at the beginning of the vegetative period, an average total daily production of about 2 g/m² is found at the tundra dry meadow in Norway. Considering the different growth periods for tops and roots and for vascular plants and cryptogams, the daily total primary production of the same site (without compensation for consumption, but for decomposition) was about 5 g/m²; about half was aboveground parts [Wielgolaski, in press (b)], ranging from about 2.5 to 6 g/m² in different years. Even somewhat higher values were calculated for wet, eutrophic alpine meadows in Norway. At Moor House daily aboveground production was about 1.6 g/m² (Forrest, 1971).

Measurements of photosynthesis and respiration by tundra plants relevant to productivity estimates have been performed by Hadley and Bliss (1964), Scott and Billings (1964) and Johnson and Kelley (1970), among others. Within the IBP-tundra group the same processes are being studied in several countries. Tieszen (1972) has provided preliminary data on wet sedge meadow at Point Barrow, Alaska. Net CO_2 incorporation by photosynthesis is esti-

mated to be 9 to 12 g/m²/day. This converts to 6 to 8 g/m²/day of dry matter which is comparable to daily production calculated by harvesting at the wet sedge meadow in Norway which usually has higher temperatures but shorter days [Wielgolaski, in press (b)]. Zalenskij et al. (1972) studied photosynthesis of tundra plants in Taimyr, U.S.S.R., and found that a deficiency of CO_2 in the atmosphere may restrict plant assimilation, especially at high light intensities. In their analyses maximum apparent photosynthesis was 6 mg CO_2 per gram dry weight per hour, which they say supports the concept of low levels of apparent photosynthesis in Arctic plants. Data on photosynthesis and respiration in Norweigian alpine tundra (Skre, in press) indicate higher maximum apparent photosynthesis in some vascular plants in moist, eutrophic communities (partly above 15 mg CO_2/g/h at 15° C and 20,000 lux) early in the growing season. There is relatively good correlation between production estimated from harvesting data. Temperatures on the day before the photosynthetic measurement seem to influence the apparent photosynthetic values, however (Nygaard, in press).

[*Editor's Note:* Material has been omitted at this point.]

REFERENCES

Alexandrova, V. D. 1970. The vegetation of the tundra zones in the USSR and data about its productivity, p. 93–114. *In* W. A. Fuller and P. G. Kevan (ed.) Proceedings of the Conference on Productivity and Conservation in Northern Circumpolar Lands, Edmonton, 1969. IUCN Publ. (new series) No. 16. 344 p.

Andreev, V. N. 1966. Peculiarities of zonal distribution of the aerial and underground phytomass on the East European Far North. Bot. Zh. 51:1410–1411. (in Russian).

Bliss, L. C. 1962. Net primary production of tundra ecosystems, p. 35–46. *In* H. Lieth (ed.) Die Stoffproduktion der Pflanzendecke. Gustav Fischer Verlag. Stuttgart. 156 p.

Bliss, L. C. 1966. Plant productivity in alpine microenvironments on Mt. Washington, New Hampshire. Ecol. Monogr. 36:125–155.

Bliss, L. C. 1970. Primary production within Arctic tundra ecosystems, p. 75–85. *In* W. A. Fuller and P. G. Kevan (ed.) Proceedings of the Conference on Productivity and Conservation in Northern Circumpolar Lands, Edmonton, 1969. IUCN Publ. (new series) No. 16. 344 p.

Bliss, L. C. 1972. Devon Island research 1971, p. 269–275. *In* F. E. Wielgolaski and Th. Rosswall (ed.) Proceedings IV. International Meeting on the Biological Productivity of Tundra, Leningrad, Oct. 1971. Tundra Biome Steering Committee, Stockholm. 320 p.

Bliss, L. C., and F. E. Wielgolaski (ed.) 1973. Primary Production and Production Processes, Tundra Biome. Tundra Biome Steering Committee, Edmonton-Oslo. 256 p.

Chepurko, N. L. 1972. The biological productivity and the cycle of nitrogen and ash elements in the dwarf shrub tundra ecosystems of the Khibini mountains

(Kola Peninsula), p. 236–247. *In* F. E. Wielgolaski and Th. Rosswall (ed.) Proceedings IV. International Meeting on the Biological Productivity of Tundra, Leningrad, Oct. 1971. Tundra Biome Steering Committee, Stockholm. 320 p.

Clymo, R. S. 1970. The growth of *Sphagnum:* methods of measurement. J. Ecol. 58:13–49.

Forrest, G. I. 1971. Structure and production of north Pennine blanket bog vegetation. J. Ecol. 59:453–480.

Gorchakovsky, P. L., and N. I. Andreyashkina. 1972. Productivity of some shrub, dwarf shrub and herbaceous communities of forest-tundra, p. 113–116. *In* F. E. Wielgolaski and Th. Rosswall (ed.) Proceedings IV. International Meeting on the Biological Productivity of Tundra, Leningrad, Oct. 1971. Tundra Biome Steering Committee, Stockholm. 320 p.

Hadley, E. B., and L. C. Bliss. 1964. Energy relationship of alpine plants on Mt. Washington, New Hampshire. Ecol. Monogr. 34:331–357.

Jenkin, J. F., and D. H. Ashton. 1970. Productivity studies on Macquarie Island vegetation, p. 851–863. *In* M. W. Holdgate (ed.) Antarctic Ecol. Vol. 2. Academic Press. London and New York. 998 p.

Johnson, P. L., and J. J. Kelley, Jr. 1970. Dynamics of carbon dioxide in an Arctic biosphere. Ecol. 51:73–81.

Kallio, P., and L. Kärenlampi. 1971. A review of the stage reached in the Kevo IBP in 1970, p. 79–91. *In* O. W. Heal (ed.) Proceedings of the Tundra Biome Working Meeting on Analysis of Ecosystems, Kevo, Finland. Tundra Biome Steering Committee, London. 297 p.

Khodachek, E. A. 1969. The plant matter of tundra phytocenoses in Western Taimyr. Bot. Zh. 54(7):1059–1073. (Translated by P. Kuchar).

Kjelvik, S., and L. Kärenlampi. In press. Plant biomass and primary production of Fennoscandian subarctic and subalpine forests and of alpine willow and heath ecosystems. *In* F. E. Wielgolaski (ed.) Fennoscandian Tundra Ecosystems. Part 1. Plants and Microorganisms. Springer Verlag, Berlin-Heidelberg-New York.

Larcher, W., L. Schmidt, G. Grabherr, and A. Cernusca. 1973. Plant biomass and production of alpine shrub heaths at Mt. Patscherkofel, Austria, p. 65–73. *In* L. C. Bliss and F. E. Wielgolaski (ed.) Primary Production and Production Processes, Tundra Biome. Tundra Biome Steering Committee, Edmonton-Oslo. 256 p.

Muc, M. 1973. Primary production of plant communities of the Truelove Lowland, Devon Island, Canada-sedge meadows, p. 3–14. *In* L. C. Bliss and F. E. Wielgolaski (ed.) Primary Production and Production Processes, Tundra Biome. Tundra Biome Steering Committee, Edmonton-Oslo. 256 p.

Nygaard, R. T. In press. Acclimatization effect in photosynthesis and respiration. *In* F. E. Wielgolaski (ed.) Fennoscandian Tundra Ecosystems. Part 1. Plants and Microorganisms. Springer Verlag, Berlin-Heidelberg-New York.

Pakarinen, P., and D. H. Vitt. 1973. Primary production of plant communities of the Truelove Lowland, Devon Island, Canada-moss communities, p. 37–46. *In* L. C. Bliss and F. E. Wielgolaski (ed.) Primary Production and Production Processes, Tundra Biome. Tundra Biome Steering Committee, Edmonton-Oslo. 256 p.

Pospelova, E. B. 1972. Vegetation of the Agapa station and productivity of the main plant communities, p. 204–208. *In* F. E. Wielgolaski and Th. Rosswall (ed.) Proceedings IV. International Meeting on the Biological Productivity of Tundra, Leningrad, Oct. 1971. Tundra Biome Steering Committee, Stockholm. 320 p.

Rosswall, T. 1972. Progress in the Swedish tundra project 1971, p. 291–194. *In* F. E. Wielgolaski and Th. Rosswall (ed.) Proceedings IV. International Meeting on the Biological Productivity of Tundra, Leningrad, Oct. 1971. Tundra Biome Steering Committee, Stockholm. 320 p.

Scott, D., and W. D. Billings. 1964. Effect of environmental factors on standing crop and productivity of an alpine tundra. Ecol. Monogr. 34:243–270.

Shamurin, V. F., T. G. Polozova, and E. A. Khodachek. 1972. Plant biomass of main plant communities at the Tareya station (Taimyr), p. 163–181. *In* F. E. Wielgolaski and Th. Rosswall (ed.) Proceedings IV. International Meeting on the Biological Productivity of Tundra, Leningrad, Oct. 1971. Tundra Biome Steering Committee, Stockholm. 320 p.

Skre, O. In press. CO_2-exchange in Norwegian tundra plants studied by infrared gas analyzer technique. *In* F. W. Wielgolaski (ed.) Fennoscandian Tundra Ecosystems. Part 1. Plants and Microorganisms. Springer Verlag, Berlin-Heidelberg-New York.

Svoboda, J. 1973. Primary production of plant communities of the Truelove Lowland, Devon Island, Canada—beach ridges, p. 15–26. *In* L. C. Bliss and F. E. Wielgolaski (ed.) Primary Production and Production Processes, Tundra Biome. Tundra Biome Steering Committee, Edmonton-Oslo. 256 p.

Tieszen, L. L. 1972. Photosynthesis in relation to primary production, p. 52–62. *In* F. E. Wielgolaski and Th. Rosswall (ed.) Proceedings IV. International Meeting on the Biological Productivity of Tundra, Leningrad, Oct. 1971. Tundra Biome Steering Committee, Stockholm. 320 p.

Vassiljevskaya, V. D., and L. A. Grishina. 1972. Organic carbon reserves in the conjugate eluvial accumulative landscapes of West Taimyr (Station Agapa), p. 215–218. *In* F. E. Wielgolaski and Th. Rosswall (ed.) Proceedings IV. International Meeting on the Biological Productivity of Tundra, Leningrad, Oct. 1971. Tundra Biome Steering Committee, Stockholm. 320 p.

Warren Wilson, J. 1957. Arctic plant growth. Adv. Sci. 13:383–388.

Wielgolaski, F. E. 1972a. Vegetation types and plant biomass in tundra. Arctic and Alpine Res. 4:291–305.

Wielgolaski, F. E. 1972b. Production, energy flow and nutrient cycling through a terrestrial ecosystem at a high altitude area in Norway, p. 283–290. *In* F. E. Wielgolaski and Th. Rosswall (ed.) Proceedings IV. International Meeting on Biological Productivity of Tundra, Leningrad, Oct. 1971. Tundra Biome Steering Committee, Stockholm. 320 p.

Wielgolaski, F. E. In press (b). Primary productivity of alpine meadow communities. *In* F. E. Wielgolaski (ed.) Fennoscandian Tundra Ecosystems. Part 1. Plants and Microorganisms. Springer Verlag, Berlin-Heidelberg-New York.

Zalenskij, O. V., V. M. Shvetsova, and V. L. Voznessenskij. 1972. Photosynthesis in some plants of Western Taimyr, p. 182–186. *In* F. E. Wielgolaski and Th. Rosswall (ed.) Proceedings IV. International Meeting on the Biological Productivity of Tundra, Leningrad, Oct. 1971. Tundra Biome Steering Committee, Stockholm. 320 p.

23

Reprinted from pages 95–97, 120, 134, 158, and 166–167 of *Preprint No. 180,*
Fort Collins, Colorado: Natural Resource Ecology Laboratory,
Colorado State University, 1976, 147pp.

ANALYSES AND SYNTHESES OF
GRASSLAND ECOSYSTEM DYNAMICS

G. M. Van Dyne, F. M. Smith, R. L. Czaplewski, and R. G. Woodmansee

[*Editor's Note:* In the original, material precedes and follows this excerpt.]

10.0 SUMMARY

This paper includes increasingly specific discussion of the world's grasslands in general, an overview of the world's IBP grassland researches, an overview of the US/IBP's grassland researches, and model experiments for a single
shortgrass prairie ecosystem. Analyses are made across sites within a nation
and between nations for comparison mainly at the primary producer level. A
partial and semi-structured set of hypotheses primarily at the ecosystem, community, or subsystem level are discussed. A model is presented briefly whose
flow functions are based on hypotheses primarily at the physiological level in
the biotic portion of the system or in an equivalent level in the abiotic portion of the system. Finally, the model is used to test hypotheses at the community or population level. Analyses are made of model outputs, thus analyses
of the syntheses.

Although milk and meat production from grasslands has increased the last 10
years throughout the world, at the same time there has been decrease in area of
grasslands. This intensification of grassland usage requires more knowledge of
functional ecology. IBP grassland research has been concerned with ecological
processes, environmental interactions, and responses to various natural and induced stresses. Examples of estimates of standing crops and productivity, both
from IBP studies and the world literature are discussed for a range of grass-

land types. To provide a context for discussion of specific IBP studies, there
is a brief review of characteristic prairies, steppes, annual grasslands, savan-
nas, and other grasslands of the world. Data on high-rainfall grasslands [see
tables 5 and 14, and figures 3 and 4] shows standing crops and annual increments
from 3000 to 6000 g . m^{-2}. Annual production in arid and semiarid grasslands is
considerably less ranging down to only a few hundred g . m^{-2}. Within IBP studies maximum aboveground live biomass shows an inverse curvilinear relationship
with latitude, a weak inverted parabolic relationship with mean annual precipitation, and a positive curvilinear relationship with mean annual temperature.
The belowground to aboveground live biomass ratio exhibits a somewhat positive
curvilinear relationship with latitude and a weak inverse curvilinear relationship with mean annual precipitation and mean annual temperature. Aboveground
net primary production exhibits a somewhat weak inverse curvilinear relationship
with latitude [see figure 3]. At about half the IBP sites temperature determines the onset of the growing season, and at the remainder soil water is the
determining factor. The growing season depends upon soil water more frequently
than on temperature [see figure 12].

Considerable new information on decomposition and nutrient cycling in
grasslands has been developed in IBP studies. For example, in the semiarid
western Canadian prairie estimated biomass of microorganisms is 300 g . m^{-2}
dry weight as compared to annual net primary production of 485 g . m^2. In a
tallgrass prairie in the United States there was more than 600 g . m^{-2} biomass
of microorganisms. In tallgrass prairie sites decomposition of organic material was more dependent upon soil temperature than on soil water. The converse
occurred in shortgrass prairie.

Belowground production is more than twice aboveground annual production in
grasslands, but standing crops of biomass belowground may be five to ten times
as much as aboveground standing crop.

Many of the world's grasslands occur in areas where there is a high degree
of variability of precipitation from year to year. For grasslands of western
North America the coefficients of variation in annual rainfall is about 30% of
the mean. Rainfall distribution from year to year is nearly normal.

The practical classification of warm-season and cool-season species follows
a general correspondence with C_3 and C_4 pathways of carbon fixation and photosynthesis. In western North America mountain grasslands there is a dominance
of cool-season species. In the tallgrass prairie and desert grassland further
south there is a dominance of warm-season species. Mixed prairie and shortgrass prairies have mixtures of warm-season and cool-season species.

Western North American grasslands intercept less than 10% of the solar radiation [see table 14]. They utilize much less, and that which they do utilize
is at a cost of exchanging water. Efficiency of energy capture ranged from
0.16% in desert grassland to a maximum of 0.97% in high mountain grassland.
Water use efficiency displays a curvilinear tendency with increasing availability of water. With sparse vegetation in arid grasslands there is inefficient
water use because much of it is evaporated directly from the bare soil surface.
These grasslands evolved under the adaptive strategy of persistence and survival. In more humid grasslands the vegetation develops a canopy capable of
intercepting more of the solar radiation.

Where domestic animals are placed on grasslands at moderate to heavy grazing rates, some 85% to 98% of the consumer community biomass is domestic herbivores; insects comprise 1% to 7%; birds make up less than 1%. In desert and
shrub steppe grasslands large herbivores are somewhat replaced by herbivorous
rodents, lagomorphs, and reptiles. In desert grasslands there may be in the
order of 3 kg . ha^{-1} wet weight of consumers as compared to 88 kg . ha^{-1} wet
weight of consumers in tallgrass prairie.

Table 5.   Some vegetation production characteristics of selected IBP
           grassland research sites (see text)

Parameter	Range	Unit
Maximum aboveground live biomass	97.0 – 1974.0	$g \cdot m^{-2}$
Maximum standing dead	13.0 – 1268.0	$g \cdot m^{-2}$
Maximum litter	107.0 – 1504.0	$g \cdot m^{-2}$
Maximum belowground biomass	139.0 – 3871.0	$g \cdot m^{-2}$
Belowground/aboveground live biomass ratio	0.34– 31.0	
Aboveground net primary production	126.3 – 3396.0	$g \cdot m^{-2} \cdot yr^{-1}$
Belowground net primary productivity	18.2 – 1465.0	$g \cdot m^{-2} \cdot yr^{-1}$
Percent of total net primary productivity reflected aboveground	10.4 – 94.1	%

Table 14.   Net primary production and efficiency of energy capture by ungrazed
            grasslands within the growing season, from the driest to the wettest
            (Sims and Singh, 1971, pp. 59-124, Range Sci. Dept. Sci. Series No.
            10, Fort Collins, Colorado: Colorado State University)

Grassland type	Total net production ($kcal \cdot m^{-2}$)	Efficiency %
Desert grassland	1177	0.16
Shortgrass prairie (southern)	1260	0.19
Shortgrass prairie (central)	2721	0.57
Low mountain	2707	0.60
High mountain	2464	0.97
Mixed prairie (central)	2446	0.46
Mixed prairie (northern 1)	2052	0.47
Mixed prairie (northern 2)	3150	0.73
Tallgrass prairie	2220	0.44

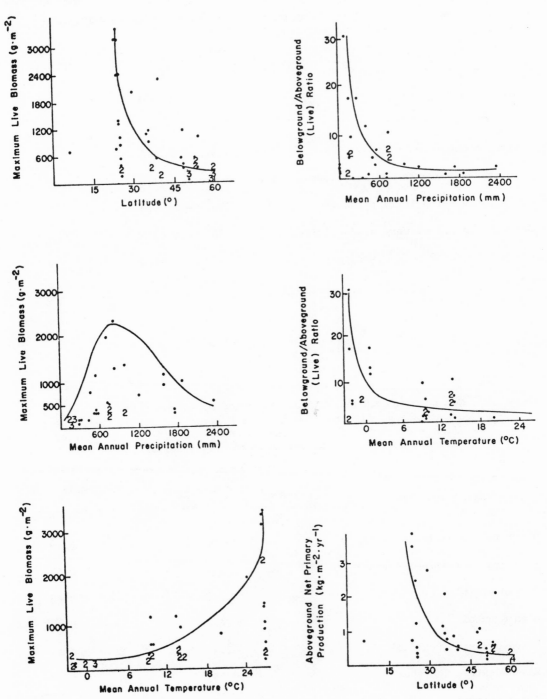

Figure 3

202

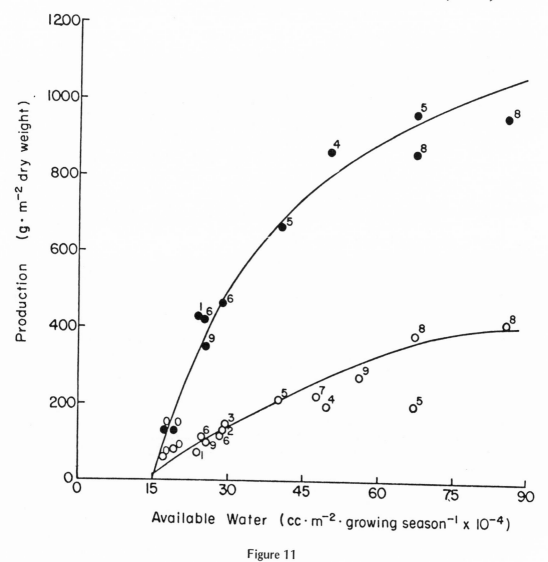

Figure 11

203

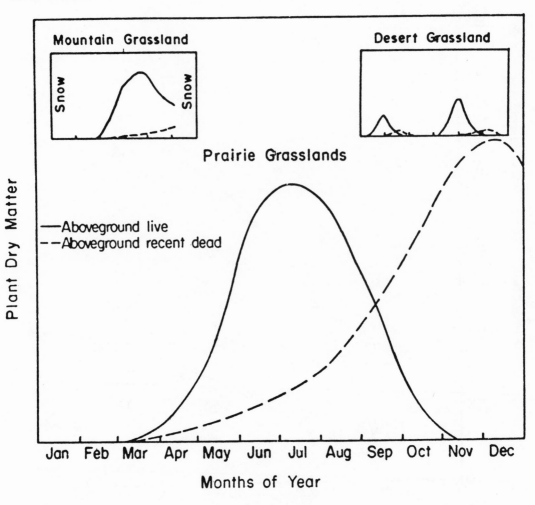

Figure 12

204

24

This material was compiled by the volume editor expressly for this Benchmark volume.

PHOTOGRAPHS, PRODUCTIVITY DATA, AND SOME ENVIRONMENTAL PARAMETERS FROM IBP SITES

The following collection of photographs from IBP sites is the result of a single request to site coordinators, directors, or principle investigators of all known IBP study groups listed in the *1972 US-IBP Directory*. We apologize to our colleagues that we did not make a second mailing. We could only accommodate up to twenty sites in this volume. The requests were mailed in 1974. By that time many sites had closed down, directors had changed, and communications had ceased.

The project was rewarding, however, since an interesting cross section of major vegetation types from all over the world was collected. The pictures reflect very well the productivity level usually attained by the types depicted.

The collection is valuable because several sites responded that had no intensive coverage in the usual IBP literature and provides productivity data in addition to the many tabulations available in this book.

No further explanation is required than that given by the respondents containing geographical coordinates, vegetation type, average temperature and precipitation levels, productivity figures, and bibliographical references. All data are the responsibility of the responding person whose signatures appear below such picture. The picture of the Solling project was provided by this volume editor. It was taken at the time when I was the site coordinator of the project. The productivity data entries were taken from the later publication by Runge. Other data for this photo were compiled from various sources, mainly from Ellenberg (1971).

REFERENCES

National Academy of Science. U.S. National Committee of the International Biological Program. 1972. *IBP Directory*. May 1972. Washington, D.C.

Ellenberg, H., ed. 1971. *Integrated Experimental Ecology. Methods and Results of Ecosystem Research in the German Solling Project*. Ecological Studies 2. New York, Heidelberg, Berlin: Springer Verlag.

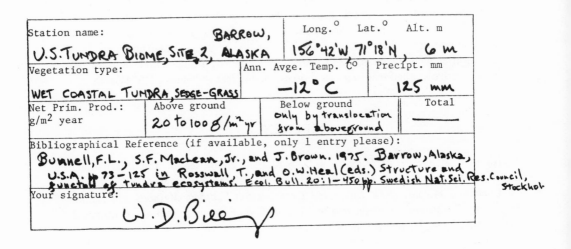

Station name:	BARROW,	Long.° Lat.° Alt. m	
U.S. TUNDRA BIOME, SITE 2, ALASKA		156°42'W 71°18'N, 6 m	
Vegetation type:		Ann. Avge. Temp. C°	Precipt. mm
WET COASTAL TUNDRA, SEDGE-GRASS		−12° C	125 mm
Net Prim. Prod.: g/m² year	Above ground 20 to 100 g/m² yr	Below ground only by translocation from aboveground	Total ———

Bibliographical Reference (if available, only 1 entry please):

Bunnell, F.L., S.F. MacLean, Jr., and J. Brown. 1975. Barrow, Alaska, U.S.A. pp 73–125 in Rosswall, T., and O.W. Heal (eds.) Structure and function of tundra ecosystems. Ecol. Bull. 20:1–450 pp. Swedish Nat. Sci. Res. Council, Stockhol.

Your signature: W. D. Billings

Station name:		Long.°	Lat.°	Alt. m
	Patscherkofel	11°20'E	47°13'N	2175 m

Vegetation type: dwarf shrub heath	Ann. Avge. Temp. C°	Precipt. mm
"Loiseleurietum 2175m"	+ 0,5°	951

Net Prim. Prod. g/m² year 110 dry matter	Above ground *biomass* 750	Below ground *biomass* 800	Total 1550 litter: 1000 ← *1100*

Bibliographical Reference (if available, only 1 entry please):

W. LARCHER, A. CERNUSCA, L. SCHMIDT, G. GRABHERR, E. NÖTZEL, N. SMEETS, 1975 ,
Mt. Patscherkofel, Austria; in: Structure and Function of Tundra Ecosystems. ROSSWALL, T. & HEAL, O.W. (eds.). Ecol. Bull. (Stockholm) 20:125-139

Your signature:

Institut für Allgemeine Botanik
der Universität Innsbruck
Sternwartestraße 15

24.6.76

Station name: PAWNEE SITE (U.S. IBP Grassland Biome Study -- central Plains Experimental Range ARS, USDA)	Long.°	Lat.°	Alt. m
	104	41	1430

Vegetation type: Shortgrass prairie	Ann. Avge. Temp. C°	Precipt. mm
	8.3°C	300

Net Prim. Prod.: g/m² year (1970 ~~1972~~ data) ~~ungrazed ame~~	Above ground	Below ground	Total
	142	458	600

Bibliographical Reference (if available, only 1 entry please):
Sims, P.L. and J.S. Singh. 1971. Herbage dynamics and net primary production in certain ungrazed and grazed grasslands in North America. p. 59-132 IN French, N.R. (editor) Preliminary analysis of structure and function in grasslands. Range Sci. Dept. Science Series No. 10. Colo. St. U. Fort Collins, Colo. USA. 387 pp.

Your signature:

George M. Van Dyne

Station name: Kawatabi		Long.° 140°15E	Lat.° 38°44N	Alt. m 600
Vegetation type: Miscanthus sinensis(tallgrass)		Ann. Avge. Temp. C° 9.8	Precipt. mm 2335	
Net Prim. Prod.:' g/m² year 1000	Above ground 800	Below ground 200		Total 1000
Bibliographical Reference (if available, only 1 entry please): Numata,M.ed.:Ecological Studies in Japanese Grasslands. Univ. of Tokyo Press, JIBP Synthesis Vol.13(1975)				
Your signature: M. Numata				

Station name: Hakkoda		Long.° 140° 55ᴇ	Lat.° 40° 40ɴ	Alt. m 600
Vegetation type: Zoysia japonia(shortgrass)		Ann. Avge. Temp. C° 6.2		Precipt. mm 1,425
Net Prim. Prod.: g/m² year 1000	Above ground 750	Below ground 250		Total 1000
Bibliographical Reference (if available, only 1 entry please): Numata,M.ed.: Ecological Studies in Japanese grasslands. Univ. of Tokyo Press, JIBP Synthesis Vol.13(1975)				
Your signature: M. Numata				

Station name: Phrygana research site (Athens-Greece)		Long.° 23°43′	Lat.° 37°58′	Alt. m 420
Vegetation type: Phrygana (Synonyms:Batha,Israel,Tomillares, Spain)		Ann. Avge. Temp. C° 18.2		Precipt. mm 416
Net Prim. Prod.: g/m² year	Above ground 412	Below ground (310		Total 722
Bibliographical Reference (if available, only 1 entry please): MARGARIS,N.S.,Structure and dynamics in a phryganic ecosystem.Dozent Thesis,University of Athens,Athens,Greece(1976)				
Your signature: (Dr.N.S.Margaris)				

211

Station name: LAKE WINGRA - MADISON WIS		Long.° W 89° 24'	Lat.° N43° 24°	Alt. m 293.6
Vegetation type: PRAIRIE-FOREST		Ann. Avge. Temp. C° 6.9		Precipt. mm 777
Net Prim. Prod.: g/m² year 1481.	Above ground 819.	Below ground 662.		Total 1481.
Bibliographical Reference (if available, only 1 entry please): BAUMANN, P.C., J.F. KITCHELL, J.J. MAGNUSON and T.B. KAYES. 1974. LAKE WINGRA, 1837-1973: A case history of human impact. Trans. Wis. Acad. Sci. Arts and Lett. 62: 57-94.				
Your signature:				

212

Station name: Solling projekt		Long.° ~ 9° 40'E	Lat.° ~51°47'N	Alt.˙m ~ 500
Vegetation type: *Luzulo-Fagetum*		Ann. Avge. Temp. C° 6.5		Precipt. mm 1100
Net Prim. Prod.: Kcal 𝑊/m² year (1969)	Above ground 5720 Kcal	Below ground		Total 6870 Kcal
Bibliographical Reference (if available, only 1 entry please): Runge, M.: Energieumsätze in den Biozönosen terrestrischer Ökosysteme. Scripta Geobotanica vol. 4, 77 p. Göttingen, E. Goltze				
Your signature: . photos by: H. Lieth				(1973)

213

Station name: Hestehaven	Long.° Lat.° Alt. m E 10°29'Min N56°18'Min 11-28		
Vegetation type: Temperate deciduous - beech	Ann. Avge. Temp. C° 7.1		Precipt. mm 660
Net Prim. Prod.: g/m² year	Above ground 1499	Below ground 375⁻	Total 1874
Bibliographical Reference (if available, only 1 entry please):			
Your signature: B. Overgaard Nielsen			

Station name: Forêt de Fontainebleau Réserves biologiques		Long.° Lat.° Alt. m 2° W 48° N 140	
Vegetation type: forêt naturelle de hêtres stade de futaie	Ann. Avge. Temp. C° 10,15		Precipt. mm 697
Net Prim. Prod.: g/m² year	Above ground 476	Below ground 80	Total 556
Bibliographical Reference (if available, only 1 entry please): G. Lemée - Recherches sur les écosystèmes de la hêtraie naturelle en forêt de Fontainebleau. Masson Ed.Paris(en prépar.)			
Your signature: Laboratoire d'Ecologie végétale Université de Paris-Sud, 91405-ORSAY, France			

Station name: ISPINA, NIEPOŁOMICE FOREST NEAR CRACOW	Long.° Lat.° Alt. m 20°22'E 50°6'N 180-185		
Vegetation type: Oak-hornbeam forest TILIO-CARPINETUM	Ann. Avge. Temp. C° 7,8	Precipt. mm 729,4	
Net Prin. Prod.: g/m² year	Above ground 1006,1	Below ground ca. 124,3	Total 1130,4
Bibliographical Reference (if available, only 1 entry please): A. Medwecka-Kornaś, A. Łomnicki, E. Bandoła-Ciołczyk. 1974. Energy Flow in the Oak-Hornbeam Forest (IBP Project "Ispina"). Bull. Acad. Polon. Sci., Sér. des Sci. Biolog. Cl. II, 22, 9: 563-567.			
Your signature: Anna Medwecka Kornaś			

Station name: THOMPSON RESEARCH CENTER /		Long.° Lat.° Alt. m 122 57'30" 47 52'33" 210	
Vegetation type: PSEUDOTSUGA MENZIESII TSUGA HETEROPHYLLA		Ann. Avge. Temp. C° 9.8	Precipt. mm 1364
Net Prim. Prod., g/m² year	Above ground 1436	Below ground 363	Total 1799
Bibliographical Reference (if available, only 1 entry please): Cole, D.W. and S. P. Gessel. 1968. Cedar River Research. Institute of Forest Products, College of Forest Resources, Univ. of Wash. Contrib No. 4. 54p.			
Your signature: Robert L. Edmonds			

Station name:		Long.°	Lat.°	Alt. m
H.J. ANDRENS EXPERIMENTAL FOREST WATERSHED NO. 10		122.20'	44 15'	430-670
Vegetation type: PSEUDOTSUGA MENZIESII TSUGA HETEROPHYLLA	Ann. Avge. Temp. C° 7.5			Precipt. mm 2313
Net Prim. Prod.: g/m² year	Above ground 193.1	Below ground 19.3		Total 212.4
Bibliographical Reference (if available, only 1 entry please): —				
Your signature: *Robert L. Edmonds*				

Station name:	Long.°	Lat.°	Alt. m
SAN JUAN DE LA PEÑA (Spain)	N 42	W 1	1230

Vegetation type:	Ann. Avge. Temp. C°	Precipt. mm
Pine-holly forest	8.0	802

Net Prim. Prod.:' g/m^2 year	Above ground	Below ground	Total
	1756	no data	no data

Bibliographical Reference (if available, only 1 entry please):

ALVERA, B., PIRINEOS 109, P. 17-29 (1973)

Your signature:

219

Station name: *Triangle Site, Eastern Deciduous Forest Biome,* ~~EDFB~~ *USIBP*	Long.° 78°W	Lat.° 36°N	Alt. m 135m
Vegetation type: *Pinus taeda plantation*	Ann. Avge. Temp. C° 13.6°C		Precipt. mm 1150
Net Prim. Prod.⟩ g/m²/year	Above ground 1190	Below ground 190	Total 1380
Bibliographical Reference (if available, only 1 entry please): *Ralston, C.W. 1973. Annual Primary Productivity in a loblolly pine plantation. In: IUFRO biomass studies. S4.01 mensuration, growth and yield. Univ. of Maine, Orono. P 105-118.*			
Your signature: *Kenneth R. Higginbotham*		photo: Fred Mowry	

Station name: LE ROUQUET	Long.$^{\circ}$ Lat.$^{\circ}$ Alt.⸱ m		
	1° 65' N 48° 36' 180		

Vegetation type: Evergreen oak forest	Ann. Avge. Temp. C$^{\circ}$ 13° 4	Precipt. mm 987

Net Prim. Prod.: g/m^2 year	Above ground 644,2	Below ground	Total

Bibliographical Reference (if available, only 1 entry please):LOSSAINT, RAPP, 1871 - Répartition de la matière organique, productivité et cycle des éléments minéraux dans des écosystèmes de climat méditerranéen. UNESCO, Productivité des écosystèmes forestiers, Actes Coll. Bruxelles (597-617)

Your signature:
P. LOSSAINT, M. RAPP

221

Station name: Parc national du Banco, Côte d'Ivoire		Long.° 4° W	Lat.° 5° N	Alt. m 50
Vegetation type: forêt naturelle équatoriale sempervirente	Ann. Avge. Temp. C° 26,2		Precipt. mm 2095	
Net Prim. Prod.; g/m² year	Above ground 1630	Below ground 70		Total 1700

Bibliographical Reference (if available, only 1 entry please):
BERNHARD-REVERSAT F.; HUTTEL C., LEMEE G.(1975)- Recherches sur l'écosystème de la forêt subéquatoriale de Basse Côte d'Ivoire. Terre et Vie, 29,169 264.

Your signature: ^   Laboratoire d'Écologie végétale
                          Université de Paris-Sud, 91405-ORSAY,France

Station name: Pasoh Forest Reserve Negeri Sembilan, West Malaysia	Long.° 102°18'E	Lat.° 2°59'N	Alt. m 80m
Vegetation type: Mixed lowland rain forest of Shorea-Diptero-carpus type.	Ann. Avge. Temp. C° ca. 25°C		Precipt. mm ca. 2000 mm

Net Prim. Prod.: g/m²yr	Above ground 2114	Below ground 553 (estimated)	Total 2667

Bibliographical Reference (if available, only 1 entry please):

Kira, T. : Community architecture and organic matter dynamics in tropical lowland rain forests of Southeast Asia. In"Tropical Trees as Living Systems, Proc. 4th Cabot Symp." in preparation, Cambridge Univ. Press.

Your signature: *Tatuo Kira*   Tatuo Kira

Part III

MODELING PRODUCTIVITY PATTERNS

Editor's Comments
on Papers 25 Through 31

Modeling of the primary productivity process have so far followed two main streams. The first is understanding and predicting the reaction of the plants under various environmental constraints. The second is constructing global and regional patterns from a limited data set of productivity values in combination with a large data pool of meaningful correlated properties.

The attempt to understand the primary productivity process as a function of plant growth and environmental properties began at the early part of this century. F. F. Blackman, mentioned by Transeau (1926), and V. H. Blackman were leading growth analysis modelers, and Paper 25 contains G. E. Blackman's 1968 report on the development of the field. A similar modeling approach is reflected in the paper by Monteith, which is included in this collection as Paper 29. V. H. Blackman (Paper 26) developed a model of plant growth equal to the law of compound interest in economics. At the same time, Mitscherlich worked on the quantification of Liebig's law of the minimum dating back to the middle of the nineteenth century. Together with the mathematician, Baule, he converted the law of the minimum to the yield law in which all growth factors may appear as percentage contributors to an a priori, set upper level of plant yield. This enabled them to predict accurately the agricultural yield in response to fertilizer application. The relations derived follow the law of diminishing return ascribed to the French agricultural economist, Talbot, in the early part of the nineteenth century about a century before Mitscherlich and Baule. The yield law has since been shown applicable to a great variety of ecosystems processes including the modeling of net primary productivity.

A few decades later with the advent of biophysics as a new endeavor in biology, researchers like Huxley, Teissier, and Bertalanffy started to develop growth models using morphological and physiological plant characters (see Bertalanffy 1951). The allometric growth models by Huxley and Teissier became the base for forest productivity calculations from stem diameter at breast height (DBH) so prominent for example, in Whittaker's dimension analysis (see page 121). Bertalanffy (1951) made the attempt to cast Rubner's rule that assimilation of substances is a function of the surface area and dissimilation a function of the volume into a mathematical equation. His successful deduction of growth patterns from surface and volume proportions resulted in one of the most significant deterministic models constructed in biology. It indicated that biology was developing from an experimental science into an exact science.

Modeling of productivity is now so abundant that no attempt is made in this volume to summarize the recent approaches. This will be subject of another Benchmark Book. What is necessary to include here are aspects of more recent models that helped to produce regional and global productivity patterns. One of the first attempts to derive global forest yield patterns from environmental data was presented by S. S. Paterson. His basic approach is the calculation of the "CVP Index." Paper 28 is part of his thesis, which was accompanied by a world map of forest yield (lumber volume.)

The papers by Monteith (Paper 29) and Lieth (Paper 31), elaborate on the usefulness of evapotranspiration for the prediction of primary productivity. This concept was used first by Lieth (1961) and later by Lieth and Box (1972) and by Rosenzweig (Paper 30) to prepare correlation models between net primary productivity and actual evapo-transpiration. Paper 30 is a good example of the use of experimental data sets to uncover environment/plant relations. It also relates to the contents of several other papers in this book. One of the attempts to elaborate the productivity pattern of the United States by Sharpe (see Paper 20) is based on Rosenzweig's model.

The final example of pattern modeling is Paper 31 by Lieth, which attempts a derivation of the global productivity pattern from synoptic temperature and precipitation averages. The approach is known as the Miami model, and the paper has become most successful. It was first presented as a US/IBP Eastern Deciduous Forest Biome report and has since been reprinted about thirty times in many different languages. The version selected for this volume originally appeared in UNESCO's *Nature and Resources* series.

REFERENCES

Bertalanffy, L. von. 1951. *Theoretische Biologie,* Vol. 2, 2nd ed. Bern: A. Francke.

Lieth, H. 1961. La producion de sustancia organica por la capa vegetal y sus prob-lemas. *Acta Sci. Venezolana* **12**:107–114.

Lieth, H., and E. Box. 1972. Evaportranspiration and primary productivity; C. W. Thornthwaite memorial model, pp. 37–46 in *Papers on Selected Topics in Climatology,* Vol. 2, J. R. Mather, ed. Elmer, N. J.: Thornthwaite Assoc.

Transeau, E. N. 1926. The accumulation of energy by plants. *Ohio J. Sci.* **26**:1–10.

ADDITIONAL READINGS

Briggs, G. E., F. Kidd, and C. West. 1920/1921. Quantitative analysis of plant growth. I. *Ann. Appl. Biol.* 7:103; II., ibid. 7:202.

Coombe, D. E. 1960. An analysis of the growth of *Trema guineensis. J. Ecol.* 48:219.

Evans, G. C., and A. P. Hughes. 1962. Plant growth and the aerial environment. On the computation of unit leaf rate. III. *New Phytol.* **61**:322.

Watson, D. J. 1952. The physiological basis of variation in yield. *Adv. Agron.* 4:101.

25

Reprinted by permission of Unesco from pages 243-244 and 259 of *Functioning of Terrestrial Ecosystems at the Primary Production Level*, edited by F. E. Eckardt, Paris: UNESCO, 1968, 516pp.

THE APPLICATION OF THE CONCEPTS OF GROWTH ANALYSIS TO THE ASSESSMENT OF PRODUCTIVITY

G. E. Blackman
Oxford University

[*Editor's Note:* In the original, material precedes these excerpts.]

In 1919, V. H. Blackman pointed out that "if the rate of assimilation per unit of leaf surface and the rate of respiration remain constant, and if the size of the leaf system bears a constant relationship to the dry weight of the whole plant, then the rate of production of new material as measured by the dry weight will be proportional to the size of the plant". In other words, growth is a cumulative and exponential process and Blackman suggested that the rate of gain per day (the efficiency index) could be employed as a comparative measure of the performance of different species under a given set of conditions. It should be noted that differences in performance would be determined by three parameters, the rate of assimilation per unit leaf surface, the respiratory losses and the size of the leaf system. Alternatively, on the basis of a diurnal time scale, the measured rate of gain in dry matter (the relative growth rate or RGR) is determined by the *net* difference between the amounts assimilated and respired—that is, the net assimilation rate per unit of photosynthetic surface—while the size of the assimilatory system is best expressed as the ratio of photosynthetic surface to the weight of the whole plant. Since for most species photosynthesis largely takes place in the leaves, the two parameters are the net assimilation rate (NAR) and the leaf area ratio (LAR).

In 1920-1921 Briggs, Kidd and West came to the important conclusion that if growth is proceeding on an exponential basis then:

The relative growth rate = net assimilation rate × leaf area ratio.

In other words, any factor which alters the relative growth rate does so by causing positive or negative changes in either the net assimilation rate or the leaf area ratio or both. Furthermore, it was clear for the first time that from sequential determinations of plant weight and leaf area, it would be possible to analyse in the field on a comparative basis the effects, either alone or in a combination of environmental factors, on the growth of different species.

Briggs et al. in their proof assumed that changes in plant weight (W) and leaf area (A) were proceeding exponentially and this being so the differential equation is:

$$\frac{1}{W}\frac{dw}{dt} = \frac{1}{A}\frac{dw}{dt} \times \frac{A}{W}.$$

Given the plant weights and leaf areas at two sampling occasions (t_1 and t_2) the mean rates for RGR and NAR can be calculated by integrating $\frac{1}{W}\frac{dw}{dt}$ and $\frac{1}{A}\frac{dw}{dt}$ to give

$$RGR = \frac{\log_e W_2 - \log_e W_1}{t_2 - t_1} \tag{1}$$

and

$$NAR = \frac{W_2 - W_1}{t_2 - t_1} \times \frac{\log_e A_2 - \log_e A_1}{A_2 - A_1} \tag{2}$$

Many years later, Williams (1946) pointed out that in the derivation of the equation for NAR, Briggs et al. had not appreciated that the integration was only feasible if A and W are linearly related. If this linearity does not hold, errors of estimation will arise and Williams suggested a graphical method of interpolation whereby errors could be minimized.

In 1960, Coombe derived a further equation for NAR based on the assumption that the weight changes with the square of the leaf area, and in 1962 Evans and Hughes and Whitehead and Myerscough simultaneously produced generalized equations based on the assumption that $W = a + bA^n$ as against the assumptions of $W = a + bA^2$ of Coombe or the original $W = a + bA$ of Briggs et al. Coombe reached the conclusion that there was little to choose between the formulae based on the second and third assumptions since the estimates of NAR did not vary by more than a few per cent provided that between sampling occasions the leaf area did not more than double. Evans and Hughes agreed that during the early vegetative phase the original formula was reasonably valid since the constant n of the generalized equation does not depart appreciably from unity but that with age n tends to rise. In this connexion Whitehead and Myerscough emphasized that the effects of stage of development could be followed from the alpha term in their equation where alpha is defined as the ratio (on a logarithmic scale) of the rates of dry-matter and leaf-area production per plant.

The question of the relationship between leaf area and plant weight again arises when it comes to calculating the mean leaf area ratio. Since the relationship does not normally depart greatly from linearity, it is generally agreed that the arithmetic mean can be taken as a reasonable estimate.

[*Editor's Note:* Material has been omitted at this point.]

REFERENCES

Blackman, V. H. 1919. The compound interest law of plant growth. *Ann. Bot. Lond.*, vol. 33, p. 353.

Briggs, G. E., F. Kidd, and C. West. 1920–1921. Quantitative analysis of plant growth. I. *Ann. Appl. Biol.*, vol. 7, p. 103: II., ibid., vol. 7, p. 202.

Coombe, D. E. 1960. An analysis of the growth of *Trema guineensis. J. Ecol.*, vol. 48, p. 219.

Evans, G. C., and A. P. Hughes. 1962. Plant growth and the aerial environment. On the computation of unit leaf rate. III. *New Phytol.*, vol. 61, p. 322.

Whitehead, F. H., and P. J. Myerscough. 1962. Growth analysis of plants. The ratio of mean relative growth rate to mean relative rate of leaf area increase. *New Phytol.*, vol. 61, p. 314.

Williams, R. F. 1946. The physiology of plant growth with special reference to the concept of net assimilation rate. *Ann. Bot. Lond.*, N.S., vol. 10, p. 41.

26

Reprinted from *Ann. Bot.* 33(131):353–360 (1919) by permission of the publisher,
Oxford University Press

The Compound Interest Law and Plant Growth.

BY

V. H. BLACKMAN.

IN many phenomena of nature we find processes in which the rate of change of some quantity is proportional to the quantity itself. Since money put out at compound interest increases in this way—the rate of increase being clearly proportional to the amount of capital at any time— Lord Kelvin called the law which such processes follow ' the compound interest law'. The rate at which a body cools follows the compound interest law, for the hotter the body relative to its surroundings the more rapidly it loses heat. Again, the variation of atmospheric pressure with height above sea-level follows this law, as does also the velocity of a chemical reaction. Wilhelmy's law, discovered as long ago as 1850, that ' the amount of chemical change in a given time is directly proportional to the quantity of reacting substance present in the system ', is simply a restatement of the compound interest law.

The importance of this law for the proper appreciation of the growth of a plant was brought home to the writer in 1917 in connexion with the results of some experiments on the growth of cucumbers carried out in association with Mr. F. Gregory at the Cheshunt Experimental Station.

It is clear that in the case of an ordinary plant the leaf area will increase as growth proceeds, and with increasing leaf area the rate of production of material by assimilation will also increase; this again will lead to a still more rapid growth, and thus to a greater leaf area and a greater production of assimilating material, and so on. If the rate of assimilation per unit area of leaf surface and the rate of respiration remain constant, and the size of the leaf system bears a constant relation to the dry weight of the whole plant, then the rate of production of new material, as measured by the dry weight, will be proportional to the size of the plant, i. e. the plant in its increase of dry weight will follow the compound interest law.

The fact that the increase in number of unicellular organisms, when not limited by external conditions, follows a regular geometric series has long been recognized. The resemblance also of the growth processes of animals and plants to an autocatalysis has been pointed out by a number of workers, as J. Loeb, W. Ostwald, Robertson, F. F. Blackman, Chodat

and his pupils.[1] The application of the compound interest law to the
growth of the higher plants, though of fundamental importance to a right
understanding of the plant's rate of increase, has, however, been over-
looked by most botanists,[2] and its recognition is sadly lacking in the textbooks both of plant physiology and general botany. Chodat in his
'Principes de Botanique' (2nd edit., p. 133) appears to be the only text-book
writer who even refers to the relation of growth to a geometric series, and
his treatment of the subject appears under the section of the book which
deals with the growth of the cell, and it is confined to pointing out the
similarity of the growth of the cell to a process resembling autocatalysis.

Apart from any question of autocatalysis it is obvious that the increase
in size of the assimilating surface of the young plant must constantly
accelerate the rate of growth, and that the consideration of this acceleration
is essential for the proper comparison of the final weight of different plants
and of the same plants grown for different periods.

When money accumulates at compound interest, the final amount
reached depends on (1) the capital originally employed, (2) the rate of
interest, (3) the time during which the money accumulates. In the case of
an annual plant the ultimate dry weight attained will depend on (1) the weight
of the seed, since that determines the size of the seedling at the time that
accumulation of new material begins ; (2) on the rate at which the material
present is employed to produce new material, i. e. the percentage increase
of dry weight per day or week or other period ; (3) the time during which
the plant is increasing in weight.

It is clear then that some simple equation is required to relate these
three factors to the final weight attained ; such an equation does not appear
to have been hitherto put forward by those few workers who have considered
the growth relations of the whole plant from this aspect. Before dealing,
however, with this attention may be drawn to the work of Noll and his pupils,
who have provided the data of the growth relations of a number of plants
during various stages.

Noll seems to have been the first to formulate the view that in the case
of an annual plant the successive dry weights taken at regular periods
follow a geometric series. In 1906 Noll read before the Niederrheinische
Gesellschaft für Natur- und Heilkunde zu Bonn a paper (which appears only

[1] There can be no doubt that the development of an increasing number of rapidly enlarging
cells which occurs in the development of most plant organs will cause a rapid acceleration of growth,
producing a curve of growth which is very similar to that of an autocatalysed reaction. A process
of autocatalytic nature does not, however, explain the rapid fall in the rate of growth in the later
stages of development of the human body, or the fall in the rate of growth of a plant organ which
has passed its 'grand period' of growth. The growth of an annual body of a plant organ of limited
size is clearly dominated by factors other than those that play their part in a simple autocatalytic
process.

[2] Since the above was written I have seen the proof of a paper (to appear in the Annals of
Applied Biology) by Dr. F. Kidd and Dr. W. West in which they point out the importance of the
compound interest law.

in title in the Proceedings of the society) entitled 'Über die Substanz-quotienten pflanzlicher Entwickelungsstadien'. In the next two years three of his pupils, Gressler,[1] Hackenberg,[2] and Kiltz,[3] published inaugural disser-tations dealing with the determination of the 'Substanzquotienten' of various plants.

From these papers it becomes clear that the 'Substanzquotient' is the factor which relates the dry weight at the end of any period of growth with the dry weight at the beginning of that period. Taking Gressler's results with *Helianthus uniflorus giganteus* we find that in five successive weeks the average dry weight in grammes of a number of plants was 0·0454, 0·147, 0·508, 1·653, 5·868, 17·33, 30·35, 46·2, 66·1, 88·9. Dividing the second by the first, the third by the second, and so on, we find that the successive weekly 'substance-quotients' were 3·25, 3·45, 3·24, 3·56, 3·0, 1·75, 1·52, 1·4, 1·3. The successive, weekly dry weights clearly exhibit at first a progression which is approximately geometrical; later, however, the rate of increase falls off, the series becoming an arithmetical one.

The 'substance-quotient' per week is obviously a clumsy and in-accurate method of expressing these results. What is required is some simple method of relating the plant's activity in the production of new material to time and to the initial weight of the seedling. Hackenberg and Kiltz merely state their results as 'substance-quotients' per week, but in calculating the average 'substance-quotient' over a period of many weeks Gressler treats his results as a discontinuous geometric series. The formula which would then apply is $W_1 = W_0 (1 + r)^t$, where $W_1 =$ the final weight, $W_0 =$ the initial weight, $r =$ the rate of interest, and $t =$ time. Gressler's results for *Helianthus* calculated in this way per week and per day are shown in columns 5 and 6 of the accompanying table.

	Seedling weight. Grm.	Final weight. Grm.	Time. Days.	Rate of Interest (Discontinuous).		Rate of Interest (Continuous).	
				per week.	per day.	per week.	per day.
H. uniflorus giganteus	0·0327	17·33	37	227·6 %	18·5 %	119·0 %	17·00 %
H. nanus	0·0348	14·805	37	214·3 %	17·7 %	114·5 %	16·36 %
H. cucumerifolius nanus	0·00106	0·401	56	110·0 %	11·2 %	74·1 %	10·59 %
H. macrophyllus giganteus	0·0241	6·772	32	243·3 %	19·3 %	123·4 %	17·63 %
H. arboreus giganteus	0·0192	14·680	40	219·6 %	18·1 %	116·1 %	16·59 %

Treatment of the results in this way would, however, only be satisfac-tory if the additional material were added *discontinuously* at the end of each day or week. It is obvious, however, that during the daylight period the plant is adding new material continuously, and during rapid growth the plant

---

[1] P. Gressler: Ueber die Substanzquotienten von *Helianthus Annuus.* Inaug. Diss., Bonn, 1-29, Tables I-V, 1907.

[2] H. Hackenberg: Ueber die Substanzquotienten von *Cannabis sativa* und *Cannabis gigantea.* Inaug. Diss., Bonn, 1-27, 1908. Also Beihefte zum Bot. Centralbl., xxiv, pp. 45-64, 1908.

[3] H. Kiltz: Versuche über den Substanzquotienten beim Tabak und den Einfluss von Lithium auf dessen Wachsthum. Inaug. Diss., Bonn, 1908 (seen only in abstract, Bot. Centralbl., cx, p. 455, 1909).

is continuously, or nearly continuously, unfolding its leaves and increasing its assimilating area. The plant's increase is thus comparable rather to money accumulating at compound interest, in which the interest is added to the principal not daily or weekly, but continuously. The simple equation which best applies to the growth of active annual plants is thus:

$$W_1 = W_0 e^{rt},$$

where, as before, $W_1 =$ the final weight, $W_0 =$ the initial weight, $r =$ the rate of interest, and $t =$ time, and e is the base of natural logarithms.[1] Some of Gressler's results have been calculated on this basis of continuous addition at compound interest, and are given in the last two columns of the table. The rate of interest required to give the same final dry weight is naturally less when it is added continuously than when it is assumed to be added discontinuously, and far less than the rate of interest per week calculated from the weekly 'substance-quotients'.

As has already been stated, it is obvious from general considerations, and also from the equation, that the final weight attained will depend on the initial weight, the rate of interest (r), and the time. The differences in the dry weight attained by two plants may thus depend on simply the initial dry weights of the seedlings; if the rate of interest is the same the final weights will then vary *directly* as the initial weights. This shows the marked effect which large seeds as compared with small seeds may have on the final weight attained. Again, if the initial weights are the same a small difference in the rate of interest (r) will soon make a marked difference in the total yield, and the difference will increase with the lengthening of the period of growth. A difference of 1 per cent. in the rate of interest will in a period of 69 days double the final weight attained.

Oats in water culture may, according to Wolff, attain a dry weight 2,359 times that of the seed. If the growing period be taken as 100 days, the rate of interest on the basis of continuous addition is 7·76 per cent. per day. If the rate of assimilation per unit area should rise by 5·8 per cent. then, allowing 10 per cent. for loss of respiration, the final weight at the end of 100 days would go up 50 per cent. Plants of *Helianthus macrophyllus giganteus* (investigated by Gressler) with a seed weight of 0·0241 grm. may in 32 days reach a dry weight of 6·77 grm., i.e. a weight 251 times that of the seed. This on calculation by the equation given above requires that r shall be 0·1763 (i. e. an average rate of 17·63 per cent.) per day. An increase

---

[1] This formula can be expressed as $\log_e \dfrac{W_1}{W_0} = rt$. In using the formula it is only necessary to find the number which expresses the relation between the final and initial dry weights of the plant; and then to find the napierian logarithm ($\log_e$) of that number, or to find the common logarithm and multiply by 2·3026. The logarithm so found when divided by the time gives the rate of interest required. Suppose a plant has doubled itself in ten days. We find that the $\log_e$ 2 is 0·69315; therefore the plant has been producing new material at the rate of 0·0693 (i. e. 6·93 per cent.) per day. If the period were 5 days the rate would be 13·8 per cent. per day; if 100 days, then 0·69 per cent. per day.

of assimilation of 2 per cent. would in this ´case increase the weight at the end of 32 days by about 20 per cent.

A marked difference in the rate of interest (r) is exhibited by different plants. The table given shows that in some species of *Helianthus* it may reach 17·6 per cent. per day, while in *H. cucumerifolius nanus* it is only 10·42 per cent. per day. In some results obtained by Stefanowska[1] with Maize in water culture the plants increased their fresh weight 27·5 times in 45 days; the rate of interest was therefore only 7·45 per cent. per day. Obviously some plants can work with far greater economy than others. Thus for every 100 grm. of dry material already present *H. macrophyllus giganteus* can produce new material at the average rate of 10·4 grm. per day; calculating from Hackenberg's results, we find that *Cannabis gigantea* may work at a rate of 13·1 per cent. per day for a short time, and (from Kiltz) that *Nicotiana Tabacum* may work at the rate of 20·5 per cent. per day. *Zea Mais*, on the other hand, as stated above, under some conditions works at the average rate of only 7·45 per cent. per day.

The rate of interest (r of the equation) is clearly a very important physiological constant. *It represents the efficiency of the plant as a producer of new material*, and gives a measure of the plant's economy in working. The rate of interest, r, may be termed the *efficiency index* of dry weight production, since not only does it indicate the plant's growth efficiency as measured by increase of dry material, but it also appears as an exponential term in the equation which expresses the relation between the initial dry weight, the final dry weight, and the period of growth. It may also be termed the 'economy constant' of the plant; it is of course comparable to the velocity constant of a chemical reaction.

It is suggested that in all water cultures, pot experiments, and similar experiments where dry weights are determined after a period of growth, the efficiency index should be calculated from the seed weight and the final weight attained, so that a measure of the plant's *average* economy of working may be obtained.[2] Such a calculation will show how far a large final weight ·is determined by a large initial weight or by a high efficiency index.

A glance at the table given above shows that the 'dwarfness' of *Helianthus cucumerifolius nanus* is due not only to the very small seed weight but also to the comparatively low efficiency of the plant, the efficiency index being only 0·1042 (or 10·42 per cent.) per day. This form of *Helianthus* is handicapped by a seed 1/300th of the weight of that of *H. uniflorus giganteus*, so that even if it had the same efficiency it could only attain 1/300 of the final weight of the latter. Other things being equal, a small seed is a permanent handicap to a plant in the production of material. *H. cucumerifolius nanus*, in order to attain after 37 days the same

---

[1] Comptes rendus de l'Acad. des Sciences, cxxxviii, 304–6, 1904.

[2] Where root parts are not available, the efficiency index can be given for the aerial parts only. The seeds should, where possible, be weighed without the testa.

weight as *H. uniflorus*, would have to work at an efficiency of 0.2621 (i. e. 26.21 per cent.) per day ; with this high economy of working the plant would double its weight in less than three days. *For the highest production of vegetative material by the single plant two factors are necessary—large seeds, and a high economy in working represented by a large efficiency index.*

The importance of these two factors in breeding cereal crop plants should be borne in mind ; it may be possible to breed for a high efficiency index. In many crop plants the matter is of course complicated by the effect of crowding on the efficiency of the individual plant, a question which requires further analyses.

The growth is naturally affected by external conditions, being higher when conditions are favourable, but even under the same conditions there are large variations in the economy of working of different plants, so the efficiency index is certainly to a large extent a characteristic of different species and varieties. It would be of great interest to determine to what these differences in efficiency are due. They may be the result of differences in the rate of assimilation per unit area of leaf surface, of differences in the rate of respiration, of differences in the thickness of the leaves, or of differences in the distribution of material to leaves on the one hand and to the axis on the other. The larger the proportion of new material that the plant can utilize in leaf production the greater, other things being equal, should be its efficiency.

It is clear, from Gressler's and Hackenberg's results with *Helianthus* and *Cannabis*, that the efficiency of the plant is greatest at first and then falls somewhat, but the fall is only slight until the formation of the inflorescence, when there is a marked diminution in the efficency index. For *Helianthus arboreus giganteus*, for example, the 'substance-quotients' for successive weeks are 3.11, 3.49, 3.71, 3.06, 2.59, 3.03, 2.0, 1.5, 1.4, 1.1, 1.3, 1.3. The sudden fall from 3.03 to 2.0 is associated with the appearance of the inflorescence.

The observations of Gressler, Hackenberg, &c., require repetition, for they worked with only a small number of plants and they give no idea of the experimental errors involved, though the differences in dry weight of the individual plants must have been considerable. It would be very valuable to have data for various agricultural and horticultural plants, showing the average efficiency index under various conditions. If the dry weight measurements were combined with a measurement of leaf area and leaf weight some insight could be obtained into the nature of the differences which exhibit themselves as differences in the efficiency index of different plants and as differences in the efficiency index of the same plant at different stages.[1]

[1] If such observations were combined with an estimation of the carbon content of the root, stem, and leaf, together with some measure of the rate of respiration, the analysis could be carried still further. It is hoped to undertake this work with some agricultural plants.

The fall in efficiency after the first few weeks of growth may perhaps be correlated with the mechanical relations connected with larger size. A doubling of the leaf area would require a stem of more than twice the weight to attain equal strength. Gregory[1] found that with larger leaf areas the ratio of stem weight to leaf weight went up.

As the efficiency of the plant is highest in its early stages favourable conditions at that time should have a marked effect on growth. If a plant, owing to such conditions, should double its size as compared with another plant, then there is no reason why that advantage should not be retained to the end of the growing season. A 'good start' means, among other things, a larger capital to work with throughout the growing season.

Gericke,[2] investigating the effect of various injuries on the growth of *Helianthus annuus*, showed that the removal of the cotyledons not only affected the total weight—as was to be expected, since a large portion of the plant's capital was thereby lost—but also markedly reduced the efficiency of the plant. Calculations from his weekly data of dry weights show that the efficiency index never rose above 14.24 per cent. per day as compared with the 18.46 per cent. of normal plants. It is not at all clear why the loss of the cotyledons should reduce the economy of working of the plant developed from the remaining portion, especially as the removal of *one* cotyledon and one of the first leaves had no such effect. It suggests that the cotyledons may contain a supply of some special material necessary for the proper development and efficiency of the plant ; a possible analogy with the growth substances of animals occurs to one. The subject certainly requires further investigation.

Kiltz (*loc. cit.*) observed in *Nicotiana* a marked decrease in dry weight at the time when the seed was matured, reaching 2.98 per cent. in the case of *N. gigantea*. This might be explained by assimilation being brought to a standstill while respiration continued. The amount to be explained in this way is, however, very large, and Chodat, Monnier and Deleano[3] have described a restoration to the soil of mineral matter up to 40 per cent. of the dry weight of the mature plant. Whatever its cause, the phenomenon is of importance in all experiments where the dry weight of annual plants which have reached the fruiting stage is in question.

SUMMARY.

Attention is drawn to the fact that the growth of an annual plant, at least in its early stages, follows approximately the 'compound interest law'. The dry weight attained by such a plant at the end of any period will depend on (1) the weight of the seed (or the seedling at its start),

[1] Experimental and Research Station, Cheshunt, Herts., Report III, p. 24, 1917.

[2] F. Gericke : Experimentelle Beiträge zur Wachstumsgeschichte von *Helianthus annuus*. Inaug. Diss., Halle, 1–43, 1909.

[3] Bull. de l'Herbier Boissier, vii. 350 and 948, 1907.

representing the initial capital with which the plant starts ; (2) the average rate at which the plant makes use of the material already present to build up new material : this represents the rate of interest on the capital (material) employed ; (3) the period of growth.

The plant is continually unfolding its leaves and increasing its assimilating power. Successive increases in the weight of the plant cannot therefore be treated as a discontinuous geometric series. as if the new material (interest) were added at the end of daily or weekly periods. New material is added continuously during daylight, and during rapid growth the plant is continuously, or nearly continuously, unfolding its leaves and increasing its assimilating rate. The growth of the plant more nearly approximates to money accumulating at compound interest where the interest is added continuously. The simple equation which best expresses the growth relations of active, annual plants is $W_1 = W_0 e^{rt}$, where $W_1 =$ the final weight, $W_0 =$ the initial weight, $r =$ rate at which the material already present is used to produce new material, and $t =$ time.

The term r is an important physiological constant, for it is a measure of the efficiency of the plant in the production of new material ; the greater r is, the higher the return which the plant obtains for its outlay of material.

The rate of interest, r, may thus be termed the '*efficiency index*' of dry weight production, for not only is it a measure of the plant's efficiency but it is also an exponential term in the equation expressing the growth of the plant. In some forms of *Helianthus* the average efficiency index for the period up to the formation of the inflorescence may reach 0·1763 (i e. 17·63 per cent.) per day.

It is suggested that in all experiments (such as water cultures, pot experiments) dealing with the production of vegetative material the efficiency index be calculated. The relative efficiency of different plants and of the same plant at different stages can thus be determined ; also the effect on the efficiency index of various external conditions.

A small difference in the ' efficiency indices ' of two plants (resulting, for example, from a slightly greater rate of assimilation or a more economical distribution of material between leaves and axis) may lead to a large difference in final weight. In oats, for example, an increase of 6 per cent. in assimilation might lead to an increase of 50 per cent. in dry weight at the end of 100 days.

The data of earlier workers show that the ' efficiency index ' is highest in the early stages of growth, and then falls slightly. In *Helianthus, Cannabis,* and *Nicotiana* it falls sharply at the beginning of the reproductive period when the inflorescence first appears.

There is evidence that annual plants at the end of their period of growth may lose considerably in dry weight.

27

THE LAW OF PLANT GROWTH

E. A. Mitscherlich

This excerpt was translated expressly for this Benchmark volume
by H. H. Lieth from "Des Gesetz des Pflanzenwachstums" in
Landw. Jarbücher **53**:167–182 (1919).

Theoretical part

. . . If we could provide all factors needed for plant growth at optimal levels, the results should be an infinitely high yield. However, it is especially impossible to change the internal growth factors. We may improve them by breeding new varieties but [we will] never succeed in converting a given species into another species of higher yield. Nature has set limits that allow improvement of yields to a maximum value only, which I shall call A. This maximum yield will naturally be determined by the constellation of all growth factors.

According to the physiological laws, any given plant yield (y) depends on the growth factors $(x_1 \ldots x_n)$ and the level (c) at which they are available to the plants according to the following relation:

$$y = A \cdot (1-e^{-c_1 x_1})(1-e^{-c_2 x_2})(1-e^{-c_3 x_3})(1-e^{-c_4 x_4})(\ldots)(\ldots) \qquad (1)^{[1]}$$

[where] e is the natural log base. In order to find and formulate the "law of physiological relations," I conducted simultaneous growth experiments in which I kept all growth factors constant except the one with which I intended to vary the yield. Let us set here also this variable factor constant and have, therefore, all growth factors constant. We assume now that all factors are calibrated such . . . that all are available in sufficient amounts for the plant yield. We assume here that:

$$c_1 x_1 = c_2 x_2 = c_3 x_3 = c_4 x_4 = \ldots = cx \qquad (2)$$

Given "n" different growth factors, equation (1) may take the following form:

$$y = A(1-e^{-cx})^n \qquad (3)$$

A law, which I want to call the "growth law." . . .

---

[1] Compare B. Baule, Landw. Jahrbücher Bd. 51, p. 339 (1917)

28

THE FOREST AREA OF THE WORLD AND ITS POTENTIAL PRODUCTIVITY

Sten Sture Paterson

[*Editor's Note:* In the original, material precedes these excerpts.]

As previously pointed out, temperature, precipitation, and light are the three climatic factors most important for the life-process of plants. From these factors it should consequently be possible to obtain a value for the "plant growth ability" of a climate. Purely theoretically, this number should have a constitutional relationship with the value of ideal site class. Thus the index should be an expression for the productivity of vegetation caused by the climate. Because of the intimate connection between climate, vegetation, and productivity, the index has been called the CVP-index after the initials of the three words. A calculation of this index value must be based on values for heat, water, and light, and on the correct application of these values in an equation.

[*Editor's Note:* Material has been omitted at this point.]

In the equation, the need of plants for warmth must be represented by means of temperature figures, which are of different kinds: annual averages, monthly averages, extreme values, or ranges. The values which have proved to be most useful here are temperature of warmest month (Tv) and annual range (Ta). The former represents the growth-promoting power of temperature — the more heat. the richer the vegetation. The latter, placed in the denominator of the fraction, is a reducer representing the inhibiting effect on vegetation caused by low values of temperature. Perhaps it would have been more natural to choose the average of the coolest month as a reducer. But that would have given too great importance to the average of the warmest month and, in the case of temperatures below zero, negative values, awkward for the subsequent calculations.

The supply of water is best expressed by annual precipitation (in mm). Its importance for vegetation increases with amount of rainfall and number of rainy days during the year. However, in the following index calculations it proved that an increase in the precipitation values did not have an unlimited effect on productivity. The effect of precipitation seems to have an upper limit as far as the amount of rainfall is concerned. If this value is exceeded, it will not cause a corresponding increase of productivity — rather, a decrease of productivity can be presumed, especially where extremely high rainfall figures prevail, as a result of desilication and other leaching of nutritive salts.

[*Editor's Note:* Material has been omitted at this point.]

By means of temperature and precipitation figures the length of the growing season (G) can be determined. This is necessary in this universal connection because of regional variations in growing season.

There are several different opinions as to the lower temperature limit for the functions of life in plants. As the discussion is still

going on, the author has preferred not to base the determination
of the length of the growing season on any of these opinions but
instead to accept the findings obtained empirically in this connection. Investigations carried out in Sweden have shown that the
growing season in a wide sense coincides with that time during
which the average temperature exceeds $+3°$ C (ÅNGSTRÖM, 1946,
pp. 35—37). The arrangement of the available meteorological material and the effort to obtain thorough conformity in the utilization
of the material and, at the same time, to get manageable numerical
material for the further calculations have here necessitated a
generalization of the above empirical principle. *The growing
season (G) has been calculated to comprise the time during which
the monthly average temperature reaches or exceeds $+3°$ C.*

In warm-temperate to tropical regions where the average temperature of the winter months never falls below $+3°$ C, the growing
season has to be calculated by another method. The relationship
between rainfall and temperature will here decide the resting
periods of vegetation which occur at a certain degree of aridity.
For the calculation of this aridity limit several formulæ are available (KÖPPEN, 1931; LANG, 1915; DE MARTONNE, 1935; WIL-
HELMY, 1944; WANG, 1941). With the exception of LANG all of
them give values very similar to each other (cf. LAUER, 1952, p.
22). Because of that and with the support of LAUER (1952, p. 24)[1]
the author has here used DE MARTONNE'S "Indice d'aridité"
calculated from the formula

$$i = \frac{P}{T + 10}$$

(P = annual precipitation in mm; T = annual average temperature in °C). A comparison between DE MARTONNE'S map works
(1935), and investigations of aridity carried out in the field
(JAEGER, 1928; PITTELKOW, 1928; SORGE, 1930) shows that
DE MARTONNE'S aridity index of 20 corresponds to PENCK'S
"dry limit" (Trockengrenze) as well as KÖPPEN'S AC-B climatic
boundaries. Thus aridity occurs at an index value of 20 or under it.
However, it should be noticed that this calculation shows the

---

[1] "Es wurde daher die Trockengrenzformel nach de Martonne (Ariditäts-
index 20) aus folgenden Gründen ausgewählt: 1. weil sie einfach ist, 2. weil
sie nach Vergleich mit den tatsächlichen Verhältnissen für die Untersuch-
ungsgebiete sich als die treffendste erwies."

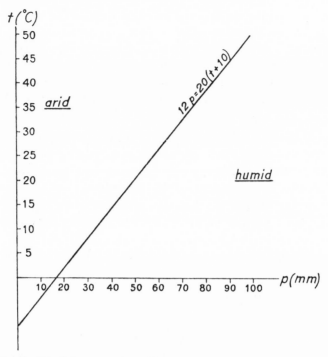

Fig. 7. Limit between humid and arid climates as presented
by Lauer in his development of de Martonne's formula for
index of aridity.

annual average of the climate and does not give an idea of the
variation of seasons. To get such an idea LAUER (1952, p. 23) has
developed DE MARTONNE'S (and KÖPPEN'S) formula to be used
for calculation of indexes for the separate months. The connection
with the annual formula has been preserved, however, by multiplying the quotient by 12. DE MARTONNE'S formula for the
calculation of the number of arid and humid months respectively
during the year thus has the following form,

$$20 = \frac{12\ p}{t + 10}$$

(p = average monthly precipitation; t = average monthly temperature). *The number of arid months constitutes the amount by
which the year should be reduced in order to establish the growing
season (G).*

The simple inverse functional relationship between p and t in the above formula facilitates the drawing of a nomogram in order to establish the boundary between aridity and humidity (Fig. 7). By means of this nomogram and with knowledge of monthly averages of precipitation (p) and temperature (t) the number of arid months can easily be settled for any place on the earth.

The third important vegetation factor, light, is more difficult to put on a numerical basis in a way useful for the equation. The author has attempted to use number of hours when the sun is above the horizon during the year as well as during the growing season — both calculated as a percentage of total number of hours of the year — but the result has not been altogether satisfactory. The fact is that the effect of light changes with the seasons, and will thereby be related to temperature conditions. The latter can therefore be supposed to express the effect of light conditions on vegetation.

The formula of the CVP-index

$$I = \frac{Tv \cdot P \cdot G}{Ta \cdot 12}$$

arrived at by the method dicussed above shows a good correlation between values of ideal site class and index values. The points of the scatter diagram are especially well gathered in the D climate region. This is essentially due to the relatively moderate variations in size of the separate climatic elements in this climatic type. Perhaps it is mainly the importance of the moderate precipitation values on the clustering of the scatter diagram that should be emphasized here. Discussing a C or A climate where precipitation — besides being high — shows examples of great differences even between two almost adjacent areas, we shall get index values reflecting these rainfall conditions. This is also shown quite clearly by the scatter diagram (cf. Fig. 8) where the points are more dispersed in the region of the C-climate (CVP $>$ 400). Dots in A-climates are not represented in this diagram because of over-high values on the CVP-index.

Thus, from a correlative viewpoint the above formula for calculation of CVP-index will give quite satisfactory values. It has one disadvantage, however, and that is the high numerical values in climatic regions of A and C type. Whereas the index values of a D climate lie around a maximum of 300, the corresponding

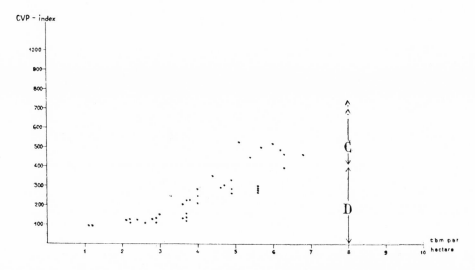

Fig. 8. Scatter diagram of the relation between CVP-index and ideal site class in C- and D-climates.

values of a C climate are about 2000 (e. g. Montgomery, Ala., 1930) and finally, in A climates reach extreme values of about 250,000 (e. g. Andagoya, Columbia, 248,800). Values of this size are cumbersome for further calculation, and therefore impracticable. Besides, they reflect a predominance of precipitation over other climatic elements which is not found in existing conditions of vegetative power. The unfavourable effect on soil caused by abundant rainfall has already been pointed out.

[*Editor's Note:* Material has been omitted at this point.]

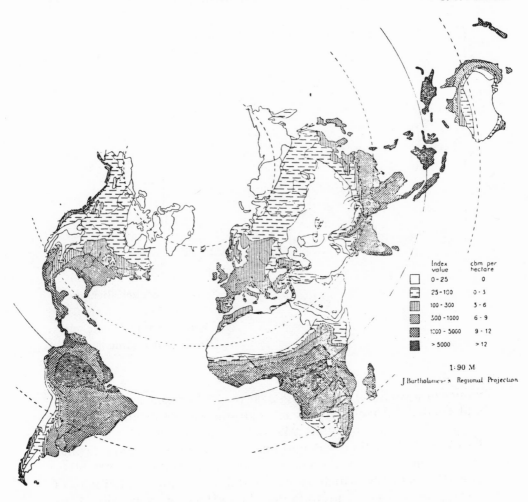

The global forest productivity pattern predicted by the CVP-index. The boundaries (Phyochores) separate regions of similar average climatic "bioeffect." The isophyte 25 represents the climatic timberline. [Source: Lieth 1962.]

REFERENCES

Ångström, A. 1946. Sweriges klimat. Stockholm.

Food and Agriculture Organization. 1948. Forest resources of the world. Washington.

——. Yearbook of forest products statistics. 1948–1951. Washington.

——. Unasylva, Vol. I–VIII. 1947–1954.

Jaeger, F. 1928. Die Gewässer Afrikas. Zeitschrift d. Gessellschaft f. Erdkunde, Berlin.

Köppen, W., and Geiger, R. 1928. Klimakarte der Erde. Gotha.

——. 1931. Grundriss der Klimakunde. Berlin.

——. 1936. Die Klimate der Erde. Handbuch der Klimatologie, I:C.

——. 1939. Klimakunde von Russland in Europa und Asien. Handbuch der Klimatologie, Band III, Teil N. Berlin.

Lang, R. 1915. Versuch einer exakten Klassifikation der Böden in klimatischer und geologischer Hinsicht. Intern. Mitt. f. Bodenkunde.

Lauer, W. 1952. Humide und aride Jahreszeiten in Afrika und Südamerika und ihre Beziehung zu den Vegetationsgürteln. Bonner Geographische Abhandlungen, Heft 9. Bonn.

Lieth, H., ed. 1962. Die Stoffproduktion der Pflanzendecke. Stuttgart: Gustav Fischer Verlag.

Martonne, E. de. 1926. Une novelle fonction climatologique: L'indice d'aridité. La Météorologie, Okt.

Milankovitch, M. 1930. Mathematische Klimalehre und astronomische Theorie der Klimaschwankungen. Handbuch der Klimatologie, Bd. I, Teil A, Berlin.

Mohnsen, K. 1953. Die Grundlagen einer Equatorialwestafrikanischen Gesamtwaldnutzung unter besonderer Berücksichtigung der Faserstoffgewinnung. Mitteilungen der Bundesanstalt für Forst- und Holzwirtschaft, No. 30.

Penck, A. 1910. Versuch einer Klimaklassifikation auf physiogeographischer Grundlage. Sitz.-ber. d. Kgl. Preuss. Akad. d. Wiss., Phys.-math. Kl.XII. Berlin.

Pittelkow, J. 1928. Die Trockengrenze Nordamerikas. Berlin.

Sorge, E. 1930. Die Trockengrenze Südamerikas. Zeitschrift der Gesellschaft für Erdkunde. Berlin.

Thornthwaite, C. W. 1948. An approach toward a rational classification of climate. The Geographical Review, Vol. 38.

Wang, T. 1941. Die Dauer der ariden, humiden und nivalen Zeiten des Jahres in China. Tübinger Geogr. und Geol. Abhandl., Reihe 2. Öhringen.

Wilhelmy, H. 1954. Die klimamorphologische und pflanzengeographische Entwicklung des Trockengebietes am Nordrand Südamerikas seit dem Pleistozän. Die Erde, Heft 3–4.

29

Reprinted from *Ann. Bot.* 29(113):17-37 (1965) by permission of the publisher,
Oxford University Press

Light Distribution and Photosynthesis in Field Crops

BY

J. L. MONTEITH

Rothamsted Experimental Station, Harpenden, Herts.

With nine figures in the Text

ABSTRACT

In a new model of light distribution in field crops a parameter s is the fraction
of light passing through unit leaf layer without interception. Radiation profiles
measured with solarimeters and photocells give values of s from 0·7 for grasses to
0·4 for species with prostrate leaves. Knowing s, leaf transmission τ and leaf-area
index L, the light distribution in a field crop may be described by a binomial
expansion of the form $\{s+(1-s)\tau\}^L$. To calculate crop photosynthesis at given
light intensity this expansion is combined with two parameters describing the
shape of the light-response curve of single leaves. Finally, the assumption that
solar radiation varies sinusoidally allows daily total photosynthesis to be estimated
from daylength and insolation.

The theory predicts about the same potential photosynthesis in a cloudy
temperate climate with long days as in a more sunny equatorial climate with
short days. When $L < 3$ photosynthesis increases as s *decreases*, i.e. as leaves
become more prostrate; but when $L > 5$, photosynthesis increases as s *increases*,
i.e. as leaves become more erect. Assuming that respiration is proportional to
leaf area, estimated dry-matter production agrees well with field measurements
on sugar-beet, sugar-cane, kale, and subterranean clover. Estimates of maximum
gross photosynthesis (for sugar-cane and maize) range from 60 to 90 g m^{-2} day^{-1}
depending on insolation.

INTRODUCTION

WHEN a healthy crop receives adequate water and nutrient, dry-matter
production is governed by solar energy available for photosynthesis.
This is a basic ecological concept, easy to demonstrate on a broad seasonal
basis when annual cycles of growth and radiation are nearly in phase. From
week to week a relationship between growth and radiation is much more
difficult to establish because it depends on parameters, such as leaf area, that
change as the crop matures; and because maximum photosynthesis of most
species is reached at a light intensity well below that of full sunlight. Black
(1963) describes a remarkably consistent dependence of growth on radiation
and leaf area for subterranean clover grown in seed boxes outdoors, but no
comparable records exist for a field crop.

Given the light-response curve of single leaves in the laboratory, the photosynthesis of a stand of the same species in the field may be estimated when it
is known how light distribution below the canopy depends on interception by
leaves (Saeki, 1963). A theoretical value of a 'potential' rate of photosynthesis,
derived by de Wit (1959), was successfully used by Alberda (1962) to interpret

248

the rate of growth of a grass sward, and by Stanhill (1962) in a similar study of lucerne. Few other ecologists have been willing to use simple physical models to describe complex biological behaviour and most analyses of the dependence of growth on weather stop at a statistical correlation without any quantitative study of cause and effect. Some of the difficulties in measurement and interpretation are apparent in this paper but one has already been overcome. There is now a meteorological method of estimating the hourly carbon dioxide exchange of a field crop, making it possible to compare and reconcile rates of photosynthesis measured in the laboratory and in the field (Monteith, 1962).

From laboratory studies on single leaves by Gaastra (1959, 1963) it is expected that the gross photosynthesis of a field crop will be

(i) relatively insensitive to daytime temperature changes during the growing season;
(ii) strongly dependent on the intensity of radiation and its distribution below the canopy; and
(iii) approximately proportional to the concentration of carbon dioxide in air surrounding the leaves.

A change in carbon dioxide from its average concentration (about 300 parts per million by volume) would be expected to produce a proportional change in photosynthesis and to this extent plant growth is 'limited' by the carbon dioxide concentration of the atmosphere. But within the crop canopy, minimum concentrations are usually between 290 and 250 p.p.m. (Tamm and Krzysch, 1961) and rarely fall to 200 p.p.m. (Lemon, 1960). This implies that the decrease of photosynthetic rate attributable to a *local* deficit of carbon dioxide will seldom reach 20 per cent and is probably about 10 per cent on average. In this sense carbon dioxide is not a seriously limiting factor in crop growth and vertical gradients of atmospheric carbon dioxide will be ignored in this paper.

There remain six parameters needed to link photosynthesis and weather. Two depend on site and season, namely mean daily insolation and daylength. Four depend on the crop, as follows. For single leaves it is necessary to know the transmission coefficient and two constants to describe how photosynthesis increases with increasing light intensity: for the crop as a whole it is necessary to specify leaf arrangement in order to calculate interception as a function of leaf area index.

The Light Response of Single Leaves

Laboratory measurements at a fixed concentration of carbon dioxide, give curves for the dependence of rate of photosynthesis p on light intensity I fitted by

$$p = (a+b/I)^{-1}. \tag{1}$$

The constants a and b acquire physical meaning when photosynthesis is regarded as the diffusion of carbon dioxide from the air around the leaf along a chain of diffusive resistances to the sites of absorption in chloroplasts. Under intense light photosynthesis approaches a maximum value of $1/a$, so a is proportional to the sum of external, stomatal, mesophyll, and carboxylation resistances (Monteith, 1963). Because the external resistance of the air surrounding the leaf is a relatively small fraction of the total resistance, values of a measured in the laboratory will be assumed valid for leaves of the same age and condition in the field. Although the theoretical derivation of equation 1 shows that b is not strictly constant, for present purposes b/I may be regarded as a photochemical resistance with b inversely proportional to the quantum efficiency in very weak light.

To work in convenient units, p will be the rate of dry-matter production in grammes of carbohydrate per square metre of leaf area per hour, and I will be the intensity of solar radiation in gramme calories per square centimetre per minute, assumed to contain 45 per cent of energy in the photosynthetically useful range 0·4 to 0·7 μ. It is probably safe to regard this fraction as constant throughout the year, at least in temperate latitudes (Black, 1960). With these units Gaastra's (1959) light-response curves yield values of a ranging from 0·5 for sugar-beet to 1·3 $m^2 hr g^{-1}$ for tomato, whereas b is almost independent of species at about 0·05 cal/g. Other measurements (Hesketh and Musgrave, 1962; Hesketh and Moss, 1963) on sugar-cane and maize gave maximum rates of photosynthesis almost twice the maximum rate for sugar-beet and yielded curve constants of $a = 0·25$, $b = 0·05$.

LIGHT DISTRIBUTION IN A CROP WITH NATURAL ILLUMINATION
The Exponential Model

On a horizontal plane at leaf-area index L measured downwards from a crop canopy, the mean intensity of radiation $I(L)$ is often assumed to follow Beers' Law for extinction in a homogeneous medium

$$I(L) = I(0)e^{-kL} \tag{2}$$

where k is a light-extinction coefficient depending on the transmission of single leaves and on their geometrical arrangement. The dependence of k on leaf and sun angle was examined in great detail by Isobe (1962) and equation 2 was used to calculate rates of photosynthesis, ignoring the distribution of light and shade (Saeki, 1960; Davidson and Philip, 1958). A weakness of this treatment is revealed by examining a horizontal leaf layer below the canopy that receives radiation at a mean intensity of, say, 0·2 cal cm^{-2} min^{-1}. At extremes the radiation could be distributed uniformly over the whole area, to give $p = 1·3$ g m^{-2} hr^{-1}; or it could be concentrated in sunflecks with an intensity of 1·0 cal cm^{-2} min^{-1} covering one-fifth of the area, when it would give $p = 0·4$ g m^{-2} hr^{-1} (using the constants for sugar-beet in equation 1). Although these extremes are rarely met in the field, they point to one possible source of error when mean light intensities, derived or measured, are used to estimate rates of photosynthesis. A further source of error is neglect of the changing spectral composition of radiation, selectively depleted of wavelengths used for photosynthesis as it filters downwards through the crop. The coefficient k, assumed constant in deriving equation 2, will therefore decrease with increasing leaf area as reported by Hayashi and Ito (1962).

Verhagen *et al.* (1963) showed that if k were allowed to vary with leaf-area index, the rate of photosynthesis predicted from equation 2 could exceed the rate when k is constant. From field measurements of growth and leaf area they infer corresponding changes of k in real crops, but the evidence is inconclusive and needs confirmation from light profiles.

A New Model of Light Absorption in a Crop

To describe the changes in intensity and quality of the light penetrating a crop with leaf-area index L, it is convenient to imagine the foliage divided into L horizontal layers, each of unit leaf-area index. It will be sufficient to specify average arrangement and orientation of the leaves by a parameter s which is the fraction of incident radiation that passes through a layer without being intercepted by any leaf or stem. At the limits, s would be zero if there was a continuous horizontal sheet of foliage; and s would be unity for leaves parallel to a collimated beam of radiation. In practice s will be the fractional area of sunflecks below the first leaf layer. Two further asumptions are needed: that horizontal spacing is nearly enough uniform for only zero or one interception per unit layer; and that s for a given crop is constant throughout the day. In practice, s may vary somewhat with solar elevation and cloudiness, and even with windspeed if the leaves move much.

The fraction of radiation intercepted by a unit leaf layer is $(1-s)$ and the fraction of this radiation to be transmitted is $(1-s)\tau$ where τ is a mean transmission coefficient. For a single leaf τ decreases with increasing angle of incidence (Tageeva and Brandt, 1961). For an assembly of randomly oriented leaves, τ will be independent of the direction of incident light, but will be less than the (maximum) transmission at normal incidence. Downward reflection in the same spectral range will help to compensate this loss of transmission and, at a later stage in the analysis, laboratory measurements of transmission by normally illuminated leaves will be used for τ.

The Decrease in Transmission with Depth

If τ_λ is the mean fractional transmission and I_λ is the corresponding radiation intensity in the part of the spectrum between λ and $\lambda+d\lambda$, the effective transmission coefficient over a spectral range is

$$\tau = \int I_\lambda \tau_\lambda \, d\lambda \bigg/ \int I_\lambda \, d\lambda \tag{3}$$

where the limits of integration will be defined after equation 11. The fraction of radiation transmitted through L leaves is then

$$\tau^L = \int I_\lambda \tau_\lambda^L \, d\lambda \bigg/ \int I_\lambda \, d\lambda. \tag{4}$$

Measuring downwards from the top of the crop canopy, the radiation below the first leaf layer is

$$I(1) = [s+(1-s)\tau]I(0). \tag{5}$$

Because of the assumption that there is no leaf overlap in unit layer, then after L layers have been penetrated the intensity is

$$I(L) = [s+(1-s)\tau]^L I(0). \tag{6}$$

In the expansion of this binomial expression, any term of the form τ^l is to be given by equation 4 with $1 \leqslant l \leqslant L$, and the form of the expansion is shown diagrammatically in Fig. 1 for $L = 3$ (i.e. $l = 1, 2, 3$). In general,

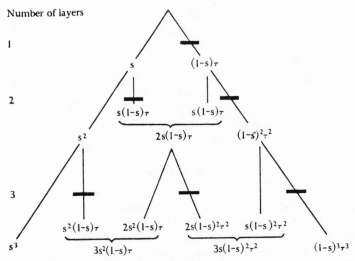

Fig. 1. Transmission and interception in a crop with three leaf layers illustrating the generation of terms in the expression $\{s+(1-s)\tau\}^3$. Bold horizontal lines represent interception.

the flux of radiation after l layers has $l+1$ components. The first component (s^l) is the fractional area of the uninterrupted beam; the second $(ls^{l-1}(1-s)\tau)$ is the fractional area of light that has suffered one interception; and the third $\frac{l(l-1)}{1.2}s^{l-2}(1-s)^2\tau^2$ is the fractional area of light that has suffered two interceptions, and so on.

Integrating over all the leaf layers, the total leaf area receiving light after interception by l higher layers is found by summing the coefficients of τ^l in a series of expressions similar to equation 6 with $L = (l+1), (l+2)$, &c. For an imaginary crop with an infinite number of leaf layers, the sum of these coefficients is $(1-s)^{-1}$ for all values of l. In a real crop with a finite number of layers, $(1-s)^{-1}$ is a limiting leaf area for each integral value of l: light of a given intensity and spectral composition cannot illuminate an area greater than $(1-s)^{-1}$. In Fig. 2 where $s = 0.6$, leaves in full sunlight $(l = 0)$

approach their maximum area (2·5) when L exceeds 7 because all the light going deeper into the crop has already been intercepted once at least. Similarly, leaves getting light after one previous interception $(l = 1)$ approach the same maximum area when L exceeds 10 because all the light penetrating below $L = 10$ has already been intercepted twice at least. In

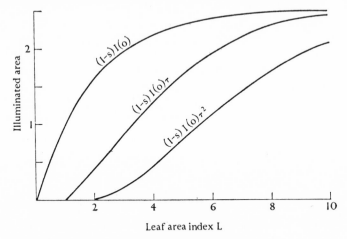

Leaf area index L

Fig. 2. Change of leaf area receiving specified intensity $(1-s)I(o)$, &c., with total leaf-area index L, where $s = o·6$. For leaves in direct sunlight the intensity is $(1-s)I(o)$; for once-shaded leaves $(1-s)I(o)\tau$; and for twice-shaded leaves $(1-s)I(o)\tau^2$.

general, the limiting leaf area for each value of l is reached at a level in the crop where all the light has previously been intercepted by $(l+1)$ leaves or more.

The intensities of equation 6 are those that would be measured by a solarimeter, and when τ and L are known, radiation records can be used to calculate values of s.

Effective Areas for Photosynthesis

Because τ is small, the effective radiation for photosynthesis has only two parts: (i) that intercepted by leaves that can see the sky. These leaves will be called 'sunlit', irrespective of whether the solar radiation they receive is direct or diffuse: (ii) that intercepted by leaves after only one transmission. These leaves will be called 'once shaded'. The areas of each are:

$$\text{Sunlit } A_0 = 1+s+s^2...s^{L-1} = (1-s^L)/(1-s). \qquad (7)$$

(Note that all leaves in the first layer are sunlit—this is part of the basic assumption.)

$$\text{Once shaded } A_1 = (1-s)\{1+2s+3s^2...(L-1)s^{L-2}\}$$
$$= \{1-s^L-(1-s)Ls^{L-1}\}/(1-s). \qquad (8)$$

TABLE I

Mean Leaf-area Indexes in given Layers (two Left-hand Columns) and Transmission Measured by Solarimeters at given Height (two Right-hand Columns)

Layer (cm)	L	Height (cm)	Transmission
	Kale 18 September 1963		
105–100	0·1	122	1·00
100–70	3·8	90	0·38
70–50	1·4	72	0·11
50–30	0·6	52	0·06
30–0	0·1	20	0·03
	Barley 23 June 1963		
		122	1·00
38–30	1·4	37	0·91
30–20	1·5	28	0·51
20–10	0·8	20	0·29
10–0	0·8	12	0·21

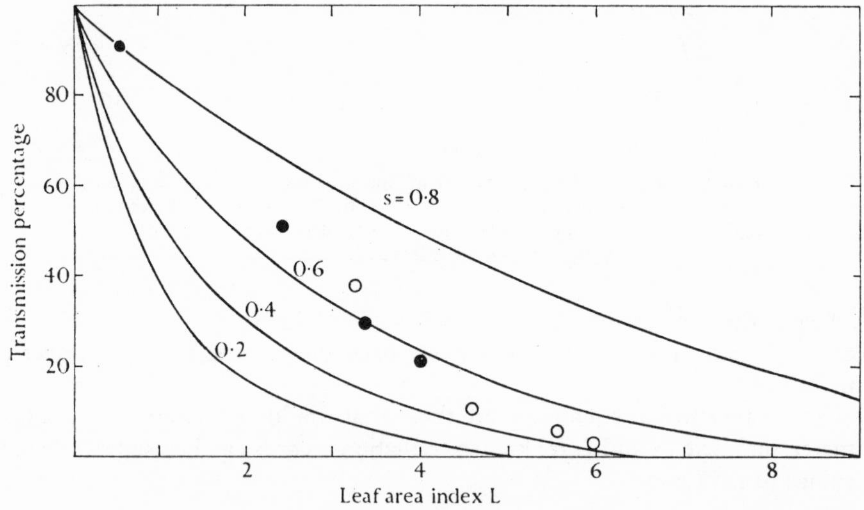

FIG. 3. Theoretical curves for extinction of *total* solar radiation with leaf-area index L at four values of s. Measurements from Table 1:

● barley   ○ kale

EXPERIMENTAL DERIVATION OF s

Transmission of Leaves and Crops

In the summer of 1963 radiation profiles in barley (*Hordeum distichum*, var. Proctor) and in kale (*Brassica olerocea acephala*, var. Improved Thousand-headed) were measured with a set of tube solarimeters described by Szeicz *et al.* (1964), and corresponding profiles of leaf area were found by sampling the crop in layers. The shape of the profile changed very little during the eight hours centred on noon, and Table 1 gives mean profiles for 8·00–16·00 GMT on two days selected for maximum leaf-area index. Knowing the

254

relation between leaf area and height, the fractional transmission of radiation was plotted against leaf area measured from the top of the canopy (Fig. 3).

Small leaf samples from both crops were sent to the I.C.I. Paints Division. There the diffuse transmission of normally incident radiation was measured between 0·4 and 0·9 μ using a spectrophotometer fitted with an integrating sphere. To complete the spectrum, τ_λ was assumed to be constant between

TABLE 2

Leaf Transmission Coefficient $\bar{\tau}_\lambda$ Weighted by Normalized Solar Spectrum y and Normalized Visible Spectrum y′ to give Total Transmission Coefficients for Solar Radiation ($\Sigma\, y\bar{\tau}_\lambda$) and Visible Radiation ($\Sigma\, y′\bar{\tau}_\lambda$)

All values in the table are multiplied by 100

Wavelength Range (mμ)		Solar Spectrum			Kale			Barley		
		y	e_λ	$y′$	$\bar{\tau}_\lambda$	$y\bar{\tau}_\lambda$	$y′\bar{\tau}_\lambda$	$\bar{\tau}_\lambda$	$y\bar{\tau}_\lambda$	$y′\bar{\tau}_\lambda$
400 to	450	4·9	0·8	..	..	..	..	1	..	..
450	500	7·7	10·7	5	..	..	..	3	0·2	0·2
500·	550	8·2	67·0	31	8·5	0·7	2·6	7	0·6	2·1
550	600	8·1	91·4	43	12·0	1·0	5·2	8	0·7	3·4
600	650	8·1	39·1	19	5·5	0·4	1·0	5	0·4	1·0
650	700	7·7	4·5	2	8·0	0·6	0·2	7	0·5	0·1
700	750	6·9	..	..	34·5	2·4	..	27	1·9	..
750	800	5·8	..	..	52·5	3·1	..	45	2·6	..
800	850	5·6	..	..	53	3·0	..	48	2·7	..
850	900	4·9	..	..	53	2·6	..	48	2·3	..
900	950	2·1	..	..	53	1·1	..	48	1·0	..
950	1,000	4·1	..	..	53	2·2	..	48	2·0	..
1,000	1,100	7·8	..	..	53	4·1	..	44	3·4	..
1,100	1,200	3·0	..	..	48	1·4	..	40	1·2	..
1,200	1,300	5·1	..	..	43	2·2	..	36	1·8	..
1,300	1,400	1·1	..	..	38	0·4	..	32	0·3	..
1,400	1,500	0·5	..	..	33	0·2	..	28	0·1	..
1,500	1,600	3·0	..	..	28	0·8	..	24	0·7	..
1,600	1,700	2·4	..	..	23	0·6	..	20	0·5	..
1,700	1,800	1·0	..	..	18	..	..	16	0·2	..
1,800	1,900	0	..	..	13	..	..	12	..	..
1,900	2,000	0·2	..	..	8	..	..	8	..	..
2,000	2,100	0·7	..	..	3	..	..	4	..	..
2,100	2,200	1·1	..	..	0	..	..	0	..	..
		100·0		100		26·8	9·0		23·0	6·8

0·9 and 1 μ, and then to decrease linearly to zero at 2 μ, an assumption which is consistent with the few spectra reproduced by Rabinowitch (1961). Table 2 gives mean values adopted for 0·05 μ bands below 1 μ and 0·1 μ bands above 1 μ. In the near infra-red region the kale leaves are about 5 per cent more transparent than the barley leaves.

The spectral distribution of solar radiation was taken from intensities calculated by Moon (1940) for air mass 2, averaged and normalized to find

$$y = I_\lambda / \Sigma\, I_\lambda$$

where I_λ is the mean intensity in a band 0·05 or 0·1 μ wide corresponding to $\bar{\tau}_\lambda$. The total transmission coefficient was then calculated from a modified form of equation 4

$$\tau^l = \Sigma \, y \bar{\tau}_\lambda^l. \tag{4 a}$$

Variation of s with Depth below the Canopy

Finally, the fraction of transmitted radiation, calculated from equation 6 for four values of s, was plotted in Fig. 3. Combining these theoretical curves with measured transmissions, the observed values of s decreased somewhat with height from 0·8 to 0·6 in barley and from 0·6 to 0·4 in kale. Three explanations are possible:

(i) Near the top of the crops, leaf area changed very rapidly with height. If the plants growing round the solarimeters were only a few centimetres taller or shorter than the sampled plants, the leaf-area index assigned to the highest solarimeters could be in error by as much as ± 1. In Fig. 3 little weight can be given to the barley point at 91 per cent or to the kale point at 38 per cent.

(ii) Near the bottom of both crops, the interception of light by stems was probably comparable with leaf interception. If the leaf-area indexes plotted in Fig. 3 were increased to allow for the intercepting area of stems, the displacement of points towards the right would increase with distance from the canopy and hence with leaf-area index. This would decrease the apparent variation of s with height.

(iii) Relaxing the condition that s is constant with height, a modified form of equation 6 can be derived that allows s to vary exponentially with L and, because the two sets of points are obviously coherent, they can be fitted by a common curve $s = 0\cdot6e^{-0\cdot15L}$. This form may express a real change of leaf posture with height or may correct for vertical changes in the ratio of leaf to stem area discussed in (ii).

Evidence from Elsewhere

To investigate the possible dependence of s on L for a wider range of crops, the results from photocell measurements by Stern and Donald (1961) are plotted in Fig. 4. Assuming that their photocells were perfectly matched to the response of the human eye, the relative photometric intensity of solar radiation (Table 2) was calculated from

$$y' = \bar{e}_\lambda \, \bar{I}_\lambda / \Sigma \, \bar{e}_\lambda \, \bar{I}_\lambda$$

where $\bar{e}_\lambda$ is a relative spectral luminosity from standard tables. Theoretical crop transmission curves calculated from equation 6, and from

$$\tau'^l = \Sigma \, y' \tau_\lambda^l \tag{4 b}$$

were then plotted in Fig. 4. The experimental curve for a sward dominated by Wimmera ryegrass (*Lolium rigidum*) defines a curve with a constant value of s ($\simeq 0\cdot75$) up to $L = 10$. In contrast, s for clover increases from about 0·3 to 0·5 as L increases from 0 to 4, suggesting that a canopy of young horizontal leaves intercepted radiation more efficiently than the foliage below.

Brougham (1958) cut swards of grass and clover to 1 inch and followed the decrease of light intensity at this height during subsequent regrowth. Plotting transmission against leaf-area index above 1 inch shows that *s* was almost constant at 0·65 below perennial ryegrass (*Lolium perenne* L.) as *L* increased from 2·5 to 8; and at 0·4 below white clover (*Trifolium repens* L.) as *L* increased from 2 to 5. Although many more measurements are needed to establish typical values of *s* for different types of crop at different stages of

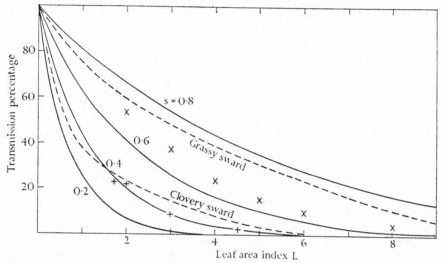

FIG. 4. Full lines are theoretical curves for extinction of *visible* solar radiation with leaf area index *L* at four values of *s*. Dashed lines are from photocell measurements by Stern and Donald (1962) on grassy and clovery swards. Symbols are from photocell measurements by Brougham (1958) on perennial ryegrass (×) and on white clover (+).

development, this analysis proceeds with the assumption that *s* is independent of leaf-area index with values near 0·7 for cereals and grasses, and near 0·4 for clover, kale, and species with prostrate leaves.

Crop Photosynthesis—Theory
Instantaneous Rates

Equation 6 gives the mean light intensity $I(l)$ incident on a horizontal plane immediately above leaf layer $(l+1)$. Leaves within this layer intercept a fraction $(1-s)$ of incident radiation and for a set of non-overlapping parallel leaves at angle θ to the incident light, $(1-s) = \sin \theta$. The radiation intensity on leaf surfaces is then $I(l) \sin \theta = I(l)(1-s)$. Assuming the same *mean* intensity for a set of randomly oriented leaves, photosynthesis by all the sunlit leaves in the $(l+1)$ layer is

$$p_0 = s^l[a+b/I(0)(1-s)]^{-1} \tag{9}$$

from equation 1. In most crops the shape of the light-response curve, and hence the values of *a* and *b*, will change with leaf age, but to make progress

a and b are assumed independent of L. The total instantaneous rate of photosynthesis of all sunlit leaves is then

$$\sum_{l=1}^{L} p_0 = (1-s)A_0[a(1-s)+b/I(0)]^{-1}$$
$$= (1-s^L)[a(1-s)+b/I(0)]^{-1}. \tag{10}$$

Similarly, the rate of photosynthesis by all leaves receiving radiation after interception in one upper layer is

$$\sum_{l=2}^{L} p_1 = (1-s)A_1[a(1-s)+b/I(0)\tau']^{-1}$$
$$= \{1-s^L-(1-s)Ls^{L-1}\}[a(1-s)+b/I(0)\tau']^{-1} \tag{11}$$

where τ' is weighted mean transmission between 0·4 and 0·7 μ. Because for many species τ' is between 0·05 and 0·1 (Kasanaga and Monsi, 1954), photosynthesis by leaves that receive radiation after two or more interceptions can be neglected. In all subsequent calculations τ' is assumed equal to 0·1 for convenience. Then total rate of photosynthesis is given by

$$P = \Sigma p_0 + \Sigma p_1. \tag{12}$$

Note that because $(1-s)A_0$ and $(1-s)A_1$ tend to unity for large values of L, they express the efficiency with which the crop intercepts radiation, whereas the terms within square brackets in equations 10 and 11 describe the performance of individual leaves.

Without modifying the general form of equation 12, the intensity I can be weighted to allow for changes in the efficiency of light-energy conversion with wavelength, but so few action spectra have been published that this refinement seems unwarranted.

Daily Totals

To integrate equation 12 over the hours of daylight, solar radiation may be assumed to vary sinusoidally between sunrise and sunset. Analysis of solar radiation measurements at Wageningen (de Vries, 1955) suggests that this approximation is good for cloudless days in all climates, and is adequate for the diurnal variation averaged over periods of a month in climates where mean cloudiness is constant throughout the day. Expressing the intensity of radiation reaching the top of the crop as

$$I(0) = I^* \sin(\pi t/h) \tag{13}$$

where h is the duration of daylight and t is time after sunrise in the same units, then the daily total radiation is

$$S = \int_0^h I(0)\, dt = 2I^*h/\pi. \tag{14}$$

The amplitude I^* is an *apparent* mid-day intensity to be found from a given radiation total S and from daylength read from tables. For example, on a cloudless mid-summer day at latitude 52° N in southern England, the maximum insolation at sea-level is about 700 cal cm^{-2} day^{-1}. With $h = 1{,}000$ minutes, $I^* = 1{\cdot}14$ cal cm^{-2} min^{-1}. Although this figure is about 10 per cent less than the observed maximum, it will always be more accurate to estimate daily photosynthesis from I^*, proportional to total radiation income, than from the true maximum intensity at mid-day. Assuming a, b, and s independent of time of day, the total photosynthesis of *sunlit* leaves is

$$2 \int_0^{h/2} \Sigma p_0 \, dt = 2(1-s)A_0 \int_0^{h/2} \{a(1-s)+b/I^* \sin(\pi t/h)\}^{-1} dt$$

$$= hA_0 \, a^{-1}\{1-f(\eta_0)\} \tag{15}$$

where $$\eta_0 = \{a(1-s)I^*-b\}/\{a(1-s)I^*+b\}. \tag{16}$$

The parameter η_0 varies from -1 when carbon dioxide diffusion is limited by very weak light to $+1$ when all leaves are light saturated.

When $a(1-s)I^* > b$, i.e. $1 > \eta_0 > 0$

$$f(\eta_0) = \frac{1}{\pi} \frac{1-\eta}{\eta^{\frac{1}{2}}} \ln\left(\frac{1+\eta^{\frac{1}{2}}}{1-\eta^{\frac{1}{2}}}\right) \tag{17 a}$$

and when $a(1-s)I^* < b$, i.e. $-1 < \eta_0 < 0$

$$f(\eta) = \frac{2}{\pi} \frac{1-\eta}{(-\eta)^{\frac{1}{2}}} \tan^{-1}(-\eta)^{\frac{1}{2}}$$

$$= \frac{2}{\pi} \frac{1+|\eta|}{|\eta|^{\frac{1}{2}}} \tan^{-1}|\eta|^{\frac{1}{2}}. \tag{17 b}$$

In the limit when $\eta = 0$ both equations 17 a and 17 b give

$$f(\eta) = 2/\pi. \tag{17 c}$$

Because $hA_0 \, a^{-1}$ is maximum possible photosynthesis with infinite radiation, in effect, when all the sunlit leaves are light-saturated, the parameter η may be regarded as a saturation factor and $\{1-f(\eta)\}$ as a saturation efficiency. In the top part of Fig. 5, η is plotted as the abscissa for a range of values of s and I^*, assuming that $a = 0{\cdot}5$ and $b = 0{\cdot}05$. Having found the appropriate value of η for specified values of s and I^*, the saturation efficiency for sunlit leaves can be read as the ordinate of the lower part of the graph. The total photosynthesis of sunlit leaves ($l = 0$) is then given by equation 15. The same procedure is repeated to find the total photosynthesis by once-shaded leaves with $l = 1$ replacing I^* by $I^*\tau'$ to find η_1, and replacing A_0 by A_1. Daily total photosynthesis is then given by

$$P = ha^{-1}[A_0\{1-f(\eta_0)\}+A_1\{1-f(\eta_1)\}]. \tag{18}$$

Application of the Theory

Photosynthesis and daylength

Although the validity of equation 18 depends on several untested assumptions, the predicted values of photosynthesis and their dependence on plant and weather factors provide a useful guide to the interpretation of field experiments. For example, some workers relate crop yields to mean daily totals of radiation, although a given amount of radiation distributed over a long day will not necessarily give the same amount of photosynthesis when received at greater intensities over a short day. Fig. 6 shows how maximum photosynthesis changes with daylength and radiation according to equation 18

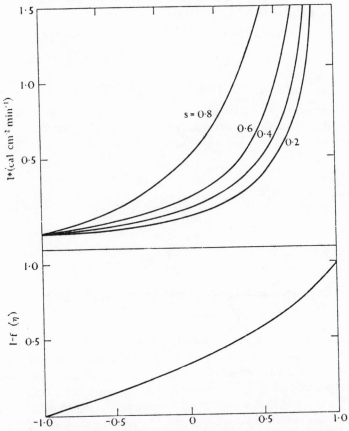

FIG. 5. Upper: variation of saturation factor η with effective intensity $I*$ and interception factor s, assuming $a = 0.5$, $b = 0.05$. Lower: variation of $1-f(\eta)$ with η.

with $s = 0.4$ and $A_0 = A_1 = (1-s)^{-1}$. The expected rate of photosynthesis is almost the same in an equatorial climate where $h = 12$ hours, and $S = 500$ cal cm^{-2} day^{-1} as it is in a cloudy temperate climate where $h = 16$ hours and $S = 300$ cal cm^{-2} day^{-1}. In subsequent calculations, the assumed daylength is 14 hours unless otherwise specified.

Photosynthesis and leaf posture

Fig. 7 shows how photosynthesis is expected to increase with leaf area for three values of s and two values of S corresponding to summer days with

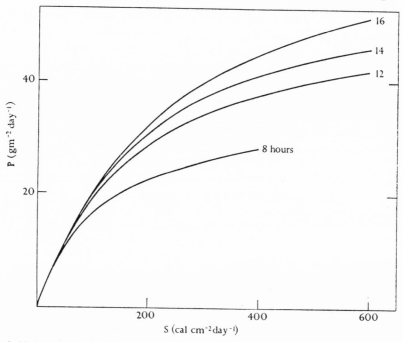

FIG. 6. Variation of gross photosynthesis P with daily insolation S, and day-length, assuming complete interception of radiation by a crop with $a = 0.5$, $b = 0.05$.

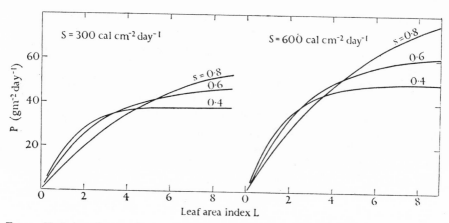

FIG. 7. Variation of gross photosynthesis P with leaf-area index L, interception factor s, and daily insolation S, assuming $h = 14$ hours, $a = 0.5$, $b = 0.05$.

and without cloud. As s increases at constant leaf area, each leaf layer intercepts less radiation; but to compensate, the radiation is received at smaller intensities and so the efficiency of energy conversion is greater. When leaf-area index is less than 3, photosynthesis increases as s *decreases* because the amount of light the crop intercepts is more important than its even distribution at a low intensity over all the leaves. When leaf-area index exceeds 5 and s is less than 0·8, interception is virtually complete. Photosynthesis then depends on the even distribution of radiation with height and so increases with s. Values of $a = $ 0·5, $b = $ 0·05 were used in Fig. 7, and for larger values of a the curves intersect at larger values of L.

Fig. 7 is relevant to the work of Watson and Witts (1959) who found that a cultivated variety of sugar-beet characterized by erect leaves produced dry matter more rapidly than three wild species with more prostrate leaves. They attributed the smaller yield of the wild plants to less uniform distribution of light over their leaves. These experimental results were used by Saeki (1960) to support a theoretical dependence of photosynthetic rate on leaf posture, but his analysis avoids a direct quantitative comparison between experiment and theory. At the maximum index of 2·8 reached in this experiment, Fig. 7 shows that the greater interception of light by the wild plants would more than compensate for their less uniform illumination. Had there been no difference other than leaf posture, the 'wild' plants would have produced more dry matter than the 'cultivated' plants over the first few weeks of growth. This illustrates the difficulty of attempting to predict the effect of leaf arrangement on photosynthesis without fully analysing the light régime.

COMPARISON OF THEORY AND FIELD MEASUREMENTS

A theoretical rate of crop dry-matter production C can be found from the calculated gross photosynthesis P by subtracting the respiratory loss R. Measurements of C reported in the literature are usually mean production rates for periods of a few weeks and are referred to the mean leaf-area index between harvests. Measurements of respiration are very few, but R/P is usually between 0·25 and 0·5 (Gaastra, 1963). So theoretical estimates of P may be regarded as consistent with field measurements of C when $(P-C)/P$ lies between 0·25 and 0·5. The balance between photosynthesis and respiration determines the optimum leaf-area index at which dry-matter production reaches a maximum, and to find this area theoretically respiration is assumed proportional to leaf area (e.g. Saeki, 1960). The term 'specific respiration' will be used here for R/L in units of g carbohydrate per m² leaf area per day.

Sugar-beet and Kale

From several experiments in which kale and sugar-beet were thinned to different values of L, Watson (1958) found that the dry-matter production of kale reached a maximum when L was between 3 and 5 whereas the production of sugar-beet increased up to the largest measured value of $L = 6$. Fig. 8 shows one set of these measurements for each crop with two theoretical curves calculated from the parameters in Table 3. For sugar-beet Gaastra's values of $a = $ 0·5, $b = $ 0·05 were used with $s = $ 0·6, a relatively large value allowing for the erect posture of the leaves. If specific respiration is now assumed to be 2·5 g m⁻² day⁻², agreement between observation and prediction

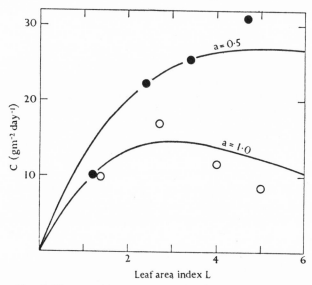

FIG. 8. Theoretical curves for net crop dry-matter production C assuming $R = 0.25L$ (see Table 3). Measurements by
Watson (1958) on sugar-beet (●) and on kale (○).

TABLE 3

*Parameters for Kale and Sugar-beet for Comparison with Measurements
by Watson (1958) Plotted in Fig. 8*

Crop	Kale		Sugar-beet	
Period	16–26 July 1954		5–16 August 1954	
S (cal cm^{-2} day^{-1})	281		265	
h (hours)	16		15	
$I*$ (cal cm^{-2} min^{-1})	0.46		0.46	
s	0.4		0.6	
a	1.0		0.5	
b	0.05		0.05	
	(g m^{-2} day^{-1})		(g m^{-2} day^{-1})	
	P	$(P-2.5L)$	P	$(P-2.5L)$
Leaf-area index 1	11.4	8.9	15.0	12.5
2	18.4	13.4	25.2	20.2
3	22.5	14.7	31.9	24.4
4	23.8	13.8	36.4	26.4
6	25.6	10.6	41.8	26.8

is very good. The ratio R/P increases from 0.17 to 0.36 per cent as L increases
from 1 to 6.

The kale produced much less dry matter than the sugar-beet and if the
specific respiration was the same for both crops, an explanation may be
sought in different values of s and a. With $s = 0.4$, to conform with results
in section 4, a resistance $a = 1.0$ is needed to give a theoretical curve con-
forming with the observation. It is not yet known whether the twofold

increase of non-photosynthetic resistance from sugar-beet to kale is characteristic of the two species or whether it reflects differences in the moisture and nutrient status of the soil during the two experiments.

The analysis emphasizes that the dependence of crop growth on leaf area may be sensitive to small physiological differences as well as to weather, so that attempts to obtain maximum yields by controlling leaf area are often frustrated (Watson and French, 1962). Even when physiological differences are trivial, the relationship between growth and leaf area will change with radiation and daylength as discussed in the section following.

Subterranean Clover

In a study of dry-matter production by subterranean clover, Black (1963) controlled leaf area by thinning plants grown in seed boxes, and he decreased the radiation by shading the plants with hessian screens. Three sets of Black's measurements are now chosen for comparison with theory because they cover a wide range of leaf area at similar temperatures. Table 4 gives the experimental conditions, and the results are plotted as points in Fig. 9. To fit theoretical curves to these points, S and h were taken from Table 4, s was taken as $0 \cdot 4$ from the analysis of light measurements by Stern and Donald (1962), and b was again assumed to be $0 \cdot 05$. It remained to choose appropriate values of a and R/L. Trial calculations for $L = 4$, $S = 290$ cal cm^{-2} day^{-1} showed that $(P-C)/P$ $(= R/P)$ exceeded $0 \cdot 5$ when $a = 0 \cdot 5$ and decreased to zero when a was increased to $1 \cdot 5$. The observed decrease of C at large values of L showed that R/L $(\simeq C/L$ when L is large) was approximately 2 g m^{-2} day^{-1}. Guided by these clues to probable values of a and R/L, the best fit between theory and measurement was found at $a = 1 \cdot 0$, $R/L = 1 \cdot 5$ g m^{-2} day^{-1}. Fig. 9 shows that the calculated curves fit the observations well when L exceeds 3 but overestimate dry-matter production at smaller values of L. The discrepancy may be attributable to the storage of dry matter in roots, neglected in these experiments because 'root weights were too variable and inconsistent to be of value'. Because rooting volume was limited by the size of the seed boxes, significant root expansion may not have been possible except in boxes where the plants were severely thinned to small values of L.

CONCLUSIONS

In the previous section the theoretical dependence of dry-matter production on leaf-area index was reconciled with field measurements for three crops and in two climates, but agreement depends on the arbitrary choice of one unknown parameter for sugar-beet (R/L) and two for both kale and clover $(a, R/L)$. The real test of this and other similar theories will come from experiments in which *all* the relevant physical and biological parameters are measured, either in the laboratory or in the field. In the laboratory more single-leaf measurements are needed to determine the shape of light response curves for different species and to follow changes of this shape with ageing and with the extraction of moisture and nutrients from the soil. Measurements of respiration from different plant organs are needed to determine how specific respiration rates change with both temperature and crop development. In the field, conventional measurements of solar radiation, dry-matter

TABLE 4

Parameters for Subterranean Clover for Comparison with Measurements by Black (1963) Plotted in Fig 9

$$a = 1 \cdot 0, \ b = 0 \cdot 05, \ s = 0 \cdot 4$$

Period	22 May–13 June 1961 (Early winter, unshaded)		10–24 November 1961 (Spring, shaded)		10–24 November 1961 (Spring, unshaded)	
S (cal cm^{-2} day^{-1})	230		290		500	
h (hours)	10		13		13	
I^* (cal cm^{-2} min^{-1})	0·60		0·50		1·00	
Mean air temperature (°C)	15		18		18	
	g m^{-2} day^{-1}		g m^{-2} day^{-1}		g m^{-2} day^{-1}	
	P	$(P-0 \cdot 15L)$	P	$(P-0 \cdot 15L)$	P	$(P-0 \cdot 15L)$
Leaf-area index 1	8·9	7·4	9·8	8·3	11·1	9·6
2	14·3	11·3	16·0	13·0	18·7	15·7
3	17·2	12·7	19·6	15·1	23·1	18·6
4	18·4	12·4	21·1	15·0	25·1	19·1
6	19·7	10·7	22·6	13·6	26·8	17·3
9	19·9	6·4	22·9	8·4	27·2	13·7

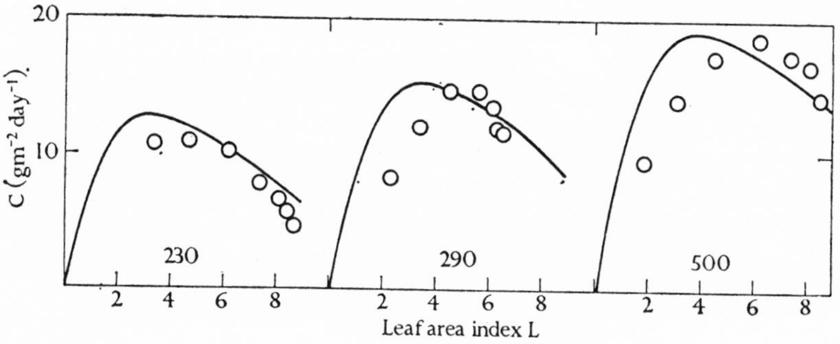

FIG. 9. Theoretical curves for net crop dry-matter production C at three values of S, assuming $R = 0 \cdot 15L$ (see Table 4). Measurements by Black (1963) on subterranean clover (○).

production, and leaf area can usefully be supplemented by studies of light interception in relation to the distribution and posture of leaves. The International Biological Programme may provide an important opportunity for standardizing the measurement of plant and weather parameters and for comparing crop yields in a wide range of climates with theoretical estimates of total dry-matter production.

To integrate daily rates of dry-matter production over the whole life of a crop at least two more parameters are needed. The first is the rate at which the photosynthetic system expands for a given rate of dry-matter production or the 'leaf-to-total growth ratio' proposed by Jackson (1963). The second is mean air temperature, often determining the length of the growing season and interacting with radiation intensity to govern rates of germination, development, and leaf expansion. The interaction has been studied in controlled environments (Blackman, 1956; Milthorpe, 1959; Newton, 1963) but

TABLE 5

Comparison of Estimated Gross Photosynthesis (P) *and Measured Yield* (C) *in* $g\,m^{-2}\,day^{-1}$

Equivalents in cwt acre^{-1} day^{-1} given in brackets

Crop	Site	Period	cal cm^{-2} day^{-1}	hours	a	P	C
Sugar-cane[1]	Hawaii 21° N	14.10.41 to 12.1.42	400	11	0·25	68 (5·4)	42 (3·3)
Maize[2]	Israel 32° N		600*	13*	0·25	90 (7·2)	42 (3·3)
Sugar-beet[3]	England 52° N	5.8.54 to 16.8.54	265	15	0·5	45 (3·6)	32 (2·5)
X	England 52° N		265	15	0·25	64 (5·1)	

[1] Borden (1942).
[2] Cited by Westlake (1963).
[3] Watson (1958).
* Assumed values.

still seems too complicated to incorporate in formal analysis. Meanwhile, in a climate where growth is unrestricted by temperature range, analyses based on radiation and daylight alone can provide estimates of maximum dry-matter production, or 'potential' photosynthesis. As a demonstration, Table 5 contains three of the largest values of dry-matter production C reported in the literature, with comparable estimates of photosynthesis P for crops with enough leaves to intercept all incident radiation.

For sugar-beet in a temperate climate the difference between estimated photosynthesis and crop yield could be attributed to respiration of $0·29P$ or 13 g m² day^{-1}. In warmer climates ratios of respiration to gross photosynthesis will exceed those in temperate climates, and for sugar-cane the difference between photosynthesis and yield could be attributed to respiration of $0·38P$ or 26 g m^{-2} day^{-1}. Assuming respiration of $0·4P$ for maize, the estimated rate of dry-matter production is 54 g m^{-2} day^{-1}. As the measured rate of 42 g m^{-2} day^{-1} represents an average over the whole life of the crop, the apparent loss of 12 g m^{-2} day^{-1} is attributable to the incomplete interception of radiation during early stages of growth.

Figs. 6 and 7 show that differences in leaf arrangement are relatively unimportant in determining maximum photosynthesis, and that a small intensity of radiation maintained over a long day may allow as much photosynthesis as a larger radiation total concentrated in a shorter day. This implies that the relatively small photosynthesis of sugar-beet in Table 5 is limited by the physiological behaviour of the leaf through the resistance *a* rather than by insufficient radiation. The last entry in the table shows that if a crop with $a = 0.25$ could be grown in a temperate summer, gross photosynthesis might reach 64 $g\,m^{-2}\,day^{-1}$, a figure that is close to the limit of 71 $g\,m^{-2}\,day^{-1}$ derived from more general reasoning by Loomis and Williams (1963).

ACKNOWLEDGEMENTS

I am most grateful to Dr. H. L. Penman for suggesting radical improvements to the first draft of this paper; to other colleagues for supplying field measurements; and to Mr. S. E. Orchard of the I.C.I. Paints Division, Slough, for providing leaf transmission spectra.

LITERATURE CITED

ALBERDA, TH., 1962. Actual and Potential Production of Agricultural Crops. *Neth. J. Agric. Sci.* **10**, 325.

BLACK, J. N., 1960. The Relationship between Illumination and Global Radiation. *Trans. Roy. Soc. S. Australia*, **83**, 83.

—— 1963. The Interrelationship of Solar Radiation and Leaf Area Index in Determining the Dry Matter Production of Subterranean Clover. *Aust. J. Agric. Res.* **14**, 20.

BLACKMAN, G. E., 1956. Influence of Light and Temperature on Leaf Growth. In *The Growth of Leaves*, p. 151. (Ed.) Milthorpe, F. London: Butterworth.

BORDEN, R. J., 1942. The Nitrogen Fertilization of the Sugar Cane Crop. *Haw. Plant. Rec.* **46**, 191.

BROUGHAM, R. W., 1958. Interception of Light by the Foliage of Pure and Mixed Stands of Pasture Plants. *Aust. J. Agric. Res.* **9**, 39.

DAVIDSON, J. L., and PHILIP, J. R., 1958. Light and pasture growth. In *Arid Zone Research XI. Proceedings of the Canberra Symposium*, p. 181. Paris: UNESCO.

GAASTRA, P., 1959. Photosynthesis of Crop Plants. *Meded. Landb-Hoogesch., Wageningen*, **59**, 1.

—— 1963. Climatic Control of Respiration and Photosynthesis. In *Environmental Control of Plant Growth*, p. 113. (Ed.) Evans, L. New York: Academic Press.

HAYASHI, K., and ITO, I., 1962. Studies on the Form of Plant in Rice Varieties: (i) Significance of the Extinction Coefficient. *Proc. Crop. Sci. Soc. Japan*, **30**, 329.

HESKETH, J. D., and MOSS, D. N., 1963. Variations in the Response of Photosynthesis to Light. *Crop Sci.* **3**, 311.

HESKETH, J. D. and MUSGRAVE, R. B., 1962 Light Studies with Individual Corn Leaves. *Crop Sci.*, **2**, 311.

ISOBE, S., 1962. Preliminary Studies in Physical Properties of Plant Communities. *Bull. Nat. Inst. Agric. Sci. Tokyo*, A. No. 9, p. 29.

JACKSON, J. E., 1963. Relationship of Relative Leaf Growth Rate to Net Assimilation Rate. *Nature, Lond.* **200**, 209.

KASANAGA, H., and MONSI, H., 1954. On the Light Transmission of Leaves. *Jap. J. Bot.* **14**, 304.

LEMON, E. R., 1960. An Aerodynamic Method for Determining the Carbon Dioxide Exchange between the Atmosphere and a Corn Field. *Agron. J.* **52**, 697.

LOOMIS, R. S., and WILLIAMS, W. A., 1963. Maximum Crop Productivity. *Crop Sci*, **3**, 67.

MILTHORPE, F. L., 1959. Studies in the Expansion of the Leaf Surface: (i) The Effect of Temperature. *J. Exp. Biol.* **10**, 233.

MONTEITH, J. L., 1962. Measurement and Interpretation of Carbon Dioxide Fluxes in the Field. *Neth. J. Agric. Sci.* **10**, 334.

—— 1963. Gas Exchange in Plant Communities. In *Environmental Control of Plant Growth*, p. 95. (Ed.) Evans, L. T. New York: Academic Press.

MOON, P., 1940. Proposed Standard Radiation Curves. *J. Franklin Inst.* **230**, 583.

NEWTON, P., 1963. Studies on the Expansion of the Leaf Surface. II. The Influence of Light Intensity and Daylength. *J. Exp. Bot.* **14**, 458.

RABINOWITCH, E. I., 1961. *Photosynthesis.* Vol. 2, part 1. New York: Wiley (Interscience).

SAEKI, T., 1960. Interrelationship between Leaf Amount, Light Distribution and Total Photosynthesis in a Plant Community. *Bot. Mag. Tokyo,* **73,** 55.

—— 1963. Light Relations in Plant Communities. In *Environmental Control of Plant Growth,* p. 79. (Ed.) Evans, L. T. New York: Academic Press.

STANHILL, G., 1962. The Effect of Environmental Factors on the Growth of Alfalfa. *Neth. J. Agric. Sci.* **10,** 247.

STERN, W. R., and DONALD, C. M., 1962. Light Relationship in Grass–Clover Swards. *Aust. J. Agric. Res.* **13,** 599.

SZEICZ, G., MONTEITH, J. L., and DOS SANTOS, J. M., 1964. Tube Solarimeter Measuring Radiation among Plants. *J. Appl. Ecol.* **1,** 169.

TAGEEVA, S. V., and BRANDT, A. B., 1961. Optical Properties of Leaves Depending on the Angle of Light Incidence. In *Progress in photobiology.* (Ed.) Christensen, B. C., *et al.,* Amsterdam: Elsevier.

TAMM, E., and KRYSCH, G., 1961. Zum verlauf des CO_2-Gehaltes der luft. *Z. Acker-u. PflBau.* **112,** 253, 377.

VERHAGEN, A. M. W., WILSON, J. H., and BRITTEN, E. J., 1963. Plant Production in Relation to Foliage Illumination. *Ann. Bot.* N.S. **27,** 641–6.

DE VRIES, D. A., 1955. Solar Radiation at Wageningen. *Meded. Landb-Hoogesch., Wageningen,* **55,** 277.

WATSON, D. J., 1958. The Dependence of Net Assimilation Rate on Leaf Area Index. *Ann. Bot.* N.S. **22,** 37.

—— and FRENCH, S. A. W., 1962. An Attempt to Increase Yield by Controlling Leaf Area Index. *Ann. Appl. Biol.* **50,** 1.

—— and WITTS, K. J., 1959. The Net Assimilation Rates of Wild and Cultivated Beets. *Ann. Bot.* N.S. **23,** 431.

WESTLAKE, D. F., 1963. Comparison of Plant Productivity. *Biol. Rev.* **38,** 385.

DE WIT, C. T., 1959. Potential Photosynthesis of Crop Surfaces. *Neth. J. Agric. Sci.* **7,** 141.

30

Reprinted from *Am. Nat.* **102**(923):67–74 (1968)

NET PRIMARY PRODUCTIVITY OF TERRESTRIAL COMMUNITIES: PREDICTION FROM CLIMATOLOGICAL DATA

MICHAEL L. ROSENZWEIG

Division of Biology, University of Pennsylvania, Philadelphia, Pennsylvania,
and
Department of Biology, Bucknell University, Lewisburg, Pennsylvania 17837

INTRODUCTION

A major research problem facing ecologists today is the quantification of energy flow in natural communities. Discussion of other ecological problems revolving around productivity would certainly benefit if some readily available quantitative estimate of productivity existed. Pianka (1966), for example, discussed the relationship of species diversity to productivity; unfortunately, he had no quantitative data to aid him. In my own work involving an analysis of the evolutionary causes of body size clines in certain terrestrial carnivores, I long ago appreciated the needs for some quantitative estimate of energy flow.

In 1947 Holdridge published a scheme which underlined the great degree to which the abiotic environment determines somehow the characteristics of the mature vegetation of terrestrial communities. This set me to wondering whether some environmental variable might exist which correlates well enough with primary terrestrial production to be usable as its predictor.

Major (1963) has noted that the actual evapotranspiration (AE) in a terrestrial environment is qualitatively related to the amount of vascular plant activity. AE may be defined as precipitation, minus runoff, minus percolation (Sellers, 1965). AE may be thought of as the reverse of rain. It is the amount of water actually entering the atmosphere from the soil and the vegetation during any period of time, that is, the evaporation plus the transpiration (Sellers, 1965). Obviously, this atmospheric entry simultaneously requires water and sufficient energy to make the phase transfer of the water possible. Thus, AE is a measure of the simultaneous availability of water and solar energy in an environment during any given period of time. Note that Holdridge's (1947) scheme depended on two variables similar to the latter, that is, precipitation and biotemperature. Further, Holdridge (1959) later showed that biotemperature and potential evapotranspiration (PE) are linearly related. PE may be defined as the amount of evapotranspiration that could occur if the soil of a large area having "vegetation typical of the surroundings" (Sellers, 1965, p. 163) were kept constantly wet, that is, at or above field capacity (Sellers, 1965). In fact, PE is the solar energy estimate used in calculating AE. Hence, it is quite relevant to wonder if AE is an environmental variable usable for the purpose of predicting production.

METHODS AND CRITERIA OF DATA COLLECTION

Thornthwaite and Mather (1957) have published a useful reference enabling an estimate of AE to be calculated from a knowledge of the latitude of a place and its mean temperature, month by month, and its precipitation, month by month. I have used such estimates of AE in this study.

Because of the fact that the special requirements of this analysis (see below) render an already scarce kind of measurement (production) scarcer, it was necessary to select as the dependent variable, net above-ground productivity—the commonest kind of production estimate. Study of other sorts of production data in the future will undoubtedly prove interesting.

Quite clearly, each production value in our collection of data (Table 1) must be accompanied by a fairly reliable estimate of the AE for the same period of time. Ideally, both productivity and AE are measured simultaneously. I know of not even one case where this happy coincidence has occurred. Fortunately, several ecologists have had the foresight to include some meteorological data along with their productivity work. Unfortunately, such data are only rarely convertible into an AE. Since this proscribes almost all work done previously, I simply made the convention of using an average annual AE garnered from some nearby weather station by the Laboratory of Climatology, Centerton, New Jersey (Thronthwaite Associates, 1962, 1964).

Due to a lack of the necessary climatological data for most tundra studies, AE could be estimated for only two constantly moist tundras where $AE = PE$.

In arid environments, where PE is greater than precipitation, over a broad area annual AE is very nearly equal to annual precipitation. Local runoff is common, however, and creates extremely variable plot-to-plot conditions (Hillel and Tadmor, 1962). These depend largely on local topographical relief. Creosote bushes are generally not inhabitants of desert slopes or gullies, but instead cover relatively broad, flat stretches of desert. Thus, point A, the creosote bush desert, was usable. Also, Pearson (1966) specifically states that there is little or no runoff from his sand dune associations. This is in agreement with the results of Hillel and Tadmor (1962). Hence, the precipitation Pearson gives for these dunes may be taken as their AE, and we obtain our point N.

Except for the one value used here, all tall grass prairie productions I have seen were obtained by a fall harvest method. Such a procedure does not take into account the amount of material that dies during the growing season. By completely denuding his prairie plots before beginning his harvests, Penfound (1964) showed the importance of such omissions in creating a gross underestimate of production. Penfound's was, therefore, the only tall grass prairie productivity used (point *D*). This work had the additional advantage of including simultaneous monthly temperatures and precipitations.

The data of Whittaker (1963, 1966) were obtained in the Great Smoky Mountains. Because of adiabatic cooling of the atmosphere, altitude is a

TABLE 1

SUMMARY OF PRODUCTION DATA USED IN REGRESSION

Code	Environment	Place	Log NAAP	Log AE	Reference
A	Creosote bush desert*	Nye Co., Nevada, U.S.	1.60	2.10	Odum, 1959
B	Arctic moist tundra†	Cape Thompson, Alaska, U.S.	2.16	2.30	Rickard, 1962
C	Alpine moist tundra‡	Mt. Washington, N. H., U.S.	2.16	2.37	Hadley and Bliss, 1964
D	Tall grass prairie*	Norman, Oklahoma, U.S.	2.75	2.79	Penfound, 1964
E	Heath Bald (Leiophyllum)§	Great Smoky Mts., Tenn., U.S.	2.66	2.58	Whittaker, 1963
F	Heath Bald (Rhododendron)§	Great Smoky Mts., Tenn., U.S.	2.61	2.58	Whittaker, 1963
G	Heath Bald (Rhododendron)§	Great Smoky Mts., Tenn., U.S.	2.69	2.58	Whittaker, 1963
H	Mixed Heath (Peregine Peak)§	Great Smoky Mts., Tenn., U.S.	2.58	2.72	Whittaker, 1963
J	Mixed Heath (Rocky Spur)§	Great Smoky Mts., Tenn., U.S.	2.80	2.69	Whittaker, 1963
K	Beech-maple forest*	Toronto, Ontario, Canada	2.98	2.75	Bray, 1964
L	Secondary tropical forest	Kade, Ghana	3.34	3.09	Nye, 1961
M	Tropical forest	Yangambi, Congo (Leopoldville)	3.46	3.12	Bartholomew, Meyer, and Laudelout, 1953
N	Cool desert sand dunes‡	Near Rexburg, Idaho, U.S.	2.24	2.34	Pearson, 1966
O	Oak-hickory forest§	Oak Ridge, Tenn., U.S.	3.08	2.92	Whittaker, 1966
P	Cheatgrass§	Hanford Reservation, Washington, U.S.	2.01	2.25	Rickard, 1962
Q	Fraser fir forest§	Great Smoky Mts., U.S.	2.75	2.61	Whittaker, 1966
R	Spruce fir forest (Mt. Mingus)§	Great Smoky Mts., Tenn., U.S.	2.97	2.68	Whittaker, 1966
S	Spruce fir forest (Mt. Collins)§	Great Smoky Mts., Tenn., U.S.	3.01	2.64	Whittaker, 1966
T	Gray beech forest§	Great Smoky Mts., Tenn., U.S.	2.96	2.69	Whittaker, 1669
U	Gray beech forest§	Great Smoky Mts., Tenn., U.S.	2.82	2.69	Whittaker, 1966
V	Hemlock—mixed forest§	Great Smoky Mts., Tenn., U.S.	3.07	2.82	Whittaker, 1966
W	Upper cove forest§	Great Smoky Mts., Tenn., U.S.	3.04	2.74	Whittaker, 1966
X	Deciduous cove forest§	Great Smoky Mts., Tenn., U.S.	3.09	2.85	Whittaker, 1966
Z	Hemlock—rhodo- dendron forest§	Great Smoky Mts., Tenn., U.S.	3.01	2.75	Whittaker, 1966

* Judged climax, assumed stable from Espenshade (1957).
† Judged stable by A. W. Johnson *et al.* (1966).
‡ Assumed stable.
§ Judged stable by original author(s).

most important cause of AE change in mountains. Whittaker recorded the altitude of each of his plots; therefore, I was able to estimate AE for his plots by interpolating their altitudes on a graph of altitude versus mean AE. The graph was drawn using mean AE data obtained at nine regional meteorological stations (Thornthwaite Associates, 1964), and coupling them with three empiral observations: Shanks' (quoted in Whittaker, 1966) observation that there is an average 4.1°C decrease in temperature for every thousand-meter increase in altitude in the Smokies; Holdridge's (1962) observation that PE = 58.93 B (where B is "biotemperature") and Major's (1963) and Thornthwaite Associates' (1964) observation that in the Smokies AE is always very nearly equal to PE. From these facts, I assumed that a 1,000-m increase in altitude was accompanied by a 4.1° decrease in biotemperature, hence by a 241.6-mm decrease in AE. The meteorological data confirmed this slope and determined the intercept.

As mentioned above, Holdridge (1947) had been successful in arranging only climax vegetation into his scheme. Prairie vegetation might be temporarily successful as a sere in an area that would later be forest. Clearly, different growth forms in the same region show widely different net productivities (Odum, 1960; Whittaker, 1963; Ovington, Heitkamp, and Lawrence, 1963). Clearly, also, one environmental value cannot be made to predict more than one productivity. One might suspect, therefore, that energy flow values in communities possessing the growth form of the local climax vegetation are the only energy flow values predictable from any general climatic variable.

Further, we have the clear and strong admonition of Whittaker (1966, p. 116) that "Unstable stands should not be compared with stable ones in the study of environmental effects on production; and unstable stands may not, unless carefully chosen for comparison, indicate reliably the effects of environment on production." Restriction of data to those from mature communities seemed an ideal strategy, too, in view of the ultimate purpose of the effort, that is, to study evolutionary pressure in widely scattered vertebrate populations which had been subjected mostly to mature vegetation for centuries past. Hence I did attempt to so restrict data.

The primary and most often used criterion for rejection or acceptance of the data was a relevant statement made by the original reporter of the data. When doubtful cases occurred, the vegetation maps of A. W. Küchler as they appear in Goode's World Atlas (Espenshade, 1957; pp. 16, 17, 52, 53) were used to determine which data were or were not acceptable. For example, the interesting and often cited data of Ovington (1963) obtained on the Cedar Creek Natural History Reservation, Minnesota, could not be used here. One plot was prairie; a second, savanna; a third, oakwood. The authors do not identify any of their plots as being climax, and Küchler denotes this area of Minnesota as being either a maple-basswood or a maple-yellow birch–hemlock–pine climax.

In the final analysis, I must disclaim any attempt or success at collecting all usable records of production. Rather, my search was for data from as wide a variety of environments as possible. This is really the only justification for including the two rather insecure but currently available estimates of production in tropical forests (L and M). Indeed, point L does not even represent a fully mature association; Nye (1961) calls it an old secondary forest. I hope that researchers in tropical forests will soon provide ecology with more secure estimates of production in this extremely interesting type of environment.

RESULTS

All data were transformed to common logarithms. Using the method of least squares, linear regression of the productivity on the AE was performed. The data and regression line appear in Figure 1. The productivity prediction equation, including 5% confidence intervals for the slope and intercept is:

$$\log_{10}\text{NAAP} = (1.66 \pm 0.27) \log_{10}\text{AE} - (1.66 \pm 0.07), \qquad (1)$$

where AE is the annual actual evapotranspiration in millimeters, and NAAP is the net annual above-ground productivity in grams per square meter.

DISCUSSION

The fact that AE is a measure of the simultaneous availability of water and energy in an environment suggests to me an explanation as to why it should be a successful predictor of production. Gross productivity may be defined as the integral of the rate of photosynthesis throughout the year. The rate of photosynthesis depends on the concentrations of its raw materials, and water and solar radiational energy are two of these. In terrestrial environments, the third, CO_2, is a more or less constant 0.029% (Sellers, 1965). Thus, the AE is a measure of the two most variable photosynthetic resources.

If AE is a good predictor of NAAP because it is a good predictor of the rate of photosynthesis, then the ratio of shoot to root and of gross-to-net production must be approximately constant in all the communities included in this study. According to Bray (1963), it is fairly accurate to assume a constant shoot-to-root ratio (roots about 20% to 50% of shoots) for these plant communities. And, according to Muller (1962), the net-to-gross statement is also fairly accurate (net about 40% to 60% of gross).

Also necessary for production is the array of biochemicals and minerals that form the photosynthetic machinery. However, unless there is a severe shortage of one or more of the essential nutrients necessary for their synthesis, this production apparatus might be expected to be synthesized in optimal amounts at just the right times by a natural plant community subject to evolutionary pressures. Perhaps plants can even evolve to sidestep some regional nutrient deficiencies. In any case, we should not expect production to be commonly limited by anything under the control of the plant organism.

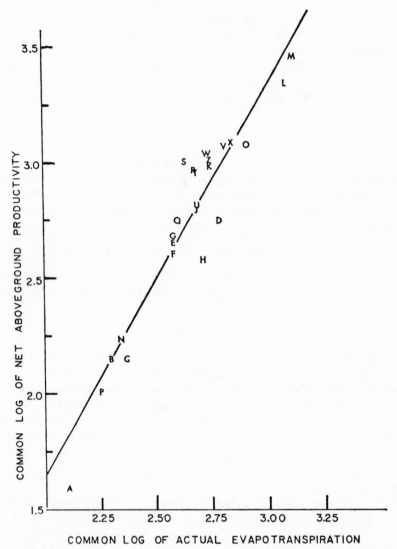

Fig. 1.—Net above-ground productivity in grams of dry matter per square meter graphed against actual evapotranspiration in millimeters. The regression line (see text) is included. See Table 1 for identities of coded points.

As I have hinted above, there are bound to be special local conditions which no general continental model can take into account. Such would be shallow soil on a mountainside, or a local deficiency of some absolutely essential nutrient like phosphorus. According to Wilson (1948), there might be local conditions of frequent fog which would raise productivity greatly by raising CO_2 concentration; AE will not predict this either.

The relationship of the productivity of successional vegetational stages to AE seems worth attention. However, this is absolutely impossible to investigate without simultaneous measurements of AE. Holdridge (1962) pointed out that the real AE (as opposed to the Thornthwaite estimated AE) is not unaffected by the character of the vegetation. Lower growths of foliage than the mature height should have a smaller AE, due presumably to decreased transpiration surface and water-holding capacity in the lower vegetation. Similarly, from his formula (Holdridge, 1962) one can conclude that higher successional stages should have greater AE than that over the mature vegetation. By using Thornthwaite estimates of AE and mature communities, I have avoided this problem. It should not be avoided forever. Interesting, indeed, would be the discovery, using real AE values, that production and real AE are closely correlated regardless of the sere. This would be strong evidence for the causality of the relationship. Further, it would be of interest to students of the evolution of efficiency in natural communities. Holdridge (1962) assumes that evaporation and transpiration are approximately equal in all types of environments. However, Hillel and Tadmor (1962) estimate that transpiration varies from 44% to about 90% of the AE in their four environments. Perhaps such variation is peculiar to deserts. It is interesting to speculate that there exists some optimal efficiency of water utilization in any environment and that it is approached as both short-term succession and long-term evolution proceed.

I cannot overemphasize the hypothetical nature of the foregoing explanation for the correlation of AE and productivity. Still, the relative usefulness of any prediction is determined by its precision. Hence, the accuracy of my explanation might be nil, but the fact that the correlation exists would remain of interest. Perhaps others can, as I have, use it in studies of the broad effects of production variation on other natural phenomena. As to my hypothesis of the cause of this correlation, that may yet prove useful, but only insofar as it can stimulate the experimental analysis that replaces it or proves it.

SUMMARY

Actual evapotranspiration (AE) is shown to be a highly significant predictor of the net annual above-ground productivity in mature terrestrial plant communities. Communities included ranged from deserts and tundra to tropical forests. It is hypothesized that the relationship of AE to productivity is due to the fact that AE measures the simultaneous availability of water and solar energy, the most important rate-limiting resources in photosynthesis.

ACKNOWLEDGMENTS

I thank Drs. R. H. MacArthur, W. J. Smith, and J. Preer, Jr., for advice and criticism. Thanks, also, are due the anonymous referees, whose criticisms were helpful in improving the manuscript. Much of the work reported on was done while I held a Cooperative-Graduate Fellowship from the National Science Foundation at the University of Pennsylvania. Material for this paper was taken, in part, from my Ph.D. dissertation.

LITERATURE CITED

Bartholomew, W. V., J. Meyer, and H. Laudelout. 1953. Mineral nutrient immobilization under forest and grass fallow in the Yangambi region (Belgian Congo). Pub. Inst. Etude Agron. Congo Belge 57.

Bray, J. R. 1963. Root production and the estimation of net productivity. Can. J. Bot. 41:65–72.

———. 1964. Primary consumption in three forest canopies. Ecology 45:165–167.

Espenshade, E. B., Jr. (ed.). 1957. Goode's world atlas. 10th ed. Rand McNally, Chicago. 272 p.

Hadley, E. B., and L. C. Bliss. 1964. Energy relationships of Alpine plants on Mt. Washington, N.H. Ecol. Monogr. 34:331–357.

Hillel, D., and N. Tadmor. 1962. Water regime and vegetation in the central Negev highlands of Israel. Ecology 43:33–41.

Holdridge, L. R. 1947. Determination of world plant formations from simple climatic data. Science 105:367–368.

———. 1959. Simple method for determining potential evapotranspiration from temperature data. Science 130:572.

———. 1962. The determination of atmospheric water movements. Ecology 43:1–9.

Johnson, A. W., L. A. Viereck, R. E. Johnson, and H. Melchior. 1966. Vegetation and flora, p. 277–354. In N. J. Wilimovsky and J. N. Wolfe [eds.], Environment of the Cape Thompson region, Alaska. U.S. Atomic Energy Comm., Div. Tech. Information, Oak Ridge, Tennessee.

Major, Jack. 1963. A climatic index to vascular plant activity. Ecology 44:485–498.

Muller, D. 1962. Wie gross ist der prozentuale Anteil der Nettoproduktion an der Bruttoproduktion? p. 26–28. In H. Leith, Die Stoffproduktion der Pflanzendecke. Gustav Fischer, Stuttgart. 156 p.

Nye, P. H. 1961. Organic matter and nutrient cycles under moist tropical forest. Plant and Soil 13:333–346.

Odum, E. 1959. Fundamentals of ecology. Saunders, Philadelphia. 546 p.

———. 1960. Organic production anh turnover in old-field succession. Ecology 41:34–49.

Ovington, J. D., D. Heitkamp, and D. B. Lawrence. 1963. Plant biomass and productivity of prairie, savanna, oakwood and maize field ecosystems in central Minnesota. Ecology 44:52–63.

Pearson, L. C. 1966. Primary productivity in a northern desert area. Oikos 15:211–218.

Penfound, W. T. 1964. Effects of denudation on the productivity of grasslands. Ecology 45:838–845.

Pianka, E. R. 1966. Latitudinal gradients in species diversity: a review of concepts. Amer. Natur. 100:33–46.

Rickard, W. H. 1962. Comparison of annual harvest yields in an arctic and a semi-desert plant community. Ecology 43:770–771.

Sellers, W. D. 1965. Physical climatology. Univ. Chicago Press, Chicago. 272 p.

Thornthwaite Associates. 1962. Average climatic water balance data of the continents, Africa. Lab. Climatol., Pub. Climatol. 15:115–287.

———. 1964. Average climatic water balance data of the continents, North America. Lab. Climatol., Pub. Climatol. 17:231–615.

Thornthwaite, C. W., and J. R. Mather. 1957. Instructions and tables for computing potential evapotranspiration and the water balance. Drexel Inst. Technol., Lab. Climatol., Pub. Climatol. 10:181–311.

Whittaker, R. H. 1963. Net production of heath balds and forest heaths in the Great Smoky Mountains. Ecology 44:176–182.

———. 1966. Forest dimensions and production in the Great Smoky Mountains. Ecology 47:103–121.

Wilson, C. C. 1948. Fog and atmospheric carbon dioxide as related to apparent photosynthetic rate of some broadleaf evergreens. Ecology 29:507–508.

31

Reprinted from *Nature and Resources* 8(2):5-10 (1972)

Modelling the primary productivity of the world

H. Lieth[1]

Two initial, central themes of the International Biological Programme (IBP) have been to achieve the best possible assessment of the world's primary productivity, and to uncover in some detail the relations between primary productivity and environmental conditions. With IBP in its end-phase, attempts are needed to summarize the incoming results of regional productivity assessments, to focus on the initial goals, and to improve earlier mapping attempts. This short notice is intended to secure co-operation from all parts of the world and to summarize the present state of knowledge, after my own current publication (1972) and that of Bazilevitch, Rodin and Rozov (1970).

The analysis of correlations between primary productivity and environmental parameters has become an important part of ecosystems modelling. Ecosystems modelling, as well as general ecological modelling, is being developed by all biomes of the United States IBP. A team at the University of North Carolina at Chapel Hill, contributing to the biome-wide studies of the Eastern Deciduous Forest Biome of the United States IBP, is developing a computer-based world model that expresses the correlation between environmental para-

1. Department of Botany, University of North Carolina, Chapel Hill, N.C. 27514 (United States).

meters and biospheric functions over large areas. As its main display module, the model uses the synagraphic mapping routine, SYMAP.

As the initial step, a computer map module of the world was developed, with which an already existing productivity map (Fig. 1) was simulated. A second map (Fig. 2), predicting the primary productivity from average annual precipitation and average annual temperature, was produced with the same map module. The maps were presented at two symposia on the world's productivity: the first, at a meeting of the German Botanical Society in Innsbruck, Austria, September 1971; and the second, at a meeting of the American Institute of Biological Sciences in Miami, Florida, October 1971.

The 'Innsbrucker' productivity map (Fig. 1) serves as the validation model for subsequent maps produced with the same map module but derived from predictive models. The original map (Lieth, 1964), of which the computer map is a simulation, was constructed from all actual productivity data available in 1962. The data base was slim at that time, and much indirect productivity calculation was necessary to draw the map. The current incoming IBP-generated figures, however, tell us that the estimate was not far off in most regions of the world. A newly made table (Table 1) from Lieth (1972) showing the production averages of various biome types of the world fits well the land productivity that can be calculated from the map. The table also gives the first breakdown and summary of annual energy fixation for the total vegetation cover of the world. Bazilevitch, Rodin and Rozov (1970) give an assessment of the world's primary productivity that is about 70 per cent larger than ours.

In order to produce some preliminary models to correlate environmental factors with productivity data, the Chapel Hill group has taken recent, mostly IBP-generated, productivity data from various parts of the world. Four north-south profiles across the Northern Hemisphere from the tundra into tropical regions were selected: one through North and South America, one

TABLE 1. Net primary productivity and energy fixation estimate for the world around 1950[1,2]

Vegetation unit	Size 10^6 km²	Net primary productivity			Combustion value Kcal/g	Annual energy fixation	
		Range g.m⁻².y⁻¹	Approx. mean	Total for area 10^9t		10^6 cal/m²	10^{18} cal per area (Col. 2)
(1)	(2)	(3)	(4)	(5)	(6)	(7)	(8)
Continental							
Forest	50.0		1,290	64.5			277.0
Tropical rain forest	17.0	1,000–3,500	2,000	34.0	4.1	8.2	139.4
Raingreen forest	7.5	600–3,500	1,500	11.3	4.2	6.3	47.2
Summergreen forest	7.0	400–2,500	1,000	7.0	4.6	4.6	32.2
Chaparral	1.5	250–1,500	800	1.2	4.9	3.9	5.9
Warm temperate mixed forest	5.0	600–2,500	1,000	5.0	4.7	4.7	23.5
Boreal forest	12.0	200–1,500	500	6.0	4.8	2.4	28.8
Woodland	7.0	200–1,000	600	4.2	4.6	2.8	19.6
Dwarf and open scrub	26.0		90	2.4			10.2
Tundra	8.0	100–400	140	1.1	4.5	0.6	4.8
Desert scrub	18.0	10–250	70	1.3	4.5	0.3	5.4
Grassland	24.0		600	15.0			60.0
Tropical grassland	15.0	200–2,000	700	10.5	4.0	2.8	42.0
Temperate grassland	9.0	100–1,500	500	4.5	4.0	2.0	18.0
Desert (extreme)	24.0		1	—			0.1
Dry desert	8.5	0–10	3	—	4.5	—	0.1
Ice desert	15.5	0–1	0	—	—	—	—
Cultivated land	14.0		650	9.1	4.1		37.8
Fresh water	4.0	100–4,000	1,250	5.0		2.7	21.4
Swamp and marsh	2.0	800–4,000	2,000	4.0	4.2	8.4	16.8
Lake and stream	2.0	100–1,500	500	1.0	4.5	2.3	4.6
TOTAL CONTINENTAL	149.0		669	100.2			426.1
Oceanic							
Reefs and estuaries	2.0	500–4,000	2,000	4.0	4.5	9.0	18.0
Continental shelf	26.0	200–600	350	9.3	4.5	1.6	42.6
Open ocean	332.0	2–400	125	41.5	4.9	0.6	199.2
Upwelling zones	0.4	400–600	500	0.2	4.9	2.5	1.0
TOTAL OCEANIC	361.0		155	55.0			260.8
TOTAL EARTH	510.0		303	155.2			686.9

1. The year 1950 has been chosen because all considerations about near-to-natural vegetation type areas become increasingly invalid from that time on.
2. Column (1): subdivisions are named according to Ellenberg and Müller-Dombois in Olson (1970) whenever possible. Column (2): basically result of the effort of three consecutive generations of geobotany students at University of North Carolina, Chapel Hill. Ajustments and compromises were made in some cases. Column (3): values were deducted from our own compilations of productivity data. Column (4): original in Whittaker and Likens (1970). Column (5): product of the positions in Columns (2) and (4). All values were rounded off to one decimal point. Column (6): original in Jordan (1971). Column (5): product of Columns (4) and (6). Column (8): product of Columns (2) and (7).
Source: Lieth (1972); more source literature is cited therein.

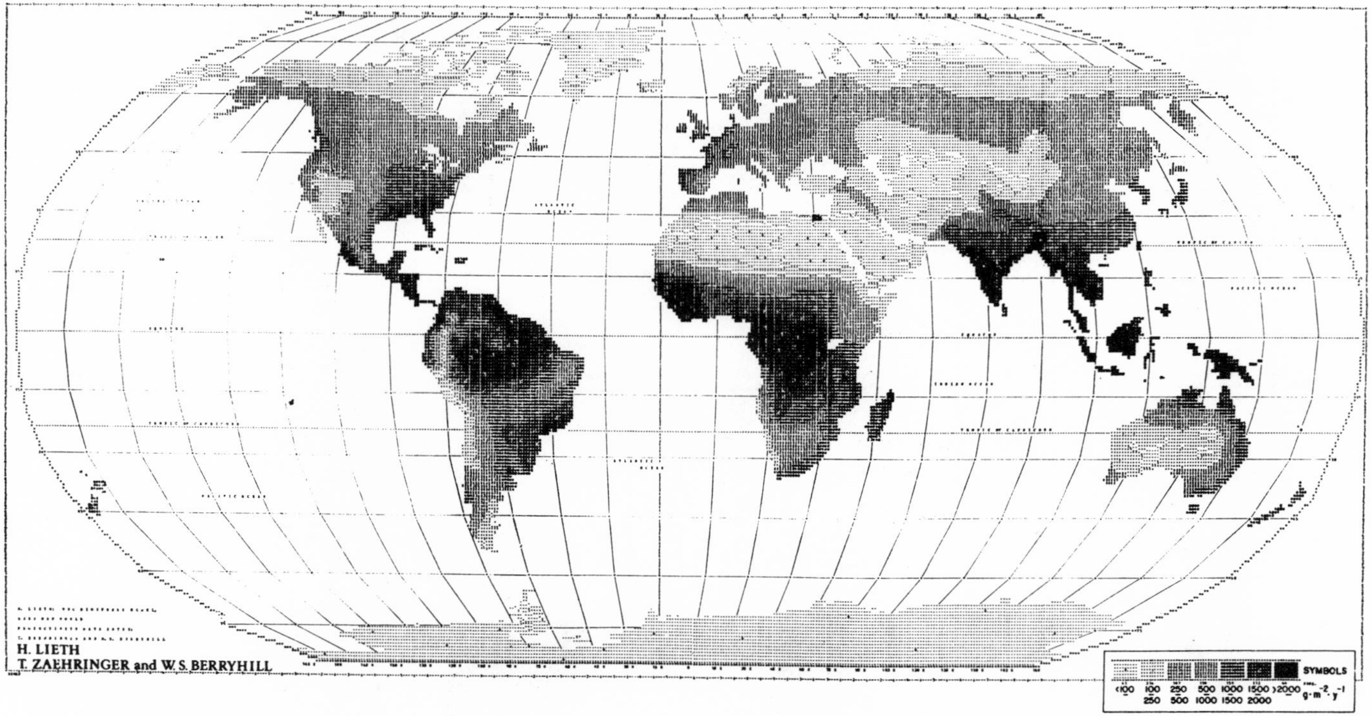

Fig. 1. 'Innsbrucker' productivity map. Computer simulated after Lieth's (1964) primary productivity map. The values shown in this map are in approximate agreement with the production rates given in Table 1 for individual vegetation types. *Source:* Lieth (1972).

H. LIETH
I. ZAEHRINGER and W.S. BERRYHILL

7

279

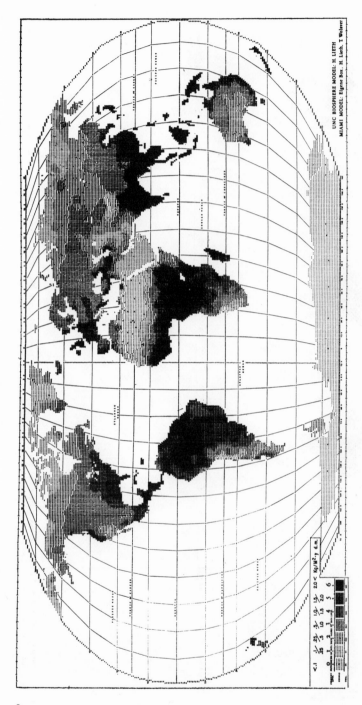

Fig. 2. 'Miami' model. Primary productivity predicted from the precipitation and temperature averages of existing meteorological stations after establishing the predictive models demonstrated in Figures 3 and 4. *Source:* Lieth (1971).

8

280

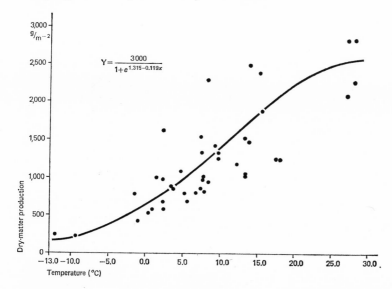

Fɪɢ. 3. Annual dry-matter production *v.* mean annual temperature.

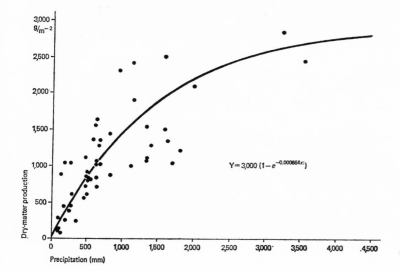

Fɪɢ. 4. Annual dry-matter production *v.* annual average precipitation.

across Europe and Africa, one through western Russia and Asia Minor, and one from Siberia through Japan to the Indo-Malayan region. These data were matched with climatic data from near-by meteorological stations, as compiled in Walter and Lieth (1961-66). The data set yielded the two curves shown in Figures 3 and 4, one predicting productivity from precipitation and the other from temperature. A table was constructed to evaluate the predicted productivity for each combination of temperature and precipitation. If temperature or precipitation predicted different productivity levels, then the lowest value was chosen (application of Liebig's law).

More than 1,000 meteorological stations were taken from the entire world, as evenly distributed as possible, and their average productivity predicted by this procedure. The values were then plotted into the same computer base map used for the Innsbruck simulation. The same productivity levels were selected for the predictive map as were used in the Innsbruck map to facilitate comparisons. The result is shown in Fig. 2, the Miami model.

Comparison shows the same general distribution of productivity patterns in the two maps. Specific areas show differences, however, and these are partly caused by the computer mapping method, and partly by wrong assumptions made while constructing the first productivity map in 1964.

The completion of all IBP and related studies should enable us to produce a more accurate map than has been possible so far. In order to further refine the Innsbruck type (validation) map, we need productivity data from as many stations of the world as possible. We would therefore appreciate it if readers of this article would send productivity data to Dr Helmut Lieth, Department of Botany, University of North Carolina, Chapel Hill, N.C. 27514 (United States). The data should

be accompanied by the following additional information: latitude and longitude to the nearest degree; approximate altitude; whether the figure contains total productivity or portions only (yield or above-ground portion); the vegetation type or crop; whether the figure is in the reporter's opinion high or low for the regional landscape; whether the productive vegetation unit uses the entire vegetation period or some portion thereof; and whether the year reported displayed any environmental extremes.

BIBLIOGRAPHY

BAZILEVITCH, N. I.; RODIN, L. E.; ROZOV, N. N. 1970. Biologicheskij Produktivnost v Geograficheskom Perspektive [Investigation of the biological productivity from a geographical viewpoint]. *Fifth meeting of the Geographical Society of the U.S.S.R., Leningrad.* (In Russian.)

JORDAN, C. F. 1971. Productivity of tropical forest and its relation to a world pattern of energy storage. *J. Ecol.,* vol. 59. p. 127-43.

LIETH, H. 1964. Versuch einer Kartographischen Darstellung der Produktivität der Pflanzendecke auf der Erde. *Geographisches Taschenbuch 1964/65.* p. 72-80. Wiesbaden, M. Steiner.

——. 1971. *Net primary productivity of the earth with special emphasis on the land areas.* Miami, Fla, American Institute of Biological Sciences Second National Congress, October 1971. (Paper presented at symposium titled, Perspectives on Primary Productivity of the Earth, organized by R. H. Whittaker and G. E. Likens.)

——. 1972. Uber die Primärproduktion der Pflanzendecke der Erde. *Angew. Bot.,* no. 1, 1972. (Symposium at meeting of the German Botanical Society in Innsbruck, Austria, September 1971.)

OLSON, J. 1970. Geographic index of world ecosystems. In: D. Reichle (ed.), *Analysis of temperate forest ecosystems.* p. 297-304. Heidelberg, Springer-Verlag.

WALTER, H.; LIETH, H. 1961-66. *Climate diagram world atlas.* Jena, G. Fischer-Verlag.

WHITTAKER, R. H.; LIKENS, G. E. 1970. Table 4.2. *Communities and ecosystems.* p. 83. New York, N.Y., Macmillan. 162 p.

Part IV

GLOBAL PRODUCTIVITY PATTERN

Editor's Comments
on Papers 32 Through 36

The global productivity pattern of the vegetation has two main features of concern to the human population: (1) food, energy, and materials as a direct result of plant production and (2) the impact of plant productivity processes on the atmosphere. Man has been aware of the first concern since prehistoric times and the second concern only since about 1800 A.D. The first discussion of both in a truly global perspective was found in Liebig's book. The passage containing the global productivity calculation follows:

> The amount of nutrient material in the air is very small, indeed, in comparison to the air mass.
> If all carbon dioxide and ammonium particles in the air were thought to be collected as strata around the earth's surface and having the density of gases at sea level, CO_2 would lie a little thicker than eight feet, ammonia gas barely two lines. Both are taken out of the air by plants and the atmosphere depoverishes on them.
> If the total surface area of the earth were one coherent meadow from which annually 100 centner hay per ha could be gained, the meadow plant would reap within 21–22 years all CO_2 from the atmosphere and

all life on earth would end. The air would discontinue to be fertile for plants; that is, to grant the necessary life conditions for their development. We know, however, perpetuity of organic life is guaranteed; man and animals live off the plant body; all organic creatures have only a temporary, relatively short residence time. The life process of the animals converts the food, which nourishes it, into the original substances. The same change undergoes all the bodies of animals and plants, alike after death: The burnable elements will be converted into carbon dioxide and ammonia [Liebig 1862, p. 262].

Liebig made an assumption of a more or less uniform productivity around the globe. The first pattern differentiation was achieved sixty years later and thereafter steadily refined. This early time period is represented in this volume by the extracts from Schroeder in Paper 32 and from Noddack in Paper 33. Several other early authors are discussed by Rodin, Bazilevich, and Rozov in Paper 34 but Schroeder's paper is probably the starting point for analytically based investigations of global productivity patterns, while Vernadsky (1926, 1934, 1940) and Noddack (1937) elaborated further on the subject. Vernadsky's papers are included in the discussion by Rodin et al., and an extract of one paper is included in *Cycles of Essential Elements*, Benchmark Papers in Ecology, edited by L. R. Pomeroy (1974).

From about 1950 on, the global productivity pattern has been under continuous investigation. The first geographical pattern analyses concerned ocean productivity since the ocean was believed by many to be a major source of food for mankind. Early pattern demonstrations appeared in tabular form while later presentations appeared as maps. Examples of the ocean productivity tables are found in Ryther (Paper 19) and as Table 1. Figure 1 is an example of a marine productivity map. Maps were presented for the ocean first and for terrestrial vegetation later.

An early attempt by an FAO sponsored program of ocean productivity mapping was included in this author's own first attempt (Lieth 1964) to present a world map of primary productivity pattern. This map is shown in Figure 2. The ocean section of this map is superseded by one by Koblentz-Mishke et al. (1970, see figure 4 in Paper 34). Other examples of land vegetation productivity patterns appear in Paper 34 (figure 1) by Rodin, Bazilevich, and Rozov, and in Paper 35 by Lieth. Several redrawings of all these maps have been made by others. The energy fixation pattern on earth by Jordan is presented in Paper 36 in tabular form. Maps and tables available at this time differ by about 30 percent. The accuracy of productivity measurements is the lowest in tropical areas. The key to future refinement of our understanding the global productivity capacity lies, therefore, in the study of tropical primary productivity.

Table 1. Ocean productivity. (Source: Sverdrup, Johnson, and Fleming 1942, p. 938; after Riley 1941. Reprinted by permission of and copyright © 1942 by Prentice-Hall, Inc.)

	Location	Production Carbon (g/m²/yr)	Method	Authority
Experimental observations	Long Island Sound	600–1000	Gross production—O₂ production in experimental bottles	Riley, 1938, 1939, present paper
	Western Atlantic, 23°–38°N	530		
	Western Atlantic, 38°–41°N	320		
	Dry Tortugas	60–430		
	Long Island Sound	400–700	Phytoplankton production—O₂ production	
		440–875	P consumption	
		100–200	N consumption	
	Western Atlantic, 23°–41°N	95–190	Chlorophyll production	
		140–365	N consumption	
	Dry Tortugas	27	P consumption	
		5	Plant pigment production	
	Western Atlantic 3°–13°N	278	Oxygen consumption	Seiwell, 1935
	Off southern California	215–430	Increase in oxygen—experimental	Sverdrup and Fleming, 1941
	Long Island Sound	384		Riley
Observations on natural environment	English Channel	84	P consumption	Atkins, 1923
		98	Changes in CO₂	
		60	Changes in O₂	
		70	Changes in P	Cooper, 1938
		88	Changes in N	
		7	Changes in Si	
		5	Changes in Ca	
	Barents Sea	170–330	P consumption	Kreps and Verjbinskaya, 1930
	Long Island Sound	138–350	P consumption	Riley
		Carbon (g/m³/day)		
Short-period experimental observations	Norwegian coast	0.14		Gran, 1927
	Scottish coast	0.16		Marshall and Orr, 1930
	Dry Tortugas	0.07	Gross production—O₂ production in experimental bottles at the surface	Riley
	Western Atlantic, 23°–41°N	0.01–0.12		
	Western Atlantic—George's Bank	0–0.88		
	Long Island Sound	0.02–0.41		
	Scripps Institution pier	0.01–0.15		Sargent, manuscript

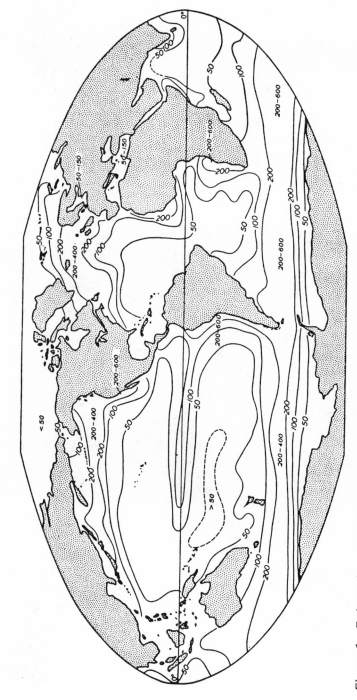

Figure 1. Estimated annual photosynthetic productivity of the oceans. Units: grams of carbon/meter²/year. (Source: Fleming 1957, p. 104)

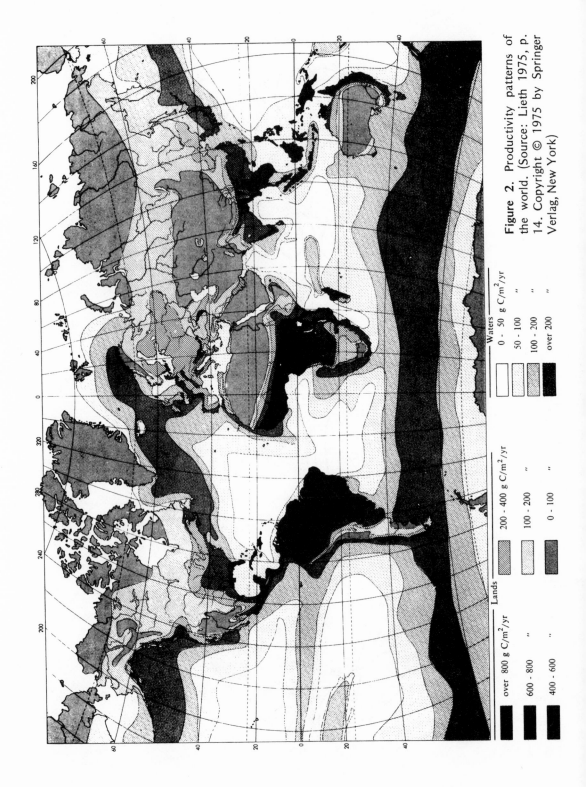

Figure 2. Productivity patterns of the world. (Source: Lieth 1975, p. 14. Copyright © 1975 by Springer Verlag, New York)

Lands

over 800 g C/m²/yr	
600 - 800 "	
400 - 600 "	
200 - 400 g C/m²/yr	
100 - 200 "	
0 - 100 "	

Waters

0 - 50 g C/m²/yr	
50 - 100 "	
100 - 200 "	
over 200 "	

REFERENCES

Atkins, W. R. G. 1923. The hydrogen ion concentration of sea water and its relation to photosynthetic changes. Part II. *Mar. Biol. Assn. U.K. J.* **12**:93–118.

Cooper, L. H. N. 1938. Phosphate in the English Channel, 1933–38, with a comparison with earlier years, 1916 and 1923–32. *Mar. Biol. Assn. U.K. J.* **23**: 181–195.

Fleming, R. H. 1957. Features of the oceans, pp. 87–107 in *Treatise on Marine Ecology and Paleoecology*, Vol. 1: Ecology, Memoir 67, Geological Society of America (reprinted in 1966), J. W. Hedgpeth, ed.

Gran, H. H. 1927. The production of plankton in the coastal waters off Bergen, March-April 1922. *Rept. Norwegian Fishery and Mar. Invest.* **3**(8):74pp.

Koblentz-Mishke, O. I., V. V. Volkovinski, and Y. G. Kabanova. 1970. New data on a magnitude of primary production in the oceans. *Dokl. Akad. Nauk USSR, Ser. Biol.* **183**(5):1189–1192.

Kreps, E., and N. Verjbinskaya. 1930. Seasonal changes in the phosphate and nitrate content and in hydrogen-ion concentration in the Barents Sea. *Conseil Perm. Internat. Explor. de la Mer J. du Conseil* **5**:326–346.

Liebig, J. von. 1862. *Der chemische Process der Ernährung der Vegetabilien*. Braunschweig: F. Viehweg & Son.

Lieth, H. 1964. Versuch einer kartographischen Erfassung der Stoffproduktion der Erde, pp. 72–80 in *Geographisches Taschenbuch 1964/65*. Wiesbaden: Max Steiner Verlag.

——. 1975. Historical survey of primary productivity research, pp. 7–16 in *Primary Productivity of the Biosphere*, H. Lieth and R. H. Whittaker, eds. New York: Springer Verlag.

Marshall, S. M., and A. P. Orr. 1927. The relation of the plankton to some chemical and physical factors in the Clyde Sea area. *Mar. Biol. Assn. U.K. J.* **14**: 837–868.

Noddack, W. 1937. Der Kohlenstoff im Haushalt der Natur. *Angew. Chem.* **50**(28): 505–510.

Pomeroy, L. R. (ed.) 1974. *Cycles of Essential Elements*, Benchmark Papers in Ecology series, volume 1. Stroudsburg, Pa.: Dowden, Hutchinson & Ross.

Riley, G. A. 1938. The measurement of phytoplankton. *Internat. Rev. Hydrobiol. Hydrogr.* **36**:373–374.

——. 1939. Plankton studies. II. The western North Atlantic, May-June 1939. *J. Mar. Res.* **2**:145–162.

——. 1941. Plankton studies. III. Long Island Sound. *Bingham. Oceanogr. Collection Bull.* **7**(art.3):93pp.

Seiwell, H. R. 1935. The distribution of oxygen in the western basin of the North Atlantic. *Papers in Phys. Oceanogr. and Meteorol.* **3**(1):86pp.

Sverdrup, H. U., and R. H. Fleming. 1941. The waters off the coast of southern California, March to July, 1937. *Scripps Inst. Oceanogr. Bull.* **4**(10):261–378.

Sverdrup, H. U., M. W. Johnson, and R. H. Fleming (eds.). 1942. *The Oceans, Their Physics, Chemistry, and General Biology*. Englewood Cliffs, N.J.: Prentice-Hall.

Vernadsky, V. I. 1926. *Biosfera*. Leningrad: Akademia Nauk.

——. 1934. *Essays in Geochemistry*. Leningrad: Akademia Nauk.

——. 1940. *Biogeochemical Essays*. Moscow, Akademia Nauk.

ADDITIONAL READINGS

Bazilevich, N. I., and L. E. Rodin. 1967. Maps of productivity and the biological cycle in the earth's principal terrestrial vegetation types. *Izvest. Geogr. Obsh., Leningrad* **99**(3):190–194.

Bazilevich, N. I., L. E. Rodin, and N. N. Rozov. 1971. Geographical aspects of biological productivity (rev. and transl.) *Sov. Geogr.* May:293–317.

Rodin, L. E., and N. I. Bazilevich. 1968. *Production and Mineral Cycling in Terrestrial Vegetation* (Engl. transl. by G. E. Fogg). Edinburgh: Oliver and Boyd.

Sverdrup, H. U., and M. W. Johnson, and R. H. Fleming. 1942. Organic production in the sea, pp. 925–945 in *The Oceans, Their Physics, Chemistry, and General Biology*, H. U. Sverdrup, M. W. Johnson, and R. H. Fleming, eds. Englewood Cliffs, N.J.: Prentice Hall.

32

THE ANNUAL PRIMARY PRODUCTION OF THE GREEN VEGETATION COVER OF THE EARTH

H. Schroeder

This excerpt was translated expressly for this Benchmark volume by H. Lieth from pages 27–28 of "Die jährliche Gesamtproduktion der grünen Pflanzendecke der Erde," in Naturwissenschaften 7:7–12, 23–29 (1919).

I summarize:

Table 15

Annual Carbon Assimilation, CO_2 Equivalent, and
Organic Matter Produced in 10^{12} Kg

	Carbon Bound			CO_2 Fixed			Organic Matter Produced		
	Average	Min	Max	Average	Min	Max	Average	Min	Max
Forest	11.0	9.0	13.0	40	32	48	23	19	28
Cultivated land	4.0	3.5	4.5	14	12	17.5	9	8	10
Steppe	1.1	0.5	2.2	4	1.6	8	2.5	1	5
Wasteland	0.2	0.1	0.5	0.9	0.4	1.8	0.5	0.2	1.0
Total	16.3	13.1	20.2	58.9	46.0	75.3	35.0	28.2	44.0

To be added are about .5 x 10^{12} Kg CO_2 for the benthos of the ocean and a yet unknown amount for the assimilating plankton.

This yields an annual CO_2 consumption of some 60 x 10^{12} Kg with a variability between 50 and 80 x 10^{12} Kg . . .

I have estimated above the amount of CO_2 assimilated by the green vegetation to be 60 x 10^{12} Kg. Of this amount, 40 Kg are contributed by forests, 14 x 10^{12} Kg by cultivated lands, 4 x 10^{12} Kg by steppes, and 1 x 10^{12} Kg by wastelands. From the forest a little less than half serves for the production of leaves and a little more than half for wood production. The life span of leaves from deciduous trees may reach from one-half to three-quarters of a year.[*] The needles of coniferous trees may last three to four years longer. According to Ebermayer it takes somewhat more than three years until the dropped leaves disappear as collectible litter. The remaining humus still contains

---

[*]This is also correct for the tropics. At least Volkens (1912) says that the trees of many species in Java drop their leaves twice within 12 to 14 months.

a major portion of the original leaf carbon. I assume, therefore, that the living leaves and the dead litter--including all stages of disintegration--together contain the triplicate amount of carbon necessary for one year's leaf production. Harvestable timber consists of trunks nearly a hundred years in age. A large portion serves man for several decades and is thus conserved. The average life time of individual trees in virgin forests is probably not shorter than individuals in cultivated forests. In addition, the decay of their carcasses requires many years. I estimate, therefore, that the entire forest--dead and alive--of the total earth contains fiftyfold the production of one year. . . .

33

CARBON IN THE HOUSEHOLD OF NATURE

W. Noddack

This excerpt was translated expressly for this Benchmark volume by H. Lieth from "Der Kohlenstoff im Haushalt der Natur," in Angew. Chem. 50(28):505-510 (1937).

The earth receives annually an energy amount by the sun in the form of radiation . . . which amounts to 1.31×10^{24} cal . . . of which 0.48 (albedo) is reflected . . . which leaves 7.47×10^{23} cal. Of this the overwhelming portion is used to maintain the surface temperature (+15° C average) which makes the earth inhabitable for organisms. Only a small amount . . . is converted by green plants into chemical energy. In order to calculate this amount we need to estimate how much chemically bound carbon the total vegetation of the earth produces annually. A balance of carbon produced by assimilation is shown in table 3.

TABLE 3

Vegetation Areas of the World

surface area of the world	510×10^6 km^2
water surface	371×10^6 km^2
land area	139×10^6 km^2

Vegetation categories	Surface area in 10^8 ha	Production of organic C in t/ha year	Total production of organic C in 10^9 t/year
Forests	44	2.00	8.8
Cultivated land	27	1.60	4.3
Steppe	31	0.60	1.9
Deserts	24	0.05	0.1
Ice covered	13	0.00	0.0
Ocean	317	0.80	25.4
Continental shelves	29	1.10	3.2
Ice covered	25	0.00	0.0

. . . For land plants we relied on data by Schroeder[1] and our own measurements[2]; for aquatic plants we used investigations by Boysen Jensen[3] and Atkins[4] . . .

The annual total production of carbon is represented by the sum of column 4 in table 3: 4.37×10^{10} t carbon per year.

---

[1] H. Schroeder, Naturwissenschaften 7,8 (1919)
[2] W. Noddack and J. Komor, Angew. Chemie 50,271 (1937)
[3] P. Boysen Jensen, Rep. Danish Mar. Biol. Stat. 22,16 (1920)
[4] W. Atkins, Sci. Progr. 106,298 (1932)

34

Reprinted from pages 15–17, 20, and 22 of *Productivity of the World's Main Ecosystems*, edited by D. E. Reichle, J. F. Franklin, and D. W. Goodall, Washington, D.C.: National Academy of Sciences, 1975, 166pp.

PRODUCTIVITY OF THE WORLD'S MAIN ECOSYSTEMS

L. E. Rodin, N. I. Bazilevich, and N. N. Rozov

[*Editor's Note:* In 1967 Bazilevich and Rodin published in Russia their first set of maps on primary productivity and biological (nutrient) cycles. This initial paper was later expanded, translated, and published under the authorship of Rodin and Bazilevich (1968). The material was revised again by Bazilevich, Rodin, and Rozov in 1971. These works formed the basis for Dr. Rodin's contribution to the Seattle Symposium. Their work was widely publicized. It contains a very detailed table (see table 3) of standing crop levels called phytomass and annual production rates. Values are given in metric tons per hectare for individual entities from the Russian soil classification system (soil-vegetation formations). The table was converted into a productivity map shown as figure 1 below. Table 3 yields a rather high level of terrestrial primary productivity compared to the other maps and tables in this book. This is due to the high productivity levels assigned to tropical soil types.

The ocean productivity map is, according to the three authors of this paper, an adaptation of the works by Koblentz-Mishke, Volkovinski, and Kabanova (1968).]

REFERENCES

Bazilevich, N. I., and L. E. Rodin. 1967. Maps of productivity and the biological cycle in the earth's principal terrestrial vegetation types. Izvestiya Geographicheskogo Obshchestva, Leningrad 99(3):190–194.

Bazilevich, N. I., L. E. Rodin, and N. N. Rozov. 1971. Geographical aspects of biological productivity. Soviet Geography: review and translation, May:293–317.

Koblentz-Mishke, O. I., V. V. Volkovinski, and Yu. G. Kabanova. 1968. New data on a magnitude of primary production in the oceans. Doklady Akad. Nauk, USSR, Ser. Biol. 183(5):1189–1192.

Rodin, L. E., and N. I. Bazilevich. 1968. Production and mineral cycling in terrestrial vegetation (English translation by G. E. Fogg). Edinburgh, Oliver and Boyd, 288p.

TABLE 3 Phytomass and Annual Primary Production of the Earth's Land Areas

Thermal Belts, Bioclimatic Regions, and Soil–Vegetation Formations	Phytomass t/ha	Phytomass 10^6 t	Production t/ha	Production 10^6 t
Polar belt, humid and semihumid regions				
Polar deserts (Arctic) on polygonal and other Arctic soils	5	353	1.0	70.6
Tundras on tundra gley soils	28	10,517	2.5	939.0
Bogs (polar) on bog permafrost soils	25	470	2.2	41.4
Floodplain formations	10	12	1.7	2.04
Mountainous polar desert formations on Arctic mountain soils	8	352	1.5	66.0
Mountainous tundra formations on mountain-tundra soils	7	2,062	0.7	206.22
TOTAL	17.1	13,766	1.6	1,325.26
Boreal belt, humid and semihumid regions				
Maritime herbaceous-forest formations on volcanic ash	100	1,350	10.0	135.0
Open forest-tundra woodland and northern taiga forest on gley-podzolic soils	125	11,000	5.0	440.0
Same, on gley-permafrost taiga soils	100	12,190	4.0	487.6
Middle taiga forest on podzolic soils	260	92,846	7.0	2,499.7
Same, on permafrost taiga soils	200	49,260	6.0	1,477.8
Southern taiga and mixed broadleaf and needle forest on turf-podzolic soils	300	95,460	7.5	2,386.5
Same, on yellowish-podzolic soils	350	13,755	10.0	393.0
Same, on turf-calcareous and turf-gley soils	350	10,535	10.0	301.0
Same, on taiga bog soils with bogs	80	6,824	4.0	341.2
Small-leaf forest on gray and gray solodized forest soils	200	4,540	8.0	181.6
Broadleaf forest on gray forest soils	370	26,307	8.0	568.8
Bogs	35	3,892	3.5	389.2
Floodplain formations	60	3,558	6.0	355.8
Mountain-taiga forests on mountain podzolic soils	170	45,832	6.0	1,617.6
Same, on mountain permafrost soils	160	48,640	5.0	1,520.0
Mountain meadows on gray mountain forest soils	300	7,560	7.5	189.0
Mountain meadows on mountain meadow soils	35	5,512.5	12.0	1,890.0
TOTAL	189.2	439,061.5	6.54	15,173.8
Subboreal belt, humid regions				
Broadleaf forests on brown forest soils	400	99,400	13	3,230.5
Same, on rendzinas	370	999	12	32.4
Herbaceous prairie on meadow chernozemlike soils (brunizems)	35	1,984.5	15	850.5
Broadleaf forests, swampy, with small bog areas	300	9,900	13	429.0
Bogs	40	64	25	40.0
Floodplain formations	90	1,782	12	237.6
Mountain forest on brown mountain forest soils	370	139,453	12	4,522.8
TOTAL	342	253,582.5	12.6	9,342.8
Subboreal belt, semiarid regions				
Steppe on typical and leached chernozems	25	1,607.5	13	1,355.9
Same, on ordinary and southern chernozems	20	2,942.0	8	1,176.8
Same, on solonets chernozems	20	500.0	8	200.0
Steppified formations on solonets	16	104.0	5	32.5
Halophytic formations on solonchak (in steppe)	12	7.2	4	2.4
Psammophytic formations on sand (in steppe)	18	108.0	8	48.0
Dry steppe on dark chestnut soils	20	1,480.0	9	666.0
Desert steppe on light chestnut soils	13	1,574.3	5	605.5
Dry and desert steppe on chestnut and solonets complexes	14	938.0	5	335.0
Same, on solonets	14	296.8	5	106.0

TABLE 3 (continued)

Thermal Belts, Bioclimatic Regions, and Soil-Vegetation Formations	Phytomass		Production	
	t/ha	10^6 t	t/ha	10^6 t
Halophytic formations on solonchak (in dry and desert steppe)	2	15.2	0.7	5.3
Psammophytic formations on sand (in dry and desert steppe)	15	213.0	6	85.2
Herbaceous bog on meadow-bog soils	15	85.5	7	39.9
Floodplain formations	80	2,280.0	12	342.0
Mountain dry steppe on mountain chestnut soils	15	1,245.0	7	581.0
Mountain steppe on mountain chernozems	25	572.5	10	229.0
Mountain meadow steppe on subalpine mountain meadow steppe soils	25	1,885.0	11	829.4
TOTAL	20.8	16,854.0	8.2	6,639.9
Subboreal belt, arid regions				
Steppified desert on brown semidesert soils	12	1,696.8	4.0	565.6
Same, on brown-soil and solonets complexes	10	407.0	3.5	142.5
Same, on solonets	9	126.0	3.2	44.8
Desert on gray-brown desert soils	4.5	617.9	1.5	205.9
Psammophytic formations on sand (in desert)	30	2,484.0	5.0	414.0
Desert on takyr soils and takyrs	3	36.3	1.0	12.1
Halophytic formations on solonchak (in desert)	1.5	32.7	0.5	10.9
Floodplain formations	80	1,088.0	13	176.8
Mountain desert on brown mountain semidesert soils	9	316.8	3	105.6
Same, on desert highland soils	7	1,437.8	1.5	308.1
TOTAL	11.7	8,234.3	2.8	1,986.3
SUBBOREAL BELT TOTAL	123.6	278,679.8	7.9	17,969.0
Subtropical belt, humid regions				
Broadleaf forest on red soils and yellow soils	450	88,965	20	3,954.0
Same, on red-colored rendzinas and terra rossa	380	6,536	16	275.2
Herbaceous prairie on reddish black soils and rubrozems	30	1,428	13	618.8
Broadleaf forest, swampy, with small bog areas	400	5,040	22	277.2
Meadow-bog and bog formations	200	5,380	130	3,497.0
Floodplain formations	250	17,050	40	2,728.0
Mountain broadleaf forest on mountain yellow soils and red soils	410	104,017	18	4,566.6
TOTAL	366.1	228,416	25.5	15,916.8
Subtropical belt, semiarid regions				
Xerophytic forest on brown soils	170	27,081	16	2,548.8
Shrub-steppe formations on gray-brown soils	35	9,331	10	2,666
Same, on gray-brown solonets soils with small solonets areas	20	150	6	45
Same, on subtropical chernozemlike and coalesced soils	25	1,187.5	8	380
Psammophytic formations on sandy soils and sand	20	102	5	25.5
Halophytic formations on solonchak soils and solonchak	1.5	10.95	0.5	3.65
Floodplain formations	250	15,725	40	2,516
Mountain xerophytic forest on brown mountain soils	120	26,832	13	2,906.8
Mountain shrub-steppe formations on gray-brown mountain soils	30	1,479	8	394.4
TOTAL	98.7	81,898.45	13.8	11,486.15
Subtropical belt, arid regions				
Steppified desert on serozems and meadow-serozem soils	12	2,376	10	1,980.0
Desert on subtropical desert soils	2	857.8	1	428.9
Psammophytic formations on sand	3	562.5	0.1	18.75
Desert on takyr soils and takyrs	1	14.7	0.5	7.35
Halophytic formations on solonchak	1	21.0	0.2	4.2
Floodplain formations	200	8,780.0	90	3,951.0
Mountain desert on mountain serozems	15	915.0	12	732.0
Same, on subtropical mountain desert soils	3	54	1	18.0
TOTAL	13.9	13,581	7.3	7,140.2
SUBTROPICAL BELT TOTAL	133.5	323,895.4	14.2	34,543.15
Tropical belt, humid regions				
Humid evergreen forest on red-yellow ferralitic soils	650	560,950	30	25,890
Same, on dark-red soils	600	11,820	27	531.9
Seasonally humid evergreen forest and secondary tall-grass savanna on red ferralitic soils	200	175,720	16	14,057.6
Same, on black tropical soils	80	1,128	15	211.5
Humid evergreen swampy forest on ferralitic gley soils	500	112,200	25	5,610
Bog formations	300	19,920	150	9,960

TABLE 3 (continued)

Thermal Belts, Bioclimatic Regions, and Soil–Vegetation Formations	Phytomass		Production	
	t/ha	10^6 t	t/ha	10^6 t
Floodplain formations	250	29,375	70	8,225
Mangrove forest	130	6,214	10	478
Humid tropical mountain forest on red-yellow ferralitic mountain soils	700	169,330	35	8,466.5
Seasonally humid tropical mountain forest on red ferralitic mountain soils	450	79,515	22	3,887.4
TOTAL	440.4	1,166,172	29.2	77,317.9
Tropical belt, semiarid regions				
Xerophytic forest on ferralitized brownish-red soils	250	115,675	17	7,865.9
Grass and shrub savanna on ferralitic red-brown soils	40	26,188	12	7,856.4
Same, on tropical black soils	30	6,150	11	2,255.0
Same, on tropical solonets soils	20	470	7	164.5
Meadow and swamp savanna on ferralitized red and meadow soils	60	5,364	14	1,251.6
Floodplain formations	200	4,420	60	1,326.0
Xerophytic mountain forest on brownish-red mountain soils	200	9,940	15	745.5
Mountain savanna on red-brown mountain soils	40	3,752	12	1,125.6
TOTAL	107.4	171,959	14.1	22,590.5
Tropical belt, arid regions				
Desertlike savanna on reddish-brown soils	15	6,435	4	1,716.0
Desert on tropical desert soils	1.5	700.95	1	467.3
Psammophytic formations on sand	1.0	282	0.1	28.2
Desert on tropical coalesced soils	1.0	21.6	0.2	4.32
Halophytic formations on solonchak	1.0	11.5	0.1	1.15
Floodplain formations	150	1,500	40.0	400.0
Mountain desert on tropical mountain desert soils	1	62.6	0.1	6.26
TOTAL	7.0	9,013.65	2.0	2,623.23
TROPICAL BELT TOTAL	243.3	1,347,144.65	18.5	102,531.63
Earth's total land area (without glaciers, streams, lakes)	180.1	2,402,547.4	12.8	171,542.86
Glaciers	0	0	0	0
Lakes and streams	0.2	40.0	5	1,000
TOTAL FOR ALL CONTINENTS	160.9	2,402,587.4	11.5	172,542.86

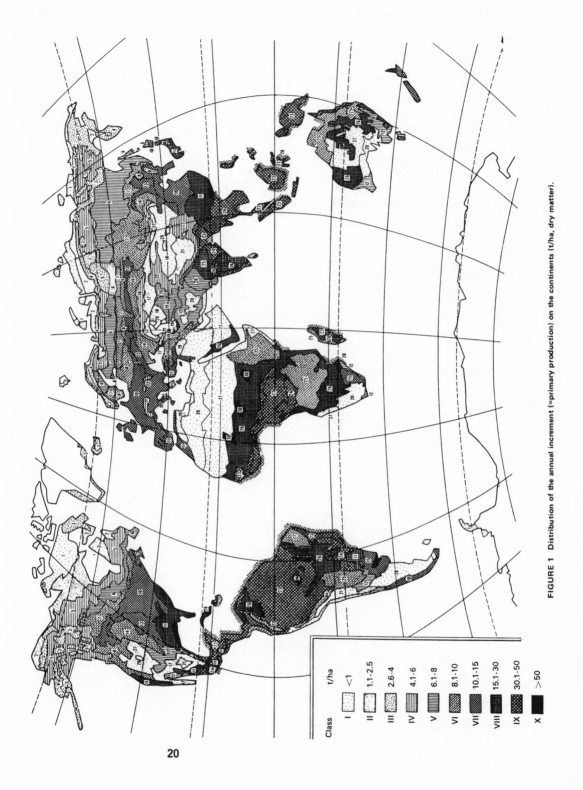

20

Class   t/ha

I   <1
II   1.1-2.5
III   2.6-4
IV   4.1-6
V   6.1-8
VI   8.1-10
VII   10.1-15
VIII   15.1-30
IX   30.1-50
X   >50

FIGURE 1   Distribution of the annual increment (=primary production) on the continents (t/ha, dry matter).

22

FIGURE 4 Distribution of the annual increment (=primary production) in the World Ocean (t/ha, dry matter).

35

Reprinted from pages 67–88 of *Unifying Concepts in Ecology*, Rept. Plenary Sessions, First Internat. Congr. Ecol., The Hague, the Netherlands, 1974, edited by W. H. van Dobben and R. H. Lowe-McConnell, The Hague: Dr. W. Junk B. V. Publishers and Wageningen: Centre for agricultural publishing and documentation, 1975, 302pp.

PRIMARY PRODUCTIVITY IN ECOSYSTEMS: COMPARATIVE ANALYSIS OF GLOBAL PATTERNS

H. Lieth

Introduction

Of the pathways of energy flow through the ecosystem: sensible heat, reflection, evapotranspiration and metabolism, metabolism is usually the narrowest but nevertheless the most important road. The entrance key to this pathway is photosynthesis. The first important observable effect is plant biomass accumulation: the net primary productivity (NPP).

This NPP varies over several orders of magnitude in different parts of the globe. The variations are of vital importance for the selfmaintenance or management of the respective ecosystems, and can in most cases be attributed to environmental conditions. The understanding of these variations and their causes is therefore of prime importance for the best use of individual types of ecosystem. It is therefore my intention to present here the global productivity patterns, to consider briefly the reasons for the differences in level, and to touch on quality differences in the organic matter produced. In order to present the material in the most compact, but usable, form I shall depend largely on models and tables. Details of methods and supporting basic data may be obtained from Lieth & Whittaker (1975).

Vegetation types and productivity levels

The earth's vegetation cover consists of a generic matrix of taxa which perform the key photosynthetic processes in very similar ways but differing widely in the quantity of production and the allocation of the products. The results of these differences substantiate the various physiognomic vegetation types as we commonly discuss them in the biological sciences. Although these vegetation types are in many ways inconvenient for the discussion of global productivity patterns it is always easier to start on the basis of a known principle. In Table 1 therefore we compare the NPP of different biomes of the world. This table presents the productive power of world vegetation in two ways: dry matter accumulation and energy fixation. It is based on a table published several times (Lieth, 1972, 1973) and corrected with new data provided by Murphy (1975), Whittaker & Likens (1975), Jones & Gore (1974) and Rodin et al. (1972). Summarizing its content we find that the total terrestrial productivity amounts to 121.7 × 10⁹ t dry

Table 1. Net primary productivity and energy fixation of major vegetation units of the world†

Vegetation unit	NPP				Annual energy fixation			
	area 10^6 km^2	range kg m^{-2} yr^{-1}	approx. mean kg m^{-2} yr^{-1}	total for area 10^9 metric tons	approx combustion value kcal g^{-1}	mean for m^2 10^6 cal m^{-2}	total for area 10^{18} cal	Authors
1	2	3	4	5	6	7	8	9
FORESTS	50			81.6			368.6	3
Tropical rainforest	17.0	1.0–3.5	2.8 (2.0)	47.4 (34.0)	4.1	11.5	195.5	3, 4
Raingreen forest	7.5	1.6–2.5	1.75 (1.5)	13.2 (11.3)	4.2	7.4	55.5	3, 4
Summergreen forest	7.0	0.4–2.5	1.0	7.0	4.6	4.6	32.2	3
Mediterranean sclerophyll forest (inclusive chaparral)	1.5	0.25–1.5	0.8	1.2	4.9	3.9	5.9	3
Warm temperate mixed forest	5.0	0.6–2.5	1.0	5.0	4.8	4.7	23.5	3
Boreal forest	12.0	0.3–1.2 (0.2–1.5)	0.65 (0.50)	7.8 (6.0)	4.6	3.0	36.0	3, 5
WOODLAND	7	0.2–1.0	0.6	4.2	4.6	2.8	19.6	3

† Previous estimates (2, 3) in ( ).

1	2	3	4	5	6	7	8	9
DWARF AND OPEN SCRUB	*26*						*11.0*	3
Tundra	8.0	0.06–1.3	0.16	2.6 / 1.3 / (1.1)	4.5	0.7	5.6	3, 5, 6
Desert scrub	18.0	0.01–0.25	0.07	1.3	4.5	0.3	5.4	3
GRASSLAND	*24*			*19.2*			*76.8*	3
Tropical grassland (including grass-dominated savannah)	15.0	0.2–2.9 (0.2–2.0)	0.8 (0.7)	12.0 (10.5)	4.0	3.2	48.0	3, 4
Temperate grassland	9.0	0.07–1.3 (0.1–1.5)	0.8 (0.5)	7.2 (4.5)	4.0	3.2	28.8	3, 5
DESERT (extreme)	*24*						*0.1*	3
Dry desert	8.5	0–0.01	0.003	—	4.5	—	0.1	3
Ice desert	15.5	0–0.001	—	—	—	—	—	3
CULTIVATED LAND	*14*	0.1–4.0	0.65	*9.1*	4.1	2.7	*37.8*	3
FRESH WATER	*4*			*5.0*			*21.4*	3
Swamps and marsh	2.0	0.8–4.0	2.0	4.0	4.2	8.4	16.8	3
Lake and stream	2.0	0.1–1.5	0.5	1.0	4.5	2.3	4.6	3

Table 1. (Contd.)†

| 1 | 2 | 3 | NPP | | Annual energy fixation | | | 9 |
Vegetation unit	area 10^6 km²	range kg m^{-2} yr^{-1}	approx. mean kg m^{-2} yr^{-1}	total for area 10^9 metric tons	approx. combustion value kcal g^{-1}	mean for m² 10^6 cal m^{-2}	total for area 10^{18} cal	Authors
TOTAL FOR CONTINENTS previous estimates in ( )	149			121.7 (100.2)			535.3 (426.1)	
Open ocean	332	0.002–0.4	0.13	41.5	4.9	0.6	199.2	1, 2
Upwelling zones	0.4	0.4–0.6	0.5	0.2	4.9	2.5	1.0	1, 2
Continental shelf	26.6	0.2–0.6	0.36	9.2	4.5	1.6	43.1	1, 2
Algae beds and reefs	0.6	0.5–4.0	2.0	1.2	4.5	9.0	3.6	1, 2
Estuaries	1.4	0.5–4.0	1.8	2.5	4.5	8.1	11.3	1, 2
TOTAL MARINE	361	0.002–4.0	0.161	55.0			258.2	2
FULL TOTAL	510			176.7			793.5	

Authors: 1. Whittaker and Likens in Lieth & Whittaker 1975*.
2. Lieth 1972.
3. Lieth 1973.

4. Murphy in Lieth & Whittaker 1975*.
5. Rodin, Bazilevich & Rozov 1972*.
6. Jones & Gore 1974*.

* Indicates paper in press.
† Previous estimates (2, 3) in ( ).

matter per year, equivalent to 2.25×10^{21} J. The marine productivity in comparison is 55×10^9 t dry matter, 1.1×10^{21} J. This new compilation gives values about 20% higher for terrestrial ecosystems. The major revision was caused by the re-evaluation of the average figures for the two tropical categories, tropical rain forest and raingreen forest.

The solar energy conversion efficiency based on the input level of 2.1×10^{24} J ($= 793.5 \times 10^{18}$ cal) annually reads with these new figures 0.16% for the total productivity of which 33% is contributed by the oceans and 67% by the terrestrial vegetation. Our previous calculations had yielded 0.13% for the global efficiency.

The increase of productivity found for tropical areas brings our calculations closer to the latest figures of Rodin et al. (1972) but our land value is still about one third smaller than theirs. It is very evident that the uncertainties still prevailing for tropical areas are the main obstacle to a better calculation. We need therefore to stress once again the urgency for more and better data sets from tropical areas (see also Brünig 1974).

The data in Table 1 may be converted into a productivity pattern map of the world. We have presented such maps in various forms (e.g. Lieth, 1964, 1972). The latest was a computer map presented at the Seattle IBP Conference in 1972 (Fig. 1). As a computer map this permits quick area-quantity calculations. This map and our previous table (Lieth, 1972) were made independently of one another but yielded similar global productivity levels (Box, 1975). The map fell short of the values in the old table by about 10% (Lieth 1972, 1973) and is about 30% short of those computed in Table 1. The reason why we have not prepared the revised 'Hague Model' map in this way will be justified in the modelling section below.

Regional productivity patterns as a base for better global patterns

A major shortcoming of Table 1 is the calculation of the mean value for each vegetation type. This was done by averaging available measurements from the respective biome types without assuring the importance or representativeness of the community chosen. It therefore seemed important to provide a more significant base for future real productivity estimates. Regional productivity maps may become far more important for future planning of land use and carrying capacity than they are now. We have undertaken such a study in North Carolina as part of the Eastern Deciduous Forest Biome (EDFB) project of the U.S. IBP. To obtain regional productivity patterns in North Carolina we used the crop statistics available and tried to assess the total net primary productivity by converting yield data into total productivity data, then multiplying the productivity data with the area covered by this crop. The averaging of the values for the major forms of land use provides a mean productivity value for the political entity listed as a subdivision in the crop yield or forestry statistics. For North Carolina it was most convenient to use the county as the basic geographic entity. North Carolina has 100 counties. With this many points it was easy to construct

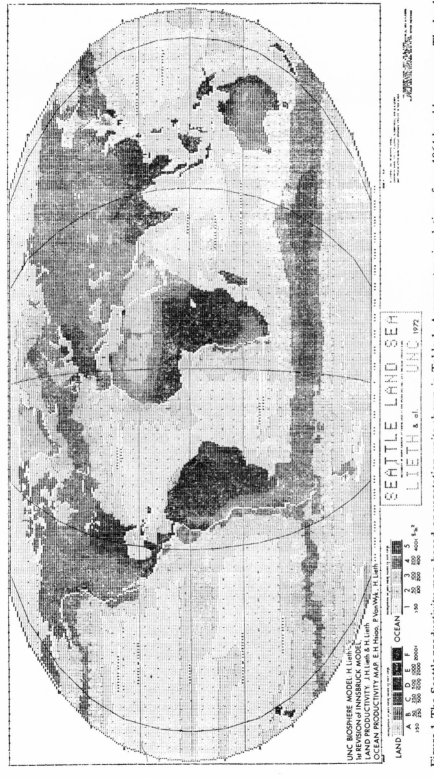

Figure 1. The Seattle productivity map based on vegetation units shown in Table 1. A computer simulation of our 1964 handdrawn map. The land portion was published as the Innsbruck productivity map in Lieth (1972). This and all other computer maps included in this paper were executed with SYMAP.

meaningful patterns for individual crops and the total productivity in the state as well. Figure 2 shows the total adjusted productivity pattern across North Carolina.

Further details including single crop production and suggestions for the improvement of the land use statistics may be obtained from Sharp et al. (1975). The final results are used here to construct models applicable on a global scale. The EDFB (Eastern Deciduous Forest Biome) of the U.S. IBP included several states wholly or partly investigated in the way described for North Carolina. The simultaneous study of Wisconsin, New York, Massachusetts, Tennessee and North Carolina now provide a north–south profile over several latitudes and altitudes. For details see Sharpe (1975), Art & Marks (1973), De Selm et al. (1971) and Stearns et al. (1971).

Together with the simultaneously run phenology projects (Hopp 1975; Reader et al. 1975) the data base was collected for a correlation study between productivity levels and length of growing season. The results are summarized in Figure 3. The average adjusted productivity values for each county are plotted in a coordinate system of length of growing period (x axis in days) and NPP (y axis in t/ha). Reader (1973) calculated for the linear correlation the equation

$$P = -1.57 + 0.0517\,S \tag{1}$$

Although constructed only for the Eastern Deciduous Forest Biome in the U.S. the model may be tested for its global usefulness by extending the equation in both directions from 0 g productivity to 365 days. If this is done we can try to use the model to convert an existing global map of the length of growing season by Wyatt & Sharp (1974) into a primary productivity map (Fig. 4).

This map is remarkable in as much as it provides us with the lowest global value that we have experienced so far with any of our models. Consistent with our past custom we will refer to this map and model combination as the 'Hague model'.

The global assessment of NPP with environmental parameters

The function of ecosystems is largely dictated by environmental forces. This is especially true for the photosynthetic process and the net primary productivity. Consequently this fact leads to the attempt to predict NPP from the ruling environmental parameters. In several recent papers (Lieth, 1973); Lieth & Box (1972) we have presented such predictions using temperature, precipitation and evapotranspiration as the environmental predictors. The combined use of average annual temperature and precipitation values for global NPP prediction was tried in the 'Miami model' (Lieth, 1973). Two equations were constructed

$$P = \frac{3000}{1 + e^{1.315 - 0.119T}} \tag{2}$$

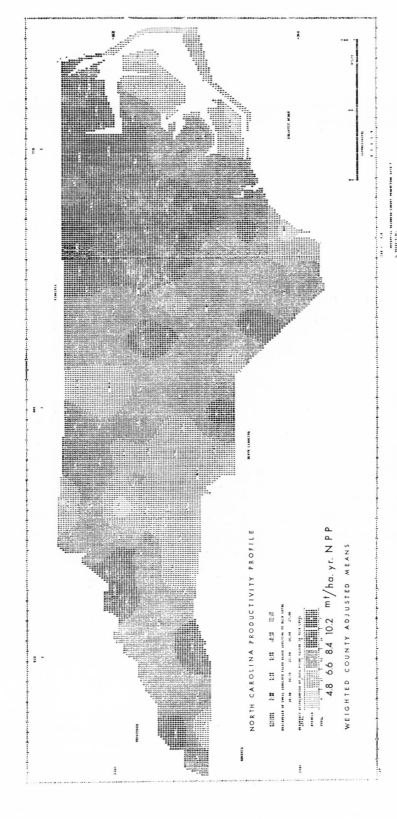

Figure 2. Productivity statistics of North Carolina. Weighted county productivity using forests, agricultural crops and other land use categories, in each case adjusting the statistical data with supporting measurements and calculations for total productivity. (These data points were used in Figure 3.)

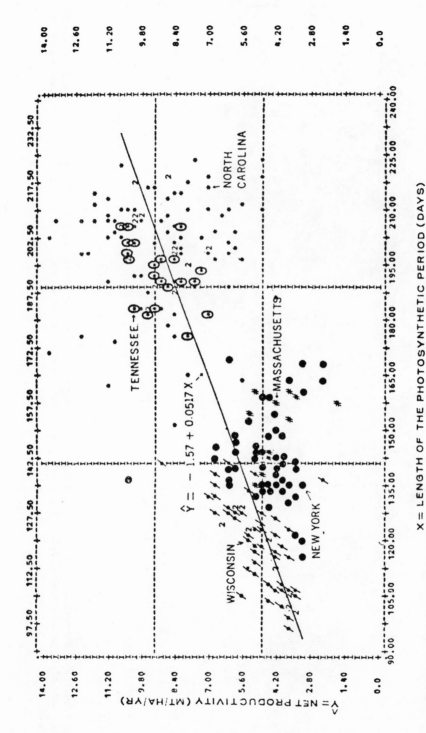

Figure 3. Net primary productivity vs. photosynthetic period. The data sets were taken from the U.S. IBP Eastern Deciduous Forest Biome memo reports for productivity profiles of Wisconsin, New York, Massachusetts, Tennessee and North Carolina.

308

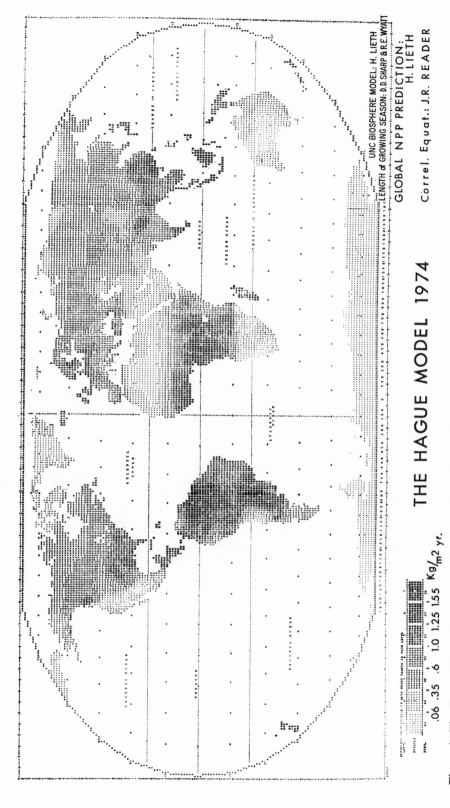

Figure 4. The Hague Model 1974: net primary productivity predicted from the length of the growing period. World growing period map based on ca. 600 points was produced by Sharp & Wyatt (1974). The predicting equation was obtained from the data set in Figure 3.

for temperature T vs. productivity, and

$$P = 3000(1 - e^{-0.000664N})$$ (3)

for precipitation N, e being the natural log base.

The models were constructed using a limited but well distributed data set, (Lieth, 1972). More than 1000 stations were then selected, spread over the continents as evenly as possible, for which both temperature and precipitation data were available and the data pairs converted into NPP by selecting the lowest value if the two models yielded different levels. The NPP values were then used to prepare a computer map similar to the two already shown.

This map, (Fig. 5), known as the Miami model, was evaluated quantitatively by Box (1975). He obtained 123.9×10^9 t of dry matter for the annual terrestrial NPP, a value which almost exactly fits the respective value of our compilation in Table 1 (121.7×10^9 t).[1]

To reinforce the demonstration of the usefulness of our environmental approach to productivity assessment we have taken evapotranspiration (E in mm) to predict NPP (P in g/m^2). This exercise was presented by Lieth & Box (1972). We constructed from our data set the equation

$$P = 3000(1 - e^{-0.0009695(E-20)})$$ (4)

With this model we converted the computer simulation of Geiger's world map of actual evapotranspiration into an NPP map. This map and model are referred to as 'Montreal 2: the C. W. Thornthwaite memorial model'. From this Box (1975) has calculated the global NPP value as 118.1×10^9 t per year.

Validations of models, maps and table

As new data sets are generated it will be desirable to test existing models with these. As new models are constructed it is necessary to test these against existing data sets. This procedure is most important either to reinforce or to reject existing patterns. Along with it we discuss the various opinions held on the global productivity level.

We have selected two examples pertinent to the topics so far discussed: (1) the validation of 'The Hague model' with recent data not used for the construction of the model, and (2) the testing of a new model by Terjung and coworkers (1973 and unpublished) against the terrestrial portion of the Seattle map.

A variety of models are available expressing in part the relation between NPP and the length of growing season. Also newer data sets provided the length of growing season in addition to other parameters. Several of the more recent approaches are compared in Figure 6. In this the standard line is the line R which is Reader's regression line for the Eastern Deciduous Forest Biome used to construct Fig. 4. We compare this first to the three other lines on the figure:

---

[1] Dr. Rodin reported in the discussion that Bodyko and Ephimova have predicted a terrestrial NPP of 134×10^9 t per year from an energy budget.

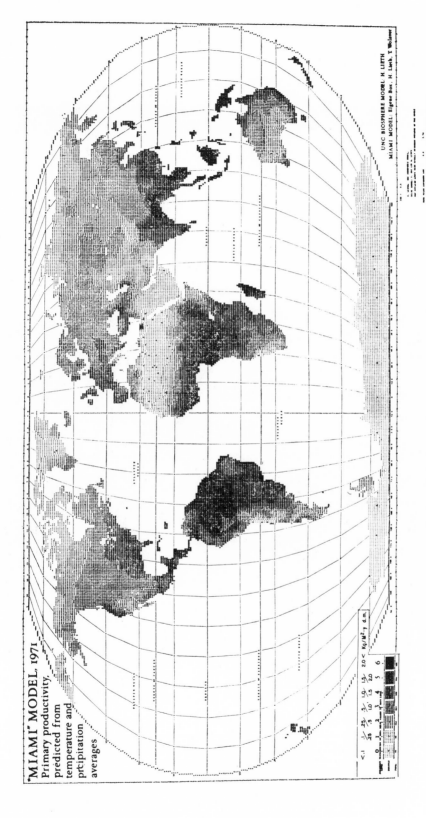

Figure 5. The Miami Model: predicting net primary productivity from precipitation and temperature. For details see Lieth (1973).

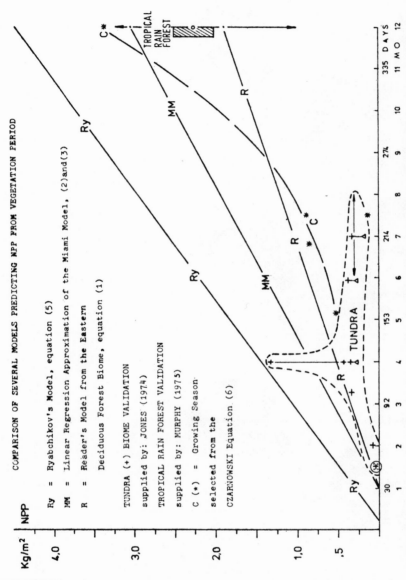

Figure 6. The comparison of several models predicting net primary productivity from the length of the vegetation period. Abscissa: vegetation period in days or months; ordinate: net primary productivity in kg/m² per year. The slopes of the lines R, MM and Ry are discussed in the text. The validation data used for tundra and tropical rain forests indicate that the global productivity level of the Miami model is probably the most accurate. This level is also supported by Table 1. For further explanation see text.

1. The line Ry which is Ryabchikov's model (1968) (from Rodin et al., 1972).

$$Kp = \frac{W\,Tv}{36\,R} \qquad (5)$$

in which Kp = productive potential, W = annual effective precipitation, Tv = length of the growing season as multiples of 10 day units, and R = annual radiation. This equation aims to assess the productive potential. It is not exactly a growing season model. It may be used here as a straight line between the two fixed end points without knowing exactly what shape the curve may have between them.

2. The line MM which uses the start and end point for the Miami model to define a straight line (which may of course be curvilinear).

3. The curve drawn through the asterisks, taken from a model developed by Czarnowski (1973). In this model

$$y = cp'\,L(1 - e^{-q}) \qquad (6)$$

in which L is an indication of the length of the growing period defined by temperature only, q the humidity index calculated from evaporation and precipitation, and p' the water vapor pressure for the growing season.

Comparing the four lines one can see the striking difference between Ryabchikov's line and all the others. The slopes of the lines R, MM and Ry have the approximate ratio $1:1.65:2.44$. Rodin et al. (1972) give for their tabular assessment of NPP the total terrestrial value of 172×10^9 t; the Miami model as well as the data set in Table 1 of this paper (derived from the same data base) provide for a NPP of about 121×10^9 t. The ratio of the slopes for MM:Ry is $1:1.47$ comparable to the ratio of the global terrestrial values derived from each data base which is $1:1.42$ (Miami model vs. the compilation by Rodin et al. Using the same approach in reverse for the Hague map based on Reader's curve we should expect 73.5×10^9 t dry matter of annual terrestrial NPP to be predicted by this model.

Czarnowski's equation has not yet been used to construct a global productivity pattern. The location of the curve suggests, however, that his assessment would come out somewhere between the growth period prediction and the Miami model.

Rodin, Bazilevich & Rozov (1972) do not specifically state that they used Ryabchikov's model to complete their very detailed productivity table of various soil-determined ecosystems, but the above mentioned discussion of the slope differences between Ry and MM strongly suggests that they have used the same data base.

The two end points of Reader's equation can be validated by two data sets recently made available to us by Jones & Gore (1974) for tundra, and Murphy (1975) for the tropical rain forest. From the values we must conclude the lower portion of Reader's equation predicts rather well (it overrates slightly); whereas the upper portion of the curve underestimates the productive power of the tropical areas. The discussion of this figure demonstrates how insecure global modeling will remain until we get more solid material from the tropical areas with which to work.

Terjung and coworkers recently started a project to predict global photosynthesis through the energy budget (Terjung & Louie, 1973; Terjung et al, in press). In their still unpublished approach they have calculated several photosynthetic models. They were kind enough to supply data for this paper and in Figure 7 we show their different models of photosynthesis vs. leaf temperature. All the models shown in this figure were used by them to calculate the annual productivity for the main growing season. Their world map based on model 3 and 4 was tested against the terrestrial portion of the Seattle map (the Innsbruck map in Lieth 1973) shown earlier in Figure 1.

Figure 8 shows the results of their comparison. The stippled areas of the map are considered to be equal in both attempts. In the white areas their models overestimate whereas in the hatched areas our model appears to overestimate. The map indicates that Terjung's map based on his model 3 + 4 yields about the same global values as our Seattle model, around 91×10^9 t of dry matter per year. The tendency of the slopes as shown in the various maps is certainly verified by Terjung and coworkers, whereas the actual degree of slope remains to be refined in either our or his map. Furthermore, investigation is needed into how sensitive their world pattern assessment is to the change from model 3 to model 2, since our experience shows that agricultural landscapes are less productive than forests in warmer regions. In particular their maize model vs. the woody species model can hardly be verified by our North Carolina data from crop statistics and plantation measurements. Should their model 1, however, become validated in the future we can well expect the NPP, presently assumed to be around the level of the Seattle or Miami model, to be raised to the level of the Ryabchikov model.

The pattern of products provided by the primary producers in different biomes

The structure and function of ecosystems depends not only on the quantity but also on the quality of the dry matter produced. We assume there is no need to justify this statement but make a brief attempt here to demonstrate major differences in materials provided by the primary producers in various biomes. We restrict our first evaluation to gross categories like carbohydrates, proteins, fatty substances and ash. Mineral component is not considered here as Rodin & Bazilevich have elaborated on this matter on various occasions. (e.g. 1966). For consumers and decomposers the organic portions are usually much more important than mineral substances provided by the plants. The following Tables 2 & 3 show how we assess such quality differences and how we may compare them in the future. This study, made as a seminar exercise at the University of North Carolina, is reported in full by Lieth (1973, 1975).

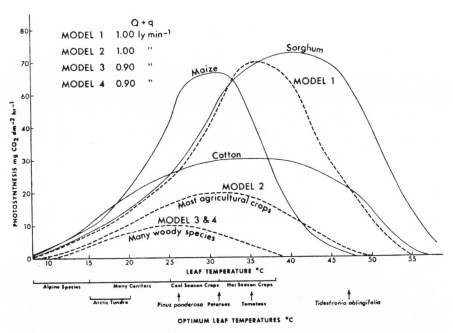

Figure 7. The various temperature-photosynthesis curves used by Terjung et al. (in press) to calculate NPP from the energy budget. Q = direct, q = indirect solar radiation, ly = cal cm^{-2}.

Table 2 shows how we assess the various qualities quantitatively. The example taken was from the grassland biome of the U.S. IBP for which we had enough details to allow a breakdown into the categories listed in the table, using the analyses provided in the Atlas of U.S. and Canadian Feeds (1971). It is clear that future studies of this kind ought to be based on our own analyses, but this approach may be justified because we first wanted to show if we can expect valuable results in terms of differences within ecosystems. The accounting procedure yields average values which we can then compare with similar computations from other ecosystems. In Table 3 we compare several ecosystem types from the temperate region. Only the grassland data are, however, assessed in the detail indicated in Table 2. The other ecosystems lack the proper data sets and need future studies in detail. The data in Table 3 are significant, however, since we do not expect changes in the figures that will overthrow the trends in the table, in which we compare a freshwater ecosystem with grassland and forest. If we analyse the trends in the separate columns we find that structural carbohydrates increase as we go from lake to forest, whereas nonstructural carbohydrates have similar levels in herbaceous communities but drop substantially in forest ecosystems. Protein has the same pattern as nonstructural carbohydrates, but deciduous forest and grassland operate with almost the same level of fat. The high

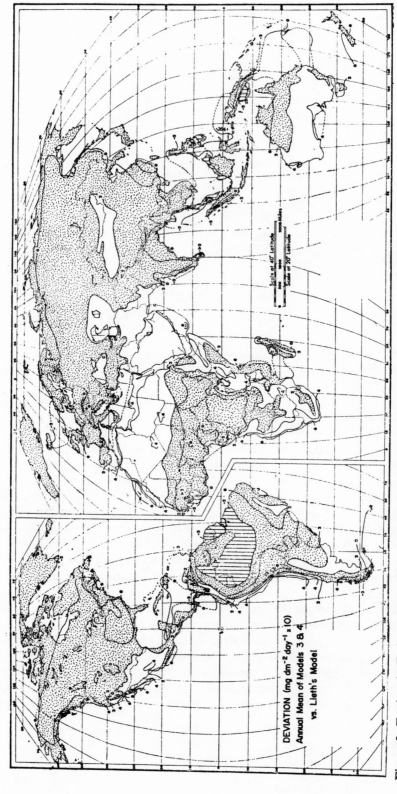

Figure 8. Testing the Terjung et al. model against the productivity map presented in Figure 1. Major deviations occur in very dry and wet areas (see text), the stippled areas are in acceptable agreement. Figures 7 and 8 are published by special permission of the authors.

316

Table 2. Chemical composition of above-ground grassland biomass; classes of chemicals distinguished in nutritional studies, entered separately for individual species. Table from Lieth (1975), compiled by J. R. Reader.

1	2	3	4	5	6	7	8	9	10	11	12	13	14
Family or tribe	Species	Peak live biomass composition %	g/m²	Ash %	g/m²	Crude fiber %	g/m²	Ether extract %	g/m²	N-free extract %	g/m²	Protein %	g/m²
Andropogoneae	Bluestem	24.2	384.3	6.6	25.4	34.3	131.8	2.3	8.8	50.2	192.9	6.6	25.4
Chlorideae	Grama	15.9	253.7	9.1	23.1	32.0	81.2	1.6	4.0	50.3	127.6	6.9	17.5
Compositae	Aster sp.	15.7	250.1	7.6	19.0	21.1	53.5	4.1	10.4	58.8	149.2	8.4	21.3
Festuceae	Fescue	9.5	151.7	7.5	11.4	31.9	48.4	2.9	4.4	48.1	72.9	9.6	14.6
Leguminosae	Lespedeza	6.7	107.2	7.4	7.2	44.9	48.1	2.1	2.2	32.8	35.2	12.8	13.7
Cactaceae	Opuntia	5.4	85.8	16.4	4.6	13.7	11.7	1.8	1.5	63.9	54.8	4.2	3.6
Stipae	Needlegrass	5.4	85.4	9.9	8.4	36.7	31.3	1.9	1.6	43.2	36.9	8.3	7.1
	Prairie												
Sporoboleae	Dropseed	4.2	67.5	10.8	7.3	31.8	21.5	1.8	1.2	43.4	29.3	12.2	8.2
Triticeae	Wheat	3.9	62.4	6.5	4.0	27.9	17.4	3.8	2.4	49.8	31.1	12.0	7.5
Liliaceae	Yucca	2.3	36.9	7.1	0.5	42.7	15.7	2.0	0.7	40.5	14.9	7.7	2.8
Chenopodiaceae	Goosefoot	1.4	22.7	6.4	1.5	19.3	8.4	7.2	2.6	52.0	19.2	15.1	5.6
Paniceae	Panicum	1.4	21.4	13.1	2.8	38.3	8.2	1.3	0.3	38.8	8.3	8.5	1.8
Boraginaceae	Stickseed	0.8	12.5	11.6	1.4	10.6	1.3	3.3	0.4	51.7	6.5	22.8	2.8
Cyperaceae	Sedge	0.6	9.5	7.7	0.7	30.2	2.8	3.2	0.3	49.3	4.7	9.6	0.9
Caryophyllaceae	Silene	0.6	9.1	16.7	1.5	19.2	1.7	—	—	—	—	10.6	0.9
Danthonieae	Danthonia	0.5	7.6	4.6	0.3	30.2	2.3	3.6	0.3	52.1	3.9	9.5	0.7
Malvaceae	Mallow	0.3	5.6	16.8	0.9	21.2	1.2	2.7	0.2	36.9	2.1	22.4	1.2
Rosaceae	Avens	0.3	5.3	3.9	0.2	21.2	1.1	17.9	0.9	36.9	1.9	17.4	0.9
Aveneae	Oats	0.3	4.8	7.2	0.3	34.6	1.6	2.3	0.1	46.7	2.2	9.2	0.4
Euphorbiaceae	Euphorbia	0.3	4.3	11.0	0.4	13.0	0.5	4.1	0.2	57.5	2.5	14.4	0.6
Onagraceae	Evening Primrose	0.3	4.0	7.9	0.3	20.0	0.8	3.0	0.1	54.7	2.2	14.4	0.6
Total		100.0	1591.8	7.6	121.2	30.8	490.5	2.7	42.6	50.2	798.3	8.7	138.1

317

Table 3. Chemical differences in primary producers of selected ecosystem types (from Lieth 1973, 1975). The numbers are given in % averages as extracted from literature data

Ecosystem	Comments	Ash	Structural carbohydrates*			Non-structural carbohydrates	Protein	Ether extract or equivalent	Total %
			Cl.	Hcl.	Lgn.†				
Coniferous forest	Biomass, mature stand	0.3	43.5	14.5	30	1.1	1.3	7.7	98.4
	productivity	4.2	44.1	8.7	18	15.5	4.0	5.8	100.3
Summergreen deciduous forest	Standing biomass	0.3	46.6	24	20	0.8	2.5	1.8	96
	productivity	4.2	37	14.4	12	22.5	6.4	2.8	99.3
Temperate grassland	Above ground	7.6	30.8			50.2	8.7	2.7	100
	peak biomass	10	28			48	11	3	100
Freshwater lake	plankton	14	18			50	17	1.5	100
	macrophytes		14–20			43–60	8–19	1.0–2.5	
	benthic algae		9–17			36–44	5–18	0.7–2	
	Typha		30–39			38–48	7–12	1.5–3.5	
	Scirpus	6.5	33			53	7	0.5	100

* Crude fiber in table 2.
† Cl. = Cellulose, Hcl. = Hemicellulose, Lgn. = Lignin.
Table compiled as a University of North Carolina seminar project (participants H. Lieth, J. R. Reader, W. Martin, P. Carlson, G. Doyle and R. Kneib.

318

terpene content is largely responsible for the values in this column being twice as high in coniferous forests as in deciduous forests.

Differences as shown in Table 3 must substantially influence the types of organisms and their performance in the consumer and decomposer groups, and with further studies we expect deeper insight into the relations between decomposer and consumer species present in a given ecosystem and the proportions of chemicals offered by the primary producers. Such investigations may enable us in the future to prepare pattern maps of protein production levels, or of other organic chemical properties of the vegetation of the world. Functional geobotany that started with the global assessment of NPP is now looking forward to a new line of work that will provide us with a better insight into evaluationary processes, management priorities and future land use improvements.

Summary

Summarizing the existing information for net primary production, we find that about 177×10^9 t dry matter are produced by the entire biosphere equivalent to about 3.35×10^{21} J (793.5×10^{18} cal), with about 122×10^9 t for the terrestrial vegetation, equivalent to 2.25×10^{21} J (535×10^{18} cal). The solar energy conversion, assuming an input of 2.1×10^{24} J (530×10^{21} cal), has a mean efficiency of 0.16%, of which 67% is from terrestrial areas and 33% from the oceans. Gross patterns of global productivity from the past decade are compared. Details in Table 1.

Correlations between net primary productivity and environmental parameters can greatly facilitate regional pattern prediction. This has also been attempted on a global scale. The procedure is to build correlation models based on a limited set of actual productivity measurements paired with environmental data, construct a regression equation and then convert a large, world covering set of environmental data into biological values. So far, we have constructed four models of this type

1. $P = -1.57 + 0.0517S$ for photosynthetic season S in days (Reader).
2. $P = 3000(1 - e^{-0.00664N})$ for precipitation N in mm (Lieth, Wolaver & Box).
3. $P = (3000/1 + e^{1.315-0.119T})$ for temperature T in °C (Lieth, Wolaver & Box).
4. $P = 3000(1 - e^{-0.0009665(E-20)})$ for actual evapotranspiration E in mm (Lieth & Box). For equation (1) P reads in t/ha, for (2)–(4) in g/m².

Each model is convertible into a map showing the global productivity pattern. The combined use of (2) and (3) is known as the Miami Model, (4) as the Thornthwaite Memorial Model. (1) was constructed from the productivity profiles of the Eastern Deciduous Forest Biome of the U.S. It uses a data set completely different from models (2)–(4). In its present form, it reads the maximum productivity possible for 365 days as about 1800 g/m² whereas the other three formulae operate with a hypothetical maximum value of 3000 g/m². With this model we have constructed a new computer map of primary productivity based on the length of vegetation period: the Hague model. This map yields the approximate global productivity level of 73.5 × 10⁹ t of terrestrial dry matter.

The primary productivity data produced in the last decade provides a starting point for future studies on a variety of scientific programs ranging from geophysics to biochemistry. From the data so far available we selected a breakdown of the organic biomass portion and compared these data for different ecosystems. Table 3 shows the differences for 3 types of ecosystem: forest, grassland and lake. Such comparisons may be used to interpret the distribution and performance of consumers and decomposers.

References

* Indicates papers and information provided before publication, used with permission of the authors.

Atlas of Nutritional data on United States and Canadian Feeds. 1971. National Academy of Science, Washington D.C.

Art, H. W. and P. L. Marks. 1973. Primary productivity profile of New York and Massachusetts. U.S.-IBP deciduous forest biome memo report 72-39. ORNL, Oak Ridge. 18 p. + tables and maps.

Box, E. 1975*. Quantitative Evaluation of the Global Productivity Models Generated by Computer. In: Primary Productivity of the Biosphere. Ecological Studies 14. H. Lieth & R. H. Whittaker, eds. Springer Verlag, New York (in press).

Brünig, E. F. 1974. Ökosysteme in den Tropen. Umschau 74: 405–410.

Czarnowski, M. S. 1973. W sprawie mapy i modelu siedliskowej zdoldosci produikcyinej Ziemi. Przeglad geograficzny 45: 295–308.

DeSelm, H. R. *et al.* 1971. Productivity profile for Tennessee. U.S.-IBP deciduous forest biome memo report 71-13. ORNL, Oak Ridge. 182 p.

Hopp, R. J. 1975. Plant Phenology Observation Networks. In: Phenology and Seasonality Modeling. Ecological Studies 8. H. Lieth, ed. Springer Verlag, New York. 25–44.

Jones, H. E. and A. J. P. Gore. 1974. Tundra Biome Synthesis Volume. Cambridge University Press, Cambridge. (in press).

Lieth, H. 1964. Versuch einer kartographischen Erfassung der Stoffproduktion der Erde. Geographisches Taschenbuch 1964/65, p. 72–80. F. Steiner Verlag, Wiesbaden.

Lieth, H. 1972. Über die Primärproduktion der Pflanzendecke der Erde. Z.f. Angew. Botanik 46: 1–37.

Lieth, H. 1973. Primary Production: Terrestrial Ecosystems. J. of Human Ecology 1: 303–332.

Lieth, H. (ed.) 1973. Chemical differences of contrasting ecosystems and their trophic levels. An exploration of a new view point in systems ecology. U.S. IBP EDF. Biome Memo Report 73-6.

Lieth, H. (ed.) 1974. Phenology and Seasonality Modeling. Springer Verlag, New York, 444 p.

Lieth, H. 1975. Some Prospects beyond Production Measurement: Comparative Analysis of some Biomass Properties on the Ecosystem Level. In: Primary Productivity of the Biosphere. Ecological Studies 14. H. Lieth & R. H. Whittaker, eds. Springer Verlag New York (in press).

Lieth, H. and E. Box. 1972. Evapotranspiration and Primary Productivity; C. W. Thornthwaite Memorial Model. In: Papers on Selected Topics in Climatology. J. R. Mather, ed. 2: 37–44, Elmer, New York.

Lieth, H. and R. H. Whittaker (eds.) 1975. Primary Productivity of the Biosphere. Ecological Studies 14. Springer Verlag, New York (in press).

Murphy, P. G. 1975. Net Primary Productivity in Tropical Terrestrial Ecosystems. In:
    Primary Productivity of the Biosphere. Ecological Studies 14. H. Lieth & R. H.
    Whittaker, eds. Springer Verlag, New York.
Reader, J. R. 1973. Phenological Investigation in Eastern North America. Thesis,
    Ph.D., University of North Carolina, Chapel Hill, N.C.
Reader, J. R., J. S. Radford and H. Lieth. 1975. Modeling Important Phytopheno-
    logical Events in Eastern North America. In: Phenology and Seasonality Modeling.
    Ecological Studies 8. H. Lieth, ed. Springer Verlag, New York. 329–342.
Rodin, L. E. and N. I. Bazilevich. 1966. Production and Mineral Cycling in Terrestrial
    Vegetation. Oliver and Boyd, Edinburgh, 228 p.
Rodin, L. E., N. I. Bazilevich and N. N. Rozov. 1972*. Productivity of the World's
    Main Ecosystems. Seen as manuscript for publication in the Seattle Symposium
    volume, National Academy of Science, Washington, D.C. (in press).
Ryabchikov, A. M. 1968. Hydrothermal Conditions and the Productivity of Plant Mass
    in the Principal Landscape Zones. Vestnik MGV, Geogr. Moscow 5: 41–48, cited
    after Rodin, Bazilevich and Rozov 1972.
Sharp, D. D., H. Lieth and D. Whigham. 1975. Assessment of Regional Productivity
    in North Carolina using Cropyield Statistics. In: Primary Productivity of the Bio-
    sphere. Ecological Studies 14. H. Lieth & R. H. Whittaker, eds. Springer Verlag,
    New York (in press).
Sharp, D. and R. E. Wyatt. 1974. Length of growing Period in Months. Map inside
    back cover. In: Phenology and Seasonality Modeling. H. Lieth, ed. Springer Verlag,
    New York.
Sharpe, D. M. 1975. Methods for Studying the Primary Productivity of Regions. In:
    Primary Productivity of the Biosphere. Ecological Studies 14. H. Lieth & R. H.
    Whittaker, eds. Springer Verlag, New York (in press).
Stearns, F., N. Kobriger, G. Cottam and E. Howell. 1971. Productivity profile of Wis-
    consin. U.S.-IBP deciduous forest biome memo report 71-14 ORNL, Oak Ridge, 82 p.
Terjung, W. H. and Stella Louie. 1973. Energy Budget and Photosynthesis of Canopy
    Leaves. Ann. Ass. Amer. Geogr. 63: 109–130.
Terjung, W. H., Stella Louie and P. A. O'Rourke. 1974. Global Photosynthesis Model.
    Seen as manuscript in preparation.
Whittaker, R. H. and G. E. Likens. 1975*. Primary production: The biosphere and
    man. In: Primary productivity of the biosphere. Ecological Studies 14. H. Lieth &
    R. H. Whittaker eds. Springer Verlag, New York (in press).

321

36

Reprinted from *Am. Sci.* **59**(4):425–433 (1971), by permission of *American Scientist*, journal of Sigma Xi, The Scientific Research Society of North America.

A WORLD PATTERN IN PLANT ENERGETICS

Carl F. Jordan

The community energetics approach to ecology (Lindemann 1942) permits the comparison of ecosystems of different structure. A comparison of rates at which an oak forest, a tomato patch, and a rice paddy convert solar energy into chemical energy, for example, could yield insights into the structure and function of these ecosystems and how the plants in them are influenced by climatic and edaphic factors. The approach, used in many agricultural and forestry studies, of measuring only economically useful plant parts, in contrast, does not give as much information concerning the nature of plant communities.

Insights into energy flow in forest ecosystems throughout the world have been gained by comparing ratios of leaf fall to litter accumulation on the forest floor (Olson 1963; Rodin and Basilevic 1968). Near the equator the ratio of leaf fall to litter accumulation

Boyhood experiences in the wilds of northern Maine led to a forestry and conservation major for Carl Jordan at the University of Michigan, where he obtained a Bachelor's degree in 1958. After four years in the Navy, he undertook graduate studies in ecology at Rutgers University. As part of the research for his Ph.D., which he received in 1966, he developed and obtained a patent for a lysimeter used in mineral cycling studies. His first position following graduate work was with the Atomic Energy Commission's terrestrial ecology project in Puerto Rico, where he studied the productivity of a tropical rain forest recovering from radiation damage, and the cycling of radioactive isotopes through the forest. In 1969 he joined the newly established terrestrial ecology project at Argonne National Laboratory, where he now holds the position of Associate Ecologist.
Gerson Rosenthal, of the University of Chicago, contributed greatly to the exploration and clarification of the ideas presented in this paper. Robert Wolfgang, of Argonne National Laboratory, gave editorial assistance. Support for the preparation of this paper was provided by the U. S. Atomic Energy Commission. Address: Radiological Physics Division, Argonne National Laboratory, Argonne, IL 60439

is high, and at higher latitudes the ratio is low. This pattern is a result of high rates of leaf production and small amounts of litter accumulation in the tropics, and low rates of leaf production and large amounts of litter accumulation at higher latitudes. Productivity of leaves and litter accumulation are related, in turn, to climatic and edaphic factors.

Few worldwide patterns such as that of the leaf fall/litter accumulation ratios have emerged from the many studies of community energetics (often called "productivity studies") now in the ecological literature. Recognizing worldwide patterns in plant production is difficult because of the large local variations in production, which are caused, in turn, by local variations in soil moisture, soil fertility, climate, and other factors.

My objective here is to find a worldwide pattern of terrestrial plant energetics that would provide a basis for calculating the productive potential of land areas of the world. The major problem in recognizing such a pattern is the elimination of confusing irregularities in the data resulting from local environmental variations. This problem is solved by forming ratios, because ratios eliminate variations in absolute values, and consequently patterns are more easily recognized.

After studying the published data it became apparent that, while absolute amounts of production depended greatly upon local conditions, energy distribution within plants followed certain worldwide environmental gradients, regardless of local conditions. By energy distribution within plants, I mean how the plant allocates the energy which is available for synthesizing parts. For example, a plant can

allocate a lot of energy to wood and little to leaves, or vice versa. Because many ecological productivity studies include both total wood and leaf and other litter production, energy distribution can be quantified by forming a ratio of amount of energy bound in wood per year to amount of energy bound in leaves and other litter.

There is a rationale in attempting to look for a world pattern in the ratios described above. Energy stored as wood in the trunk and large branches of trees is bound in parts which remain intact for most of the life of the plant. Energy stored as leaves and other litter, including fruits, flowers, bark, and twigs, is energy that is quickly available to herbivores and decomposers. Under certain conditions, it seems possible that the environment would select in favor of plants which use a large proportion of the energy available as relatively permanent tissue, and under other conditions, it would select in favor of those which use only a small proportion as such tissue. Later I will discuss some of the possible selective mechanisms.

The ratios

The ratios discussed here are of amounts of energy bound in long-lived tissue per year to amount of energy bound in short-lived tissue per year. They are calculated as follows:

$$\frac{(P_w)\,(C_w)}{(P_l)\,(C_l)}$$

where P_w = rate of wood production, C_w = caloric concentration (calories per gram, dry weight) in wood, P_l = rate of litter production, and C_l = caloric concentration in the litter. The ratio indicates how energy utilized by the plant for tissue synthesis is allocated. A high ratio indicates that a relatively high proportion of energy is bound in parts which remain intact for most of the life of the plant, including trunks and large roots of woody trees and rhizomes of certain perennial herbs and grasses. A low ratio indicates that a high proportion is bound in parts that are short-lived relative to the life of the plant, including leaves, fruits, and flowers, and in some cases bark and twigs.

Because the caloric concentrations in leaves and wood are similar (Golley 1961) and because leaves make up most of the litter, it is possible to cancel C_w and C_l in the above ratio, and to approximate it by the ratio of wood production to litter production. The data required for the latter ratio are more commonly found in the literature than those required for the former and, therefore, the latter are presented in this paper.

By comparing the ratios of wood to litter production we should be able to find a worldwide pattern in plant energetics. Because patterns in nature are a result of adaptations to environment, according to evolutionary dogma, recognizing major environmental patterns should help in finding patterns in biological data.

Two major environmental factors which form somewhat regular patterns are solar radiation and precipitation. Maps of annual amount of solar radiation reaching the earth's surface (Budyko 1956; Kondratyev 1969) show that annual solar radiation generally decreases as latitude increases. Precipitation patterns are somewhat more complex, but still are predictable in relation to prevailing winds, mountains, and land masses.

All other environmental factors fall into two categories: (1) They are a result of one or both of the two major factors and therefore have the same approximate world patterns as the major factors. Examples are temperature and length of growing season, which are functions of the amount of solar radiation and, therefore, have the same general pattern of world distribution as solar radiation. (2) They form patterns so complex that correlation with ecological productivity is impossible because of the limited number of productivity studies. An example is geological formations.

While annual amounts of solar radiation reaching the earth decrease with increasing latitude, the amount of solar radiation that can be used by plants may follow a slightly different pattern, and it is this pattern that determines the pattern of energetics to be shown here. An example of the difference is as follows: The British Isles are at a latitude of about 55° N, and presumably have many areas that receive less annual solar radiation than areas in the north-central part of the United States, at a latitude of about 45° N. Because of the warming influence of the Gulf Stream, however, the British Isles may have a longer growing season, and, as a result, there would be more usable solar radiation there than in the north-central United States. The amount of solar radiation impinging on the plant *during the growing season* determines the plant's response, not simply the amount *per year*.

To calculate the amount of usable solar radiation during the growing season at the locations of studies cited in this paper, I used maps and tables published by Trewartha (1954) and Kondratyev (1969). Trewartha presents a world map showing the average length of the growing seasons. Kondratyev gives the total possible radiation for each month for each 5° of latitude. Using the growing-season data and the latitude for each community from a world atlas, the total solar radiation during the growing season was calculated from the tables of Kondratyev.

Before ratio patterns and environmental patterns can be compared, two biological factors must be considered. One factor is that the ratio of wood production to litter production is a function of age of community as well as a function of environment. Later it will become evident that even-aged forests (forests composed of trees all about the same age) which are either young or old have ratios lower than those of both intermediate even-aged forests and forests that have trees of all ages. Rate of wood production in young forests could be limited by the cambial area of the trees, causing the ratio in these communities to be low. In communities where most of the biomass occurs in old trees, the ratio also is low. As trees reach old age, their rate of wood production decreases faster than their rate of leaf production.

The other biological factor that influences the world pattern of wood/litter production ratios is that vascular plants have two basically different structures. In trees and shrubs, energy used as wood is stored mainly above ground. In herbs and grasses, energy used as wood or woody-like tissue is stored below ground. Trees and grasses growing in the same region can be subjected to the same environmental forces and yet have different ratios because of their basically different structure. For this reason, a pattern of ratios in forest communities is considered separately from a pattern in grass and herb communities. In areas where forests and grasslands are intermixed, forests often replace grassland through the process of succession.

Methods of measuring litter fall and wood production have not been standardized; consequently some variation in the ratios results from variations in methods. The difference in ratios of forests caused by nonstandard litter collections is probably small, because leaves comprise by far the greatest bulk of the litter that falls from trees, and all authors included leaves in their litter fall term.

Rate of wood production is given in more than half the forest productivity papers, while the annual increment in total standing biomass, including an increase in leaves attached to the tree, is shown in the rest. Errors in ratios arising from the use of change in standing crop per year, instead of annual wood production, are small, because, as Kira and Shidei (1967) have reported, leaf biomass changes very little in most trees after an age of about 20 years, while wood biomass increases almost till the tree dies. It appears that, in forest studies where shrub data were not included in the term wood production, the shrub component comprised an insignificant part of total production, and therefore errors from this source probably are small. In the calculation of ratios for grassland and herb communities, wood production is assumed to be the growth of rhizomes and roots, and litter production, the growth of aboveground parts.

Only studies which present production data broken down into the categories used here could be included in the sources for the calculations presented in this paper. Herb and grass community studies which lack root and

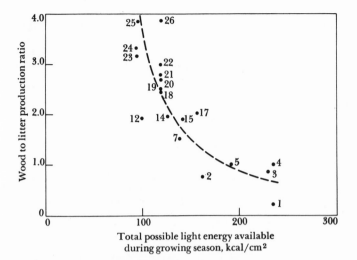

Figure 1. Ratio of wood production to litter production in forest communities as a function of amount of light energy available during the growing season. The numbers refer to communities in Table 1.

rhizome production data, essential for calculating the wood to litter production ratios, and forest productivity studies that failed to include leaf and litter fall, data also essential for calculating the ratios, could not be used.

The pattern

Available solar energy gradient. The ratio of wood production to litter production in forests generally increases along a gradient of decreasing solar energy available during the growing season as shown in Figure 1. The ratios in this figure are from Table 1, and the available light energy was calculated as explained above.

Four of the communities plotted in Figure 1 are of special interest because their ratios differ greatly from other forests of the same latitude. Community 14 is on Long Island, New York, where, because of the ocean's warming influence during the winter, the growing season is longer than at inland locations. Consistent with the hypothesis, the ratio is one of the lowest for temperate zone forests. Communities 23 and 24 are at a relatively high altitude where the growing season is short and, in keeping with the hypothesis, their ratios are relatively high. Community 17, which has the highest ratio for tropical communities, is noteworthy because its growing season is not governed by temperature but by a winter dry season (Kendrew 1953).

Not all communities in Table 1 are included in Figure 1. Some are excluded because they are young or old even-aged communities or because the ratio was affected by precipitation, a factor in the world pattern considered later. Communities 6 and 9 have ratios relatively low for their location because they are young even-aged stands. Community 13 has a low ratio because it is an old even-aged stand. In communities 10, 11, and 16, the rate of fall of dead trees is approximately the same as the term "real primary production," suggesting that there is little annual increase in total plant biomass in the forest, a condition that would prevail in an old even-aged forest. Community 8 has a low ratio for a temperate forest, but it is located where precipitation is light. A low ratio is correlated with low precipitation as well as with high amounts of available light.

Only community 27 has a ratio that drastically departs from the general pattern. The leaf fall data in this study may be in error, because the author's estimate of leaf fall was not based on direct measurement but on correlations made by other authors from trees at other locations.

The ratios of wood production to litter production in grass, herb, and sedge communities, like the ratios in forest communities, increase along a gradient of decreasing available solar energy, as indicated by type and loca-

tion of community. The data in Table 2 show that the ratio in a tundra community is higher than ratios in temperate grassland communities. Aging apparently has the same effect on the ratios of grass and herb communities as on forest communities. The ratios are low when the community is young (Table 2, community 4) and old (communities 1 and 2).

Precipitation gradient. The ratio of wood production to litter production decreases as precipitation decreases. The ratios for mesic temperate forests is between about 2.0 and 4.0 (Table 1). For savannah forests, the ratio is about 1.5 (Table 1, communities 7 and 8). For the prairie community, the one ratio available is 0.87 (Table 2, community 5). In an Arizona desert shrub community, annual wood production was 6.3 grams per square meter per year, and litter production was 133 grams per square meter per year (Chew and Chew 1965), giving a ratio of 0.05.

With a worldwide pattern of plant energetics established, it is easier to recognize worldwide patterns in absolute production values. Although data in the wood production column of Table 1 varies, there seems to be no relation between productivity and available solar energy (as indicated by latitude and altitude). The variations in wood production apparently are caused by local variations including soil moisture and fertility. Extremes in soil moisture and fertility could cause a reduced rate of growth in the species dominant throughout the region, or they could result in the local dominance of other species better adapted to the extreme conditions. On a worldwide scale, the absolute rate of wood production in mesic forested areas would be almost constant if variations in such local factors as soil moisture and fertility were eliminated.

In contrast, litter fall values in Table 1 decrease with increasing latitude and/or altitude. Litter fall decreases from approximately 2,000 grams/sq. meter/year in the tropics to just over 200 grams/sq. meter/year in north temperate latitudes or on mountain tops. In grassland communities (Table 2) absolute rates of wood production appear to be generally lower than in forest communities (Table 1), as might be expected considering the structure of the two groups of plants.

Table 1. Rate of wood production and rate of litter production in various forests, and the ratio between the two rates

Community number	Type of community	Location	Age (in years)	Wood production g/m²/yr	Litter fall g/m²/yr	Wood production/ litter production	Author
1	Tropical rain forest	Thailand	mature	533	2322	0.23	Kira et al., 1967
2	Broadleaf evergreen forest	Japan	?	920	1140	0.80	Kimura, cited in Kira et al., 1967
3	Tropical rain forest	Puerto Rico	intermediate	486	547	0.88	Jordan, in press
4	Tropical rain forests	Average of several	?	1650	1600	1.03	Rodin & Basilevic, 1968
5	Subtropical forests	Average of several	?	1250	1200	1.04	Rodin & Basilevic, 1968
6	Ash plantation	Denmark	12	410	330	1.46	Boysen & Jensen, cited in Kira et al., 1967
7	Dry savannah	Turkmenistan, Russia	?	440	290	1.52	Rodin & Basilevic, 1968
8	Oak forest	Steppe zone, Russia	?	550	350	1.57	Mina, cited in Rodin & Basilevic, 1968
9	Beech plantation	Denmark	8	480	270	1.77	Möller et al., cited in Kira et al., 1967
10	Fir	Mid taiga, Russia	?	450	250	1.80	Rodin & Basilevic, 1968
11	Fir	Southern taiga, Russia	?	550	300	1.83	Rodin & Basilevic, 1968
12	Beech forest	Germany	?	850	450	1.88	Ebermayer, cited in Rodin & Basilevic, 1968
13	Beech plantation	Denmark	85	740	390	1.89	Möller et al., cited in Kira et al., 1967
14	Oak-pine forest	Long Island, New York	40	783	406	1.92	Whittaker & Woodwell, 1969
15	Pine forest	Virginia	17	940	490	1.93	Madgwick, 1968
16	Fir forest	Northern taiga, Russia	?	300	150	2.00	Manakov, cited in Rodin & Basilevic, 1968
17	Tropical seasonal forest	Ivory Coast	?	900	440	2.04	Müller & Nielsen, cited in Kira et al., 1967
18	Beech plantation	Denmark	25	960	390	2.46	Möller et al., cited in Kira et al., 1967
19	Beech plantation	Denmark	46	960	390	2.46	Möller et al., cited in Kira et al., 1967
20	10 Angiosperm forests	England	20–50	369	137[1]	2.70	Ovington, 1956
21	10 Angiosperm forests	Europe	?	770	280	2.75	Bray & Gorham, 1964
22	18 Cold temperate forests	varied	varied	890	290	3.07	Bray & Gorham, 1964
23	10 Angiosperm forests	Mts. of Tennessee	varied	1014[2]	320	3.16	Whittaker, 1966
24	13 Gymnosperm forests	Mts. of Tennessee	varied	889	267	3.32	Whittaker, 1966
25	7 Gymnosperm forests	Europe	varied	1050	270	3.88	Bray & Gorham, 1964
26	22 Gymnosperm forests	England	20–50	892	228[1]	3.90	Ovington, 1956
27	Tulip tree stand	Mts. of Tennessee	?	2407[2]	410	5.87	Whittaker, 1966

1. Calculated from non-leaf/leaf production ratios given by Bray and Gorham.
2. Above-ground portion only.

In contrast, litter fall seems to be similar in the two types of communities. No statistical test would show a significant difference between the litter fall columns of Tables 1 and 2.

To summarize the features of the world pattern of absolute production values we can say that, along a gradient of decreasing available solar energy, wood production is constant but litter production decreases; along a gradient of decreasing precipitation, litter production is constant but wood production decreases.

World patterns of plant efficiency (calories of solar energy required to produce plant material with an energy equivalent of one calorie) can be recognized by using data already presented. Figure 1 shows that communities 1 through 5 receive approximately twice the solar energy during the growing season as communities 18–26, yet there is no significant difference in rates of wood production between the two groups of communities (Table 1). This means that communities at higher latitudes and altitudes produce wood more efficiently than communities at lower ones.

Efficiencies of litter production, in contrast, might be fairly constant throughout the world, or perhaps even

Table 2. Rate of woody tissue production and rate of litter production in various perennial herb and grass ecosystems, and the ratio between the two rates.

Community number	Type of community	Location	Age (in years)	Wood production g/m²/yr	Litter fall g/m²/yr	Wood production/ litter production	Author
1	Old field, upland	Michigan	30	0	312	0	Weigert & Evans, 1964
2	Old field, swale	Michigan	30	0	1003	0	Weigert & Evans, 1964
3	Perennial herbs	Japan	?	294	1484	0.19	Iwaki et al., 1966
4	Perennial grass	Georgia	8	148	461	0.32	Golley, 1965
5	Tallgrass prairie	Missouri	?	452	520	0.87	Kucera et al., 1967 & Dahlman & Kucera, 1965
6	Old field	Michigan	14	1023	668	1.53	Golley, 1960
7	Mesic alpine tundra	Wyoming	?	302	162	1.86	Scott & Billings, 1964

decrease at higher latitudes and altitudes. Litter production in communities 1–5 is roughly 3 to 5 times greater than in communities 18–26 (Table 1) yet, as mentioned above, they have available only about twice the solar radiation as the latter communities. Where precipitation becomes severely limiting to plant growth, efficiency of solar energy utilization would be low regardless of amounts of solar energy.

Ecological aspects of the pattern

In the previous section, I pointed out that efficiency of wood production increases as amounts of available solar energy decrease, while efficiency of litter production appears to change very little. This increase in efficiency of wood production, coupled with an almost constant efficiency of litter production, causes the pattern of increasing wood to litter production ratios along gradients of decreasing available solar energy. One explanation for this pattern might be that fast growth (that is, efficient wood production) is or was a greater competitive advantage in areas with relatively little light, as in the taiga, than in areas with relatively high levels of available light, as in the moist lowland tropics.

Another adaptive mechanism that could result in high ratios at high latitudes and altitudes is the maximizing of the rate of wood production. This would be an adaptive advantage for plants in highly stressful environments, such as those of high altitude and latitude, if larger plants were more resistant to these stresses than small

plants. A larger tree has larger roots and possibly more energy reserves to withstand low light levels, defoliation, or an unusually short growing season, events more common at high latitudes than in the warm lowland tropics.

Since the wood to litter production ratio decreases along a gradient of decreasing precipitation, in temperate zone communities, a low ratio is usually the result of a low rate of wood production, not a high rate of litter production (see Tables 1 and 2). Plants of prairie and steppe communities have a low rate of wood production because of the adaptive advantage of small plant size in dry areas: the smaller the plant, the less water it requires. Low rates of wood production are found in woody plants of dry areas because these plants have been selected for the ability to survive while maintaining very slow growth rates. The desert shrub community of Chew and Chew (1965), for example, produced woody material at the rate of only 6.3 grams/m²/yr.

I have already suggested that wood production in mesic forests of the world is fairly uniform, excluding local variations caused by local environmental conditions. But we can ask: Does uniform wood production result from equal rates of production by all species or from the average production of all species in each community being equal? Either could be hypothesized from data cited in this paper, which represent total production from all species in various communities.

Not all species have equal rates of wood production. The predominant

and largest tree species in all mesic forests of the world may have equal rates, and it is possible that their production outweighs the production of other species to the extent that the pattern of plant energetics is determined by their production. For example, spruce trees in the taiga and oaks in southern temperate regions could have similar rates of production (on a unit area of forest floor basis), and their rates might be so large in comparison to shrub and other tree production that variations in the latter production would not affect the pattern.

On the other hand, should the more or less uniform wood production throughout mesic forests of the world result from the average production of all species in each community being equal, some sort of community evolution would be implied. The implication arises from the improbability of equal production occurring by chance. While the possibility of co-evolution of at least two species is evidenced by the positively mutualistic association of algae and fungi in lichens, no evidence exists that entire communities can evolve or have evolved as some sort of supra-organism.

Other aspects of the ratios of wood production to litter production, although not directly related to the overall pattern presented so far, are of ecological interest. One is the change in ratio during succession. In mesic environments, abandoned farms or cut-over forests initially are invaded by grasses and herbs. Later, trees replace the grass and herbs. During this succession, the ratio of wood production to litter production increases.

Margalef (1957) has discussed changes in plant energetics during succession and clarified their meaning.

Succession may be defined as a gradual, irreversible change in the structure of a mixed population in the direction of a replacement of systems slightly structured and having a rapid dynamics, made up of relatively small organisms having a high productivity/biomass relationship, adapted to the rapid utilization of the resources of the medium, by other, more stable communities made up of larger organisms with a greater thermodynamic output, adapted to an efficient utilization of the resources, and having a lower productivity/biomass relationship.

The inverse of Margalef's productivity/biomass relationship reflects changes in ecosystem energetics during succession similar to the wood production/litter production ratio discussed here. A grassland with a high total leaf and wood production/total living biomass relationship (Margalef's ratio) has a low ratio of wood production to leaf and litter production. A forest with a low productivity/biomass relationship has a high wood production/litter production ratio.

A second aspect of these ratios is a consideration of the maximum ratio possible. Within limits, the greater the proportion of available energy bound as wood to that bound in leaves and other litter, the faster a woody plant grows, considering plant growth to be the rate of increase of biomass over a period of years. If a plant has bound too much energy in wood, it would have insufficient energy to support production of photosynthetic apparatus, and eventually would cease to grow and would die from insufficient energy for metabolism. The highest ratio of wood production to litter production listed in Table 1, with the exception of community 27 discussed earlier, is 3.90. This figure may represent the balance of production between wood and litter that results in the fastest growth of woody plants.

Another aspect of the ratio of wood to litter production is also of ecological interest. It has been hypothesized that evolutionary selection has modi-

fied the manner in which plants distribute energy between long-lived and short-lived tissue. This idea can be extended to include annual plants. Annuals devote all their available energy to materials required only through the growing season, except for the seeds. The annual plant's seeds may represent its adaptation to stress. While perennial plants and trees have adapted to stress by producing woody tissue at a high rate, or by putting woody tissue underground, annual plants have adapted to stress by producing an especially resistant seed. Seeds of annuals always must endure stressful cold or dry seasons on or in the soil; the seeds of many perennials, on the other hand, are protected by the vegetative stage of the life cycle until just before the seed germinates. Exceptions, including oaks and members of the rose family, have seeds especially adapted to over-winter.

The difference in energy utilization between annuals and perennials would account for the decrease in productivity of above-ground biomass in an abandoned field after the first year of abandonment noted by Odum (1960). During succession, as perennials replace annuals, less energy is used in succulent above-ground tissue and more is stored in woody underground tissue, and above-ground production decreases.

Another explanation is required for the high productivity of annuals in New Jersey reported by Botkin and Malone (1968). Their measurement of both above-ground and below-ground production showed the total to be 2,340 grams/sq. meter/growing season. Because annual plants do not have to produce tissue capable of withstanding stress by drought or cold, perhaps they can produce relatively more nonresistant tissue. Thus less energy may be required to produce a gram of nonwoody tissue than a gram of woody tissue.

Caloric concentrations

Golley (1961) was first to point out that there was a general increase in caloric concentration (calories per gram of dry weight) in plants with increasing latitude, but also noted some exceptions. Too few studies were available at that time to recognize the various factors involved in a world pattern of caloric concentra-

tions. Data now available permit the formation of a pattern that encompasses all values: this pattern is identical to the world pattern of ratios of wood production to litter production.

It is easier to see patterns in the caloric data than in the ratio data (Tables 1 and 2), for two reasons. (1) Sampling and analytic techniques for determining productivity of biomass are not as standardized as techniques for measuring caloric values. The less the variation in data due to technique, the easier it is to see the variations caused by environmental factors. (2) Caloric values do not change much during the life of an individual plant, but changes in the wood/litter production ratio during the life of individual plants complicate the world pattern of these ratios.

Caloric concentrations in plants increase along a gradient of decreasing amounts of available solar radiation, as can be seen in Figure 2. Caloric concentrations in this figure are from Tables 3 and 4, and light energy available during the growing season was calculated as described above. Only caloric values of nondomestic vegetation and plants native to the respective areas were included in Tables 3 and 4. Communities 8, 9, and 14, Table 3, were not plotted in Figure 2 because they are influenced by low amounts of precipitation, and community 4, Table 4, was not plotted because of the difficulty of determining available solar radiation at the site of that study.

Just as in the world pattern of wood to litter production ratios, it is the amount of solar energy available for the plant to use, not total annual solar radiation, that is the important environmental factor. Plants in an alpine shrub community (Table 3, community 18) and in an area with a cold continental climate (Table 3, communities 13, 15–17) have higher concentrations than plants at a higher latitude but in a location with a maritime climate and a longer growing season (Table 3, communities 6, 7, 12). The higher concentration in English gymnosperms (community 12) than in Minnesota angiosperms (communities 10 and 11) results from differences in the two plant groups to be discussed below.

The pattern of caloric concentration

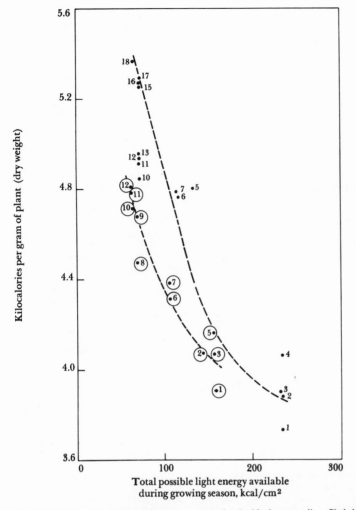

Figure 2. Caloric concentrations in plants as a function of amount of light energy available during the growing season. Uncircled points and numbers are tree data from Table 3 and are associated with the upper line. Circled points and numbers are grass and herb data from Table 4 and are associated with the lower line.

environmental factors. The adaptive advantage, in a region of limited light, of a plant that concentrates energy in its wood and leaf tissue is not readily recognized. High caloric concentration in wood and leaves would not necessarily make them more resistant to physical destruction or chemical decay. Resins of northern gymnosperms have a high caloric content, but gums and resins are also synthesized by tropical trees. It may not be the structural tissue or the presence or absence of resins that are responsible for higher caloric values but the energy concentration in the carbohydrates used for respiration. It seems possible that a high energy concentration in fuel could result in greater efficiency of energy use by the plant, because the plant would have to process relatively less fuel material in order to obtain a given amount of energy for metabolic purposes. Another advantage of a high energy concentration in carbohydrates would be that more energy can be stored per gram of root.

The reason that caloric concentrations decrease along a gradient of decreasing precipitation is also difficult to determine. If metabolic efficiency and/or concentration of energy in storage tissue are the factors involved, then at the lower end of the precipitation gradient high values of these factors confer less of an advantage than at the higher end of the precipitation gradient. High metabolic efficiency and high storage concentration might be less of an advantage to plants in areas of low precipitation because, in these areas, plants are often smaller and generally have less biomass to support per unit area of leaf surface than in areas of high precipitation.

Gymnosperm vs angiosperm

There is an aspect of the world pattern of wood production to litter production ratios and of the pattern of caloric concentrations that is of interest to plant geographers, because it helps explain the world distribution of two great plant groups. That aspect is the difference in ratios and caloric concentrations between gymnosperms (such as pine and fir trees) and angiosperms (broad-leafed trees like oak). A comparison of communities 20–26 in Table 1 indicates that within a general region there may be a higher ratio in gymnosperms than in angiosperms,

along a gradient of precipitation is similar to the pattern of ratios along a precipitation gradient: both decrease with decreasing precipitation. Plants in communities 2 and 4, Table 4 (both dry area communities), have lower caloric concentrations than plants in the temperate mesic communities of Table 3. In tundra regions, the same pattern occurs: in Table 3, plants in communities 8, 9, and 14 have lower concentrations than community 18. At the site of the former three communities, "precipitation is light" (Aleksiuk, 1970)— 10–20 inches of annual precipitation fall in the general region, according to a world precipitation map presented

by Trewartha (1954). At the latter site, annual precipitation is 73 inches per year (Hadley and Bliss 1964).

It is evident from Figure 2 that, within a given locality, grasses and herbs have lower caloric concentrations than trees. This pattern coincides with the pattern of ratios (Tables 1 and 2) in that both the ratios and caloric concentrations are lower in herbs and grasses than in trees.

Because the world pattern of caloric concentrations coincides with world patterns of available light and precipitation, the caloric pattern must reflect evolutionary adaptations to those

Table 3. Average energy values of trees in natural ecological communities, listed in order of increasing value.

Community number	Community or species	Location	No. of samples	Calories/gram leaves	entire plant	Author
1	Tropical moist forest	Panama	4	3732		Golley, 1969
2	Gallery forest	Panama	4	3879		Golley, 1969
3	Tropical rain forest	Puerto Rico	15		3897	Golley, 1961
4	Pre-montane tropical forest	Panama	4	4060		Golley, 1969
5	Oak forests	France & Spain	?		47–4900	Lieth, 1968
6	Angiosperm forests	England	8	4759		Ovington & Heitkamp, 1960
7	Scotch pine	England	14		4787	Golley, 1961
8	Poplar	N.W. Territory, Canada	2	4700	4800	Aleksiuk, 1970
9	Willow	N.W. Territory, Canada	2	4600	4800	Aleksiuk, 1970
10	Savanna	Minnesota	composite	4846		Ovington & Lawrence, 1967
11	Oakwood	Minnesota	composite	4916		Ovington & Lawrence, 1967
12	Gymnosperms	England	17	4926		Ovington & Heitkamp, 1960
13	Oak	Minnesota	"few"	4930		Gorham & Sanger, 1967
14	Alder	N.W. Territory, Canada	2	4800	5130	Aleksiuk, 1970
15	Cedar	Minnesota	"few"	5250		Gorham & Sanger, 1967
16	Larch	Minnesota	"few"	5260		Gorham & Sanger, 1967
17	Pine	Minnesota	"few"	5290		Gorham & Sanger, 1967
18	Alpine shrub	New Hampshire	30	5367		Hadley & Bliss, 1964

but there are exceptions to this pattern, possibly caused by gymnosperm communities that are young (community 15) or old (communities 10, 11, and 16). There are no ambiguities in the differences between gymnosperms and angiosperms in Table 3. In Minnesota, gymnosperms (communities 15–17) have higher values than angiosperms (communities 10, 11, and 13), and in England the same pattern is clear (communities 7 and 12 are higher than 6).

In previous sections, adaptive ad-

vantages of high ratios and high caloric concentrations have been hypothesized for plants in regions of limited solar radiation. If gymnosperms generally have higher ratios and caloric concentrations than angiosperms, it would suggest why gymnosperms replace angiosperms along a gradient of decreasing solar radiation from the tropics to the taiga, or with increasing altitude in the mountains of mesic regions. The less the available solar radiation within a general region, the greater the advantage of gymnosperms over angiosperms.

Along the gradient between taiga and tundra, small plant size apparently replaces efficiency of wood production as the most important adaptive mechanism. Also apparent is that no small gymnosperm species capable of surviving in the tundra have evolved, but some small angiosperms have become adapted to tundra conditions.

Conclusions

A pattern of plant productivity and caloric concentrations that encompasses naturally occurring terrestrial

Table 4. Average energy values of perennial herbs, grasses, and sedges in natural ecological communities, listed in order of increasing values. Cultivated plants, or plants not native to areas are not included, because they deviate from the natural distribution pattern of caloric values.

Community number	Community or species	Location	No. of samples	Calories/gram leaves	entire plant	Author
1	Perennial grass	Georgia	143		3905	Golley, 1961
2	Prairie grass	Missouri	composite	4071		Kucera et al., 1967
3	*Spartina* marsh	Georgia	14		4072	Golley, 1961
4	Desert grass	Utah	24	4080		Cook et al., 1959
5	Old field herbs	Georgia	35		4177	Golley, 1961
6	Perennial forbs	Michigan	composite	4315		Weigert & Evans, 1964
7	Perennial grass	Michigan	composite	4384		Weigert & Evans, 1964
8	Prairie herbs	Minnesota	composite	4471		Ovington & Lawrence, 1967
9	"Ground flora"	Minnesota	"few"	4680		Gorham & Sanger, 1967
10	Alpine meadow	New Hampshire	3		4711	Golley, 1961
11	Alpine *Juncus* dwarf heath	New Hampshire	2		4790	Golley, 1961
12	Alpine sedges & herbs	New Hampshire	40	4796		Hadley & Bliss, 1964

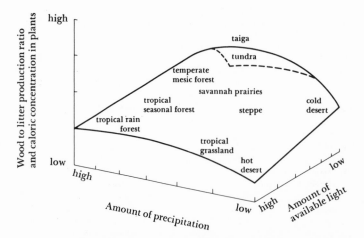

Figure 3. Production ratios and caloric concentrations as a function of amount of light available during the growing season and amount of precipitation. Any point on the three-dimensional plane represents a combination of light and precipitation on the horizontal axes, and ratio and caloric concentration on the vertical axis. The type of community generally occurring at various combinations of light and precipitation are indicated at the appropriate location on the plane.

plant communities appears to be correlated with worldwide gradients of available solar radiation and precipitation. The relationship between the pattern of productivity, caloric concentrations, and the environmental factors is summarized in Figure 3.

One objective of science is to seek order in the universe; when we find order, we gain the ability to predict. When we find order in plant production on a worldwide scale, we are able to predict the productive capability of the world. The importance of finding the pattern of plant energetics is that it provides a basis on which the productive capability of the continents can be calculated.

One of the objectives of the International Biological Program, in which biologists from many countries are participating, is "to estimate existing and potential plant and animal production in the major climatic regions, particularly in relation to human welfare" (U.S. IBP). The pattern presented is a step in the realization of this objective.

Literature cited

Aleksiuk, M. 1970. The seasonal food regime of arctic beavers. *Ecology* 51:264–70.

Botkin, D. B., and C. R. Malone. 1968. Efficiency of net primary production based on light intercepted during the growing season. *Ecology* 49:438–44.

Bray, J. R., and E. Gorham. 1964. Litter production in forests of the world. *Advances in Ecol. Res.* 2:101–57.

Budyko, M. I. 1956. *The Heat Balance of the Earth's Surface.* (Transl. from the Russian.) Office of Climatology, U.S. Dept. of Commerce, Washington, D.C. PB 131692, 259 pp.

Chew, R. M., and A. E. Chew. 1965. The primary productivity of a desert-shrub (*Larrea tridentata*) community. *Ecological Monographs* 35:355–75.

Cook, C. W., L. A. Stoddert, and L. E. Harris. 1959. The chemical content in various portions of the current growth of salt desert shrubs and grasses during winter. *Ecology* 40:644–50.

Dahlman, R. C., and C. L. Kucera. 1965. Root productivity and turnover in native prairie. *Ecology* 46:84–89.

Golley, F. B. 1960. Energy dynamics of a food chain of an old field community. *Ecological Monographs* 30:187–206.

Golley, F. B. 1961. Energy values of ecological materials. *Ecology* 42:581–84.

Golley, F. B. 1965. Structure and function of an old-field broomsedge community. *Ecological Monographs* 35:113–37.

Golley, F. B. 1969. Caloric value of wet tropical forest vegetation. *Ecology* 50:517–19.

Gorham, E., and J. Sanger. 1967. Caloric values of organic material in woodland, swamp, and lake soils. *Ecology* 48:492–94.

Hadley, E. B., and L. C. Bliss. 1964. Energy relationships of alpine plants on Mt. Washington, New Hampshire. *Ecological Monographs* 34:331–57.

Iwaki, H., M. Monsi, and B. Midorikawa. 1966. Dry matter production of some herb communities in Japan. The Eleventh Pacific Science Congress, Tokyo. August–September 1966. 1–15.

Jordan, C. F. Productivity of a tropical rain forest. *J. Ecology,* in press.

Kendrew, W. G. 1953. *The Climates of the Continents.* London: Oxford Press.

Kira, T., H. Ogawa, K. Yoda, and K. Ogina. 1967. Comparative ecological studies on three main types of forest vegetation in Thailand. IV. Dry matter production, with special reference to the Khao Chong rain forest. *Nature and Life in Southeast Asia* 5: 149–74.

Kira, T., and T. Shidei. 1967. Primary production and turnover of organic matter in different forest ecosystems of the western Pacific. *Japanese J. Ecology* 17:70–87.

Kondratyev, K. YA. 1969. *Radiation in the Atmosphere.* International Geophysics Series 12, Academic Press. 222 pp.

Kucera, C. L., R. C. Dahlman, and M. R. Koelling. 1967. Total net productivity and turnover on an energy basis for tallgrass prairie. *Ecology* 48:536–41.

Lieth, H. 1968. The measurement of calorific values of biological material and the determination of ecological efficiency. In *Functioning of Terrestrial Ecosystems at the Primary Production Level.* Proc. Copenhagen Symposium. UNESCO, pp. 233–41.

Lindeman, R. L. 1942. The trophic-dynamic aspect of ecology. *Ecology* 23:399–418.

Madgwick, H. A. I. 1968. Seasonal changes in biomass and annual production of an old-field *Pinus virginiana* stand. *Ecology* 49: 149–52.

Margalef, D. R. 1957. La Teoria de la informacion en ecologia. *Mems. R. Acad. Ciencias y Artes Barcelona* 32:373–499 (trans. in *Gen. Systems* 3:36–71, 1958).

Odum, E. P. 1960. Organic production and turnover in an old field succession. *Ecology* 41:34–49.

Olson, J. S. 1963. Energy storage and the balance of producers and decomposers in ecological systems. *Ecology* 44:322–31.

Ovington, J. D. 1956. The form, weights, and productivity of tree species grown in close stands. *New Phytologist* 55:289–304.

Ovington, J. D., and D. Heitkamp. 1960. The accumulation of energy in forest plantations in Britain. *J. Ecology* 48:639–46.

Ovington, J. D., and D. B. Lawrence. 1967. Comparative chlorophyll and energy studies of prairie, savanna, oakwood, and maize field ecosystems. *Ecology* 48:515–24.

Rodin, L. E., and N. I. Basilevic. 1968. World distribution of plant biomass. In *Functioning of Terrestrial Ecosystems at the Primary Production Level.* Proc. Copenhagen Symposium. UNESCO, pp. 45–52.

Scott, D., and W. D. Billings. 1964. Effects of environmental factors on standing crop and productivity of an alpine tundra. *Ecological Monographs* 34:243–70.

Trewartha, G. T. 1954. *An Introduction to Climate.* McGraw-Hill. 395 pp.

U. S. IBP. *Man's Survival in a Changing World.* U. S. International Biological Program, National Academy of Sciences—National Research Council, Washington, D. C.

Weigert, R. G., and F. C. Evans. 1964. Primary production and disappearance of dead vegetation on an old field in southeastern Michigan. *Ecology* 45:49–63.

Whittaker, R. H. 1966. Forest dimensions and production in the Great Smoky Mountains. *Ecology* 47:103–21.

Whittaker, R. H., and G. M. Woodwell. 1969. Structure production, and diversity of the oak-pine forest at Brookhaven, New York. *J. Ecology* 57:155–74.

AUTHOR CITATION INDEX

SUBJECT INDEX

Names of taxa from tables are not included in this index.

About the Editor

HELMUT H. F. LIETH is professor of ecology at the University of Osnabrueck, F.R.G. since January 1, 1978, and adjunct professor at the department of botany, University of North Carolina at Chapel Hill. He taught from 1967–1978 at the University of North Carolina at Chapel Hill, in 1967 at the University of Hawaii, Manoa Campus, and from 1955–1966 at the University of Stuttgart-Hohenheim, F. R. G. He received his undergraduate education at the Philosophical Theological College in Bamberg and at the University of Cologne where he finished his graduate studies in 1953 with a Ph.D. in botany and worked as an assistant until he transferred to the University of Stuttgart-Hohenheim.

Dr. Lieth traveled extensively to all continents and has held several guest professorships and research fellowships in North and South America, and Europe as well. He served for many years in various capacities for national and international committees of the International Biological Program.

Dr. Lieth is on the editorial board of *Vegetatio*, the Journal of Biogeography, and Radiation and Environmental Biophysics. Among his numerous publications are four books, two of which deal with the subject of primary productivity.